199
Topics in Current Chemistry

Springer-Verlag Berlin Heidelberg GmbH

Fullerenes and Related Structures

Volume Editor: A. Hirsch

With contributions by
C. Bellavia-Lund, F. Diederich, A. Hirsch, W. K. Hsu,
J. C. Hummelen, H. W. Kroto, M. Prato, Y. Rubin,
M. Terrones, C. Thilgen, D. R. M. Walton, F. Wudl

 Springer

This series presents critical reviews of the present position and future trends in modern chemical research. It is addressed to all research and industrial chemists who wish to keep abreast of advances in the topics covered.

As a rule, contributions are specially commissioned. The editors and publishers will, however, always be pleased to receive suggestions and supplementary information. Papers are accepted for "Topics in Current Chemistry" in English.

In references Topics in Current Chemistry is abbreviated Top. Curr. Chem. and is cited as a journal.

Springer WWW home page: http://www.springer.de
Visit the TCC home page at http://www.springer.de/

ISSN 0340-1022
ISBN 978-3-662-14729-0   ISBN 978-3-540-68117-5 (eBook)
DOI 10.1007/978-3-540-68117-5

Library of Congress Catalog Card Number 74-644622

© Springer-Verlag Berlin Heidelberg 1999
Originally published by Springer-Verlag Berlin Heidelberg New York in 1999.

The use of general descriptive names, registered names, trademarks, etc. in this publication does not imply, even in the absence of a specific statement, that such names are exempt from the relevant protective laws and regulations and therefore free for general use.

Cover design: Friedhelm Steinen-Broo, Barcelona; MEDIO, Berlin
Typesetting: Fotosatz-Service Köhler GmbH, 97084 Würzburg

SPIN: 10649319   66/3020 – 5 4 3 2 1 0 – Printed on acid-free paper

Preface

The development of the chemistry of the fullerenes was and continues to be a most exciting challenge in chemical research of the 1990s. It is not only the symmetrical structure and the beauty of these molecular allotropes of carbon but also their unprecedented properties that keep a large number of chemists, physisists and material scientists working with these spherical architectures. It came as no surprise that, the discovery of the fullerenes by Curl, Kroto and Smalley was rewarded with the Nobel prize in 1996.

The prototype of all fullerenes is Buckminsterfullerene C_{60}. Since it is the most abundant fullerene obtained from macroscopic preparation procedures such as the classical Krätschmer-Huffman method, its chemical and physical properties were developed in very quick sucession soon after it became available in 1990. The icosahedral football shaped Buckminsterfullerene C_{60} is now the most intensively studied molecule of all. Many principles of the chemistry of C_{60} are known. These allowone to tailor design new fullerene derivates with specific properties useful for biological applications or as new materials. The main types of fullerene derivative are exohedral addition products, endohedral fullerenes, heterofullerenes and cluster opened systems. Whereas many examples of the first two groups are already known, sophisticated methods for the synthesis of the latter two groups have started to emerge only recently.

It was also more recently that the chemistry of the less abundant and less symmetrical higher fullerenes like C_{70} and C_{76} started to develop. Aspects such as the inherent chirality of some of these carbon cages or the search for the principles of the regioselectivities of addition reactions to their conjugated π-system make these investigations most attractive.

The macromolecular analogues of the fullerenes are the carbon nanotubes. The development of nanotube research started soon after the bulk production of C_{60} and their identification in soot deposits formed during plasma arc experiments. Next to their remarkable and at the same time variable shape the carbon nanotubes have unique mechanical and electronic properties which are very sensitive to their geometries and dimensions.

The present volume combines reports and the current status of the principles of fullerene reactivity, of cluster modified fullerenes, of higher fullerenes and nanotubes and reviews potential applications of fullerene materials. This volume is also meant to inspire interested fullerene researchers to find elegant solutions and improvements for the perpectives described here. Fullerene and nanotube chemistry is still a young field and many facinating ideas await realization.

Andreas Hirsch, Erlangen   August 1998

Contents

Contents of Volume 196
Carbon Rich Compounds I

Contents of Volume 197
Dendrimers

Contents of Volume 198
Design of Organic Solids

Volume Editor: E. Weber

ISBN 3-540-64645-0

Principles of Fullerene Reactivity

Andreas Hirsch

Institut für Organische Chemie, Universität Erlangen-Nürnberg, Henkestr. 42,
D-91054 Erlangen, Germany. *E-mail: hirsch@organik.uni-erlangen.de*

The fullerenes have been established as new and versatile building blocks in organic chemistry. A large number of fascinating fullerene derivatives, especially of the icosahedral buckminsterfullerene C_{60}, have been synthesized. The chemistry of C_{60} continues to be good for many surprises. However, based on present knowledge a series of reactivity principles can be deduced which makes derivatization of this all carbon cluster more and more predictable. In this article first the geometric and electronic properties of the parent molecule are analyzed. The bent structure of the carbon network C_{60} and the filling of its molecular orbital with 60 π-electrons dictate the chemical reactivity. A very important aspect that was introduced with the investigation of fullerene chemistry is the shape dependence of reactivity.

Keywords: Fullerenes, Reactivity, Regiochemistry, Stereochemistry.

1
Introduction

Since the fullerenes [1] became available in macroscopic quantities in 1990 [2] a large number of chemical transformations [3–6] have been developed allowing for the synthesis of many different types of derivatives. The fullerenes are now established as versatile building blocks in organic chemistry, introducing new chemical, geometric, electronic and photophysical properties. The most intensively studied fullerene is the football shaped buckminsterfullerene C_{60}. It is the spherical shape which makes C_{60} and its higher homologues unique. The fullerenes can be considered as three-dimensional analogues of benzene and other planar aromatics. In contrast to such classical aromatics, however, the conjugated π-system of the fullerenes has no boundaries. In other words they contain no hydrogens that can be replaced via substitution reactions. Two main types of primary chemical transformations are possible: addition reactions and redox reactions. This simple topology consideration alone makes it evident that the reactivity of the fullerenes is significantly different from that of classical planar "aromatics".

Addition and redox reactions lead to covalent exohedral adducts and salts, respectively. Subsequent transformations of specifically activated adducts pave the way to other classes of fullerene derivatives (Fig. 1). These are heterofullerenes, defined degradation products or partial structures, open cage species and endohedral fullerenes.

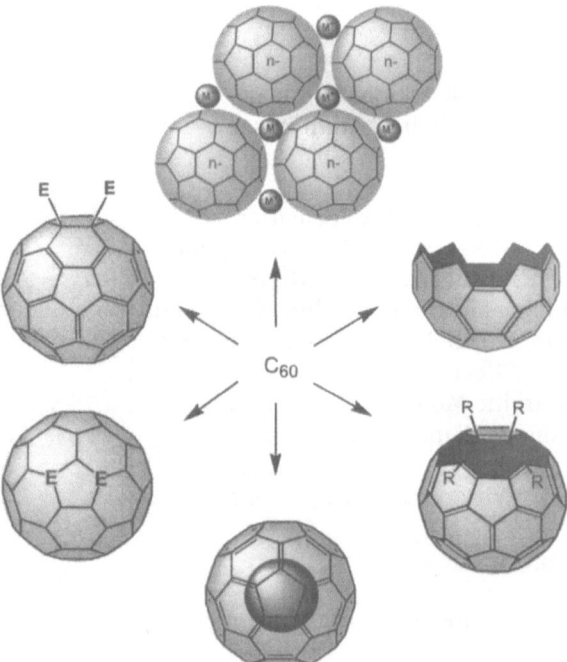

Fig. 1. Possible derivatizations of C_{60}

One example for a challenging yet unrealized synthesis goal is the introduction of a window into the fullerene framework large enough to allow atoms, ions or small molecules to enter the cage followed by a cluster closure reaction. Such a reaction sequence would provide an elegant access to endohedral fullerenes.

Alternatively, some of the prototypes of fullerene derivatives shown in Fig. 1 can be obtained directly during the fullerene formation out of graphite in the presence of foreign elements, by particle implantation methods or by total synthesis approaches. Most of the endohedral metallofullerenes are currently generated during the fullerene formation whereas partial structures are basically provided from total syntheses. For total synthesis approaches of fullerenes and their derivatives see [7].

For all the prototypes of fullerene derivatives depicted in Fig. 1 several, in some cases a large number, of examples have already been realized (Fig. 1). Among the salts the superconductors M_3C_{60} (M is, for example, an alkali metal) are the most prominent representatives [8, 9]. The first C_{60} derivative with an uncovered orifice was the ketolactam 1 [10]. The endohedral fullerene N@C_{60} (2) [11] contains atomic nitrogen located in the center of the cluster. The encapsulated N-atom does not form endohedral covalent bonds with the C-atoms of the fullerene shell [12].

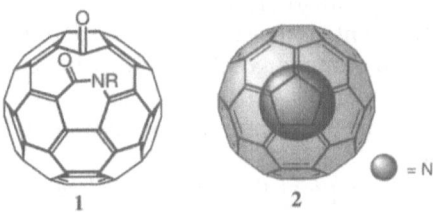

Heterofullerenes like 3 [13–16] were synthesized via specific addition/elimination sequences starting from C_{60} or C_{70}. As an example for a partial structure, semibuckminsterfullerene 4 was prepared exclusively using total synthesis approaches [17].

The largest group of fullerene derivatives are exohedral adducts. Next to monoadducts, stereochemically defined multiple adducts, for example trisadduct 5, having an inherent chiral addition pattern were synthesized [18].

Addition reactions contain the largest synthetic potential in fullerene chemistry but they can also be used as a probe for screening the chemical properties of fullerene surfaces. Analysis of the nature of addition reactions, of geometric and electronic structures of educts and products allows one to deduce reactivity principles which are the central subject of this article. One striking conclusion is the following: the fullerenes drew attention to a mostly overlooked chemistry criterion – the shape dependence of reactivity. This contribution focuses on the reactivity principles of C_{60} as the most important and most intensively investigated fullerene. The addition chemistry of the higher fullerenes follows similar principles.

2
Structure and Electronic Properties of C_{60}

The icosahedral C_{60} ([60-I_h]fullerene) consists of 12 pentagons and 20 hexagons (Fig. 2). This building principle obeys Euler's theorem predicting that exactly 12 pentagons are required for the closure of a carbon network consisting of n hexagons. [60-I_h]fullerene is the first stable fullerene because it is the smallest possible to obey the isolated pentagon rule (IPR) [19, 20]. The IPR predicts fullerene structures with all the pentagons isolated by hexagons to be stabilized against structures with adjacent pentagons. A destabilization caused by adjacent pentagons is (i) due to pentalene-type 8π-electron systems, leading to resonance destabilization and (ii) due to an increase of strain energy, as a consequence of enforced bond angles.

Due to the spherical shape of the unsaturated carbon network the C-atoms are pyramidalized. This has several consequences.

1. The deviation from planarity introduces a large amount of strain energy. Thermodynamically, C_{60} is therefore considerably less stable than the planar graphite. The heat of formation of C_{60} was determined experimentally by calorimetry to be 10.16 kcal/mol per C-atom compared to graphite as reference [21]. It was estimated by Haddon that the strain energy within C_{60} makes up about 80% of its heat of formation [22].

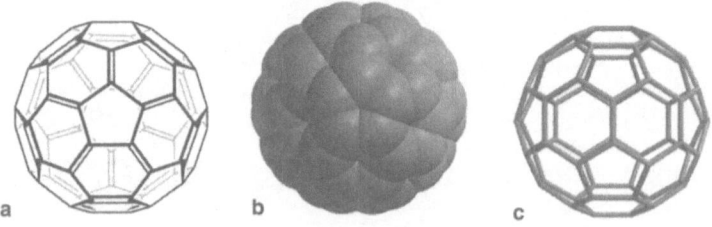

Fig. 2 a–c. Representations of C_{60}: **a** lowest energy VB structure; **b** space filling model; **c** tube model

2. The conjugated C-atoms of a fullerene respond to the deviation from planarity by rehybridization of the sp^2 σ and π orbitals, since pure p character of π orbitals is only possible in strictly planar situations [22].

The electronic structure of non-planar conjugated organic molecules was analyzed by Haddon using the π orbital axis vector (POAV) analysis. For C_{60} an average σ bond hybridization of $sp^{2.278}$ and a fractional s character of 0.085 (PAOV1) or 0.081 (POAV2) was found [23–27]. As a consequence, of the π-orbitals extend further beyond the outer surface than into the interior of C_{60}. This consideration implies, moreover, C_{60} to be a fairly electronegative molecule [28, 29] since due to the rehybridization low lying π^* orbitals also exhibit considerable s character.

Another structural aspect is of importance and plays a central role for the chemical behavior of C_{60}. The bonds at the junctions of two hexagons ([6,6]-bonds) are shorter than the bonds at the junctions of a hexagon and a pentagon ([5,6]-bonds). This was demonstrated by a series of theoretical [30–34] and experimental investigations [35–38] (Table 1). As a consequence, in the lowest energy Kekulé structure of C_{60} the double bonds are placed at the junctions of the hexagons ([6,6] double bonds) and there are no double bonds in the pentagonal rings. Topologically, each hexagon in C_{60} exhibits a cyclohexatriene and each pentagon a [5]radialene character (Fig. 3).

Table 1. Calculated and measured bond distances in C_{60} in Å

Method	[5,6] Bonds	[6,6] Bonds	Reference
HF(STO-3G)	1.465	1.376	30
HF($7s3p/4s2p$)	1.453	1.369	31
LDF($11s6p$)	1.43	1.39	32
HF	1.448	1.37	33
MP2	1.446	1.406	34
NMR	1.448	1.370	35
Neutron diffraction	1.444	1.391	36
Electron diffraction	1.458	1.401	37
X-ray	1.467	1.355	38

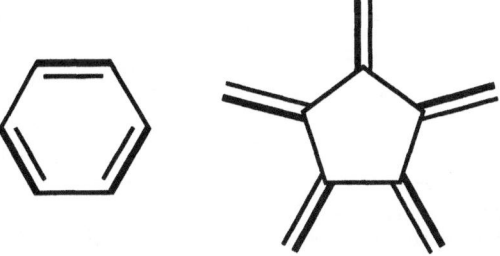

Fig. 3. Cyclohexatriene and radialene subunits of C_{60}

What is the reason for the bond length alternation in C_{60}? The bond length alternation in C_{60} cannot be explained with strain arguments for two reasons. First, the strain energy is predominantly due to the pyramidalization of the C-atoms and this does not depend on whether there is bond length alternation or not. Second, in C_{60}^{6-} the bond length alternation is considerably reduced and in C_{60}^{12-} it almost vanishes [39]. The difference in bond lengths between [6,6]-bonds and [5,6]-bonds is rather directly connected with the symmetry of C_{60} and the occupation of its molecular π-orbitals. The shortening of one of the types of bonds is equivalent to a partial localization of the π-orbitals into localized bonds. This can easily be rationalized considering a structureless shell model [40–42] and by analyzing the nature of the frontier orbitals within the fullerene framework as demonstrated for the first time in this review. The electronic levels for the 60 π-electrons of C_{60} can be deduced starting with a spherical approximation [42]. The corresponding spherical harmonics are used to describe the harmonic wave functions of this electron gas with their angular momentum quantum numbers. The irreducible representations of the icosahedral group can be found using group theory by lowering the symmetry from full rotational symmetry to icosahedral symmetry treated as a perturbation (Table 2).

Considering the Pauli principle it can be seen that 60 π-electrons will completely fill angular momentum states up through $\ell = 4$, leaving 10 electrons in the $\ell = 5$ level which can accommodate a total of 22 electrons. For $\ell = 4$ the filled

Table 2. Filled shell and subshell π-electron configurations for C_{60} [42]

l^a	Shell	Electrons/state	n_c^b	HOMO in I_h symmetry^c
0	s	2	2	a_g^2
1	p	6	8	f_{1u}^6
2	d	10	18	h_g^{10}
3	f	14	24	f_{2u}^6
			26	g_u^8
			32	$f_{2u}^6 g_u^8$
4	g	18	40	g_g^8
			42	h_g^{10}
			50	$g_g^8 h_g^{10}$
5	h	22	56	f_{1u}^6 or f_{2u}^6
			60	h_u^{10}
			62	$f_{1u}^6 f_{2u}^6$
			66	$f_{1u}^6 h_u^{10}$ or $f_{2u}^6 h_u^{10}$
			72	$f_{1u}^6 f_{2u}^6 h_u^{10}$

[a] Angular momentum for a spherical shell of p-electrons.
[b] Number of p-electrons for closed shell ground state configurations in icosahedral symmetry.
[c] Symmetries of all the levels of the l value corresponding to the HOMO; the superscript on the symmetry label indicates the degeneracy of the level.

states in icosahedral symmetry which can take up 18 electrons are given by the irreducible representations g_g^8 and h_g^{10}. Since the angular momenta are symmetrically distributed, no distortion from spherical or icosahedral symmetry is expected in cases where all states are completely filled. Therefore, no significant bond length alternation is expected for C_{60}^{10+} (50 π-electrons) with the $\ell = 4$ states filled. The complete filling of the $\ell = 5$ state would lead to an accumulation of 72 π-electrons, assuming that all the $\ell = 5$ states are filled, before any $\ell = 6$ level becomes occupied. In icosahedral symmetry the $\ell = 5$ states split into the $H_u + T_{1u} + T_{2u}$ irreducible representations. The lowest energy level is the H_u level, which in neutral C_{60} is completely filled by the ten available electrons. What are the consequences of the filling of the H_u level with 10 electrons on the geometry of C_{60}? A comparison of the representations of the sets of σ orbitals along the [6,6]-bonds with that of the π-electrons shows that when filling the h_u shell, the occupied π-orbitals form a set which is completely equivalent to the set of localized σ orbitals along the [6,6]-bonds [41]. Therefore, the molecular orbitals can be weakly localized at these sides by a unitary transformation (Fig. 4).

Localization then increases and the total energy decreases when the [6,6]-bonds shorten. This distortion exactly corresponds to the internal structural degree of freedom that the C_{60} molecule has without breaking the I_h symmetry. Adding 12 electrons into the t_{lu} and t_{2u} orbitals would lead to a closed shell situation again which does not favor bond length alternation. The t_{1g} orbitals derived from the $\ell = 6$ shell, however, are lower in energy than the t_{2u} orbitals from the $\ell = 5$ shell. This implies that the t_{1g} orbitals are filled before the t_{2u} orbitals.

Using the picture of the graphical representations of the calculated h_u, t_{lu}, and t_{1g} orbitals makes these group theoretical considerations transparent and easy to

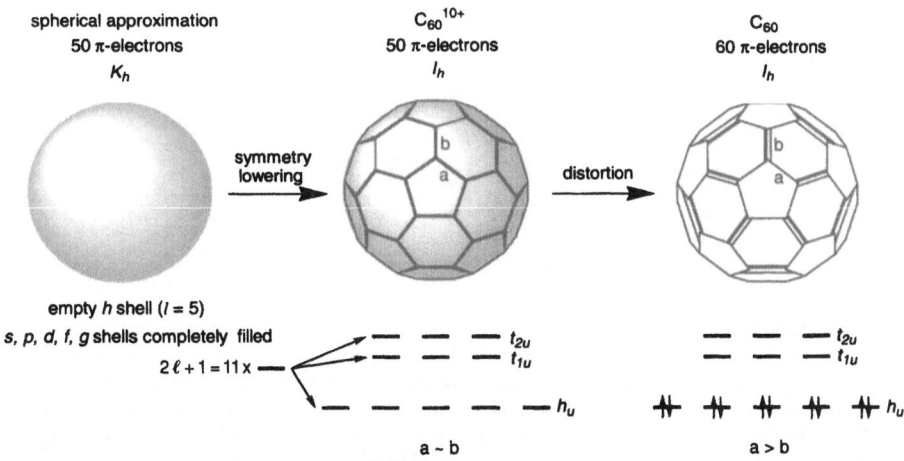

Fig. 4. Incomplete filling of the h shell deduced from the spherical approximation and the resulting bond length alternation in neutral C_{60}

understand in a qualitative manner. As can be seen from Fig. 5, the bonding interactions in the five-fold degenerate h_u orbitals are located preferably at the [6,6]-sites, whereas at the [5,6]-sites there are mainly antibonding interactions (nodes). A different situation comes out for the t_{1u} and t_{1g} orbitals, where the nodes are located at the [6,6]-sites and the bonding interactions at the [5,6]-

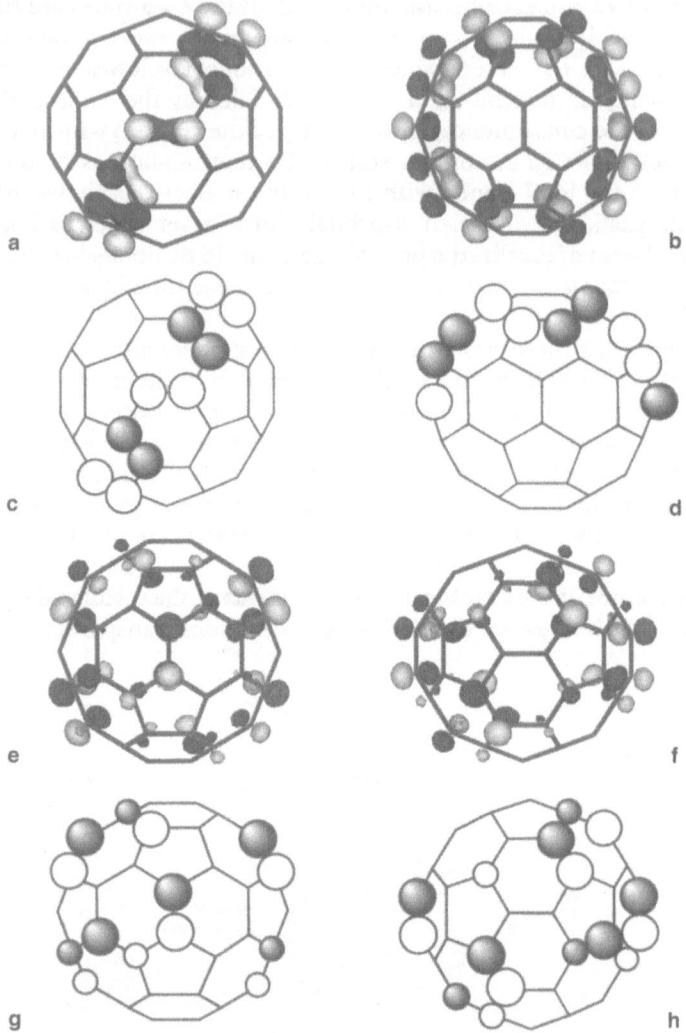

Fig. 5 a–h. Coefficients of C_{60} frontier orbitals: **a** one of the five degenerate PM3 calculated HOMOs seen from the front; **b** the perpendicular side perspective; **c,d** front and side views of schematic representation of the HOMO showing only the exohedral part of the orbital projected to the corresponding C-atom for clarity; **e** one of the three degenerate PM3 calculated LUMOs seen from the front; **f** the perpendicular side perspective; **g, h** schematic representation of the LUMO. The *larger circles* in the qualitative schematic presentations denote large and the smaller circles medium coefficients. Very small coefficients are omitted for clarity. The three LUMOs of C_{60} have the same geometry and are oriented perpendicular to each other

sites. As a consequence filling the h_u orbitals causes a shorting of the [6,6]-bonds leading to an increase in bond energy and at the same time it favors longer [5,6]-bonds reducing antibonding interactions. Upon filling the t_{1u} and t_{1g} orbitals the opposite behavior is predicted.

Indeed, calculations [39] on K_6C_{60} show that the bond lengths become more equalized with the [6,6]-bonds being 1.42 Å and the [5,6]-bonds 1.45 Å. The same calculations on C_{60}^{12-} ($Li_{12}C_{60}$) predict a further elongation of the [6,6]-bonds (1.45 Å) and already a shortening of the [5,6]-bonds (1.44 Å) [39]. In conclusion, it is the filling of the h_u, t_{1u}, and t_{1g} orbitals which is responsible for the degree of bond length alternation. This is closely related to the situation in the annulenes where, depending on the filling of the molecular orbitals with either 4n or 4n + 2 π electrons, a distortion from the ideal D_{nh} takes place or not. In contrast to the latter, however, the distortion in neutral C_{60} exactly corresponds to the internal structural degree of freedom and does not lead to a symmetry lowering.

The t_{1u} orbitals of C_{60} are low lying, suggesting that it is a fairly electronegative molecule comparable to benzoquinone. This theoretical picture is confirmed by electrochemical investigations [3, 43 – 45] and by series of reduction reactions with electropositive metals [3, 8, 9] and organic donor molecules [3, 46 – 49]. For example, successive reversible one-electron reductions up to the hexaanion can be determined by cyclic voltammetry [43 – 45]. As mentioned above this electron affinity is mainly due to the pyramidalization of the C-atom leading to π^* orbitals with significant s character. The reduction of C_{60} can also be considered as a strain relief process assisted by the pyramidalized C-atoms, since carbanions are formed and carbanions are known to prefer pyramidalized geometries. On the other hand the h_u orbitals are also low in energy which implies that C_{60} is difficult to oxidize which was corroborated by cyclic voltammetry [50] and other investigations [3]. This can also be understood in terms of strain associated with the rigid spherical geometry of C_{60} because upon oxidation carbocations are formed, preferring inaccessible planar geometries leading to an increase of strain energy.

The fullerenes show evidence for ring currents on the length scale of the individual rings (segregated ring currents) (Fig. 6) [51 – 53]. The strength of the

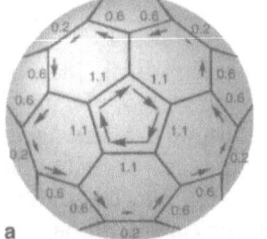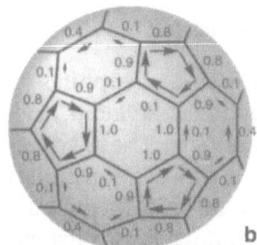

a   b

Fig. 6 a, b. π-electron ring currents in C_{60} for a magnetic field oriented perpendicular to: **a** a five-membered ring (5-MR); **b** a six-membered ring (6-MR), calculated with the London theory. The currents in the 5-MRs are paramagnetic (clockwise), whereas those in the 6-MRs are mostly diamagnetic. The current strength is given with respect to that in benzene [51]

ring currents is comparable to those in benzenoid hydrocarbons. Of particular importance is the finding of paramagnetic ring currents in the pentagons. These currents are found to be sufficiently large that their sum over the twelve pentagons almost exactly canceled the diamagnetic currents in the 20 6-membered rings.

The cancellation of ring currents in neutral C_{60} is a direct consequence of the fullerene topology, with the diamagnetic term of the magnetic susceptibility largest in the hexagons and its paramagnetic Van Vleck term predominant in the pentagons in such a way that the overall effect of the π-electrons on the magnetic susceptibility is negligible. These findings explain the vanishingly small ring-current contribution to the total magnetic susceptibility of C_{60} determined experimentally. A powerful tool for measuring the magnetic shielding in the interior of a fullerene is ^{3}He NMR spectroscopy on endohedral fullerenes encapsulating the nucleus ^{3}He [54, 55]. Whereas the chemical shift of He in ^{3}He@C_{60} is comparatively small ($\sigma = -6$) a large additional shielding by more than 40 ppm is observed in ^{3}He@K_6C_{60} with C_{60} being in the hexaanionic state with the t_{1u} orbitals occupied [53]. This is consistent with the symmetry of the t_{1u} orbitals, where the nodes are preferably located in the hexagons and the bonding interactions in the pentagons. This orbital structure allows for strong diamagnetic ring currents within the pentagons in addition to those already present in the hexagons.

The specific chemical properties of C_{60} are a direct consequence of its spherical structure with the bond length alternation in the neutral state and the resulting electrophilicity. This will be demonstrated in detail in the following paragraphs. Moreover, it will be shown that many fullerene derivatives retain typical properties of the parent molecule, like bond length alternation and electrophilicity. Also for derivatives it is the unique spherical structure of the fullerene core and the resulting stereo-electronic properties which govern the reactivity.

3
Outside Reactivity

3.1
Typical Addition Reactions of C_{60}

3.1.1
Nucleophilic and Radical Additions

C_{60} readily reacts with nucleophiles [56–87] and radicals [88–100] to give, for example, hydrogenated, alkylated, arylated, alkynylated, silylated, aminated and phosphorylated adducts. These types of additions are among the earliest observed reactions in fullerene chemistry. In the case of reactions with nucleophiles the initially formed intermediates $Nu_nC_{60}^{n-}$ can be stabilized by (1) the addition of electrophiles E^+, for example, H^+, or carbocations to give $C_{60}(ENu)_n$ [59], (2) the addition of neutral electrophiles E-X like alkyl halogenides to give $C_{60}(ENu)_n$ [75], (3) an S_{Ni} or internal addition reaction to give methanofullerenes [61] and cyclohexenofullerenes [86], respectively or (4) by an oxidation (air) to give, for example, $C_{60}Nu_2$ [63a] (Scheme 1).

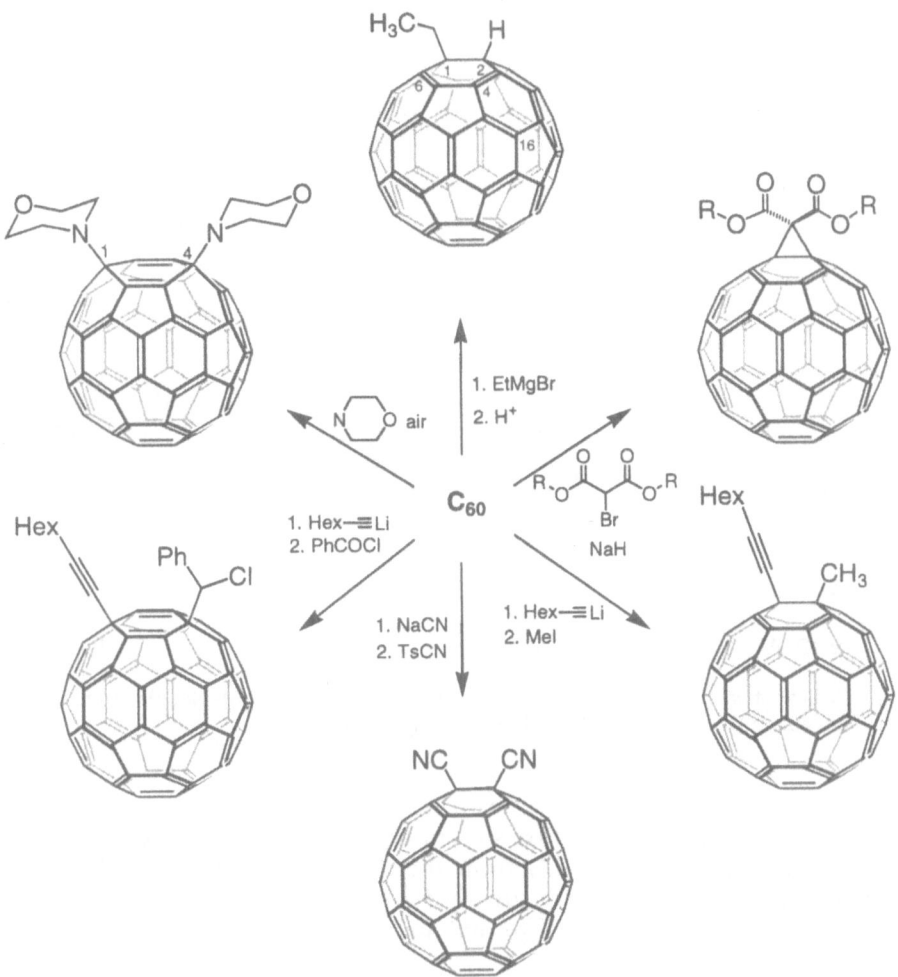

Scheme 1. Typical examples for nucleophilic additions to C$_{60}$

Although a large number of isomers are in principle possible for additions of segregated addends, the preferred mode of addition is 1,2. For a combination of sterically demanding addends 1,4 additions [63, 66, 72, 75, 80] and even 1,6 additions [87] (to the positions 1 and 16) can take place alternatively or exclusively (Fig. 7).

The addition of free radicals like R$_3$C˙, R$_3$Si˙, R$_3$Sn˙ and RS˙ generated chemically, photochemically, or thermally leads also to substituted dihydro- or polyhydrofullerenes, with the same preferences of addition modes [88–93]. As far as the resulting addition patterns of products and the presumed mechanisms are concerned, nucleophilic additions and radical additions are closely related and in some cases it is difficult to decide which mechanism actually

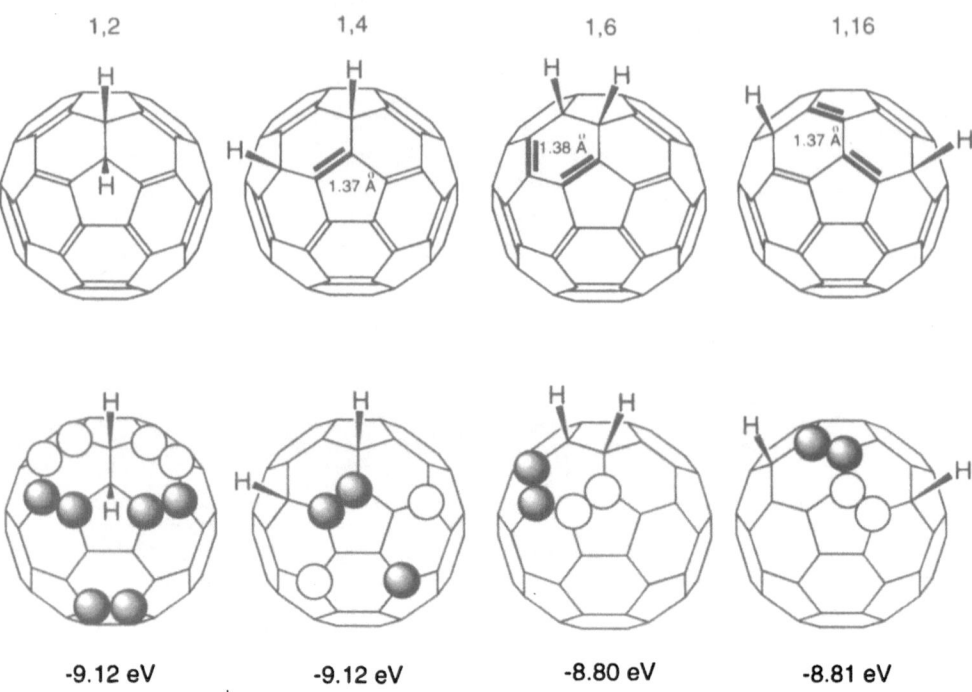

Fig. 7. Lowest energy VB structures, PM3 calculated lengths of [5,6]-"double" bonds, HOMO coefficients and HOMO energies of different dihydro[60]fullerenes

operates. For example, the first step in the reaction of C_{60} with amines is a single-electron-transfer (SET) from the amine to the fullerene [101]. The resulting aminofullerenes are finally formed via a complex, in most cases unknown, sequence of radical recombination, deprotonation, and redox reactions.

The regiochemistry of such addition reactions is governed by an important principle of C_{60} chemistry, namely the minimization of [5,6]-double bonds in the lowest energy Kekulé structure. If the size of two addends in a substituted dihydro[60]fullerene does not cause sterical constraints, 1,2-dihydro[60]fullerenes are formed as the thermodynamically most stable structures after addition of the two segregated addends to a [6,6]-double bond. This addition pattern represents the only cluster closed dihydro[60]fullerene, whose lowest energy Kekulé structure does not have double bonds placed in a five-membered ring. This VB consideration is confirmed by a variety of X-ray single crystal structure investigations and computations showing that in 1,2-adducts the bond lengths alternation between [6,6]-bonds and [5,6]-bonds is totally preserved, with significant deviations from the values of the parent C_{60} only in the direct neighborhood of the addends [102]. Moreover, like in parent C_{60} the bonding interactions in the occupied frontier orbitals are located predominantly at [6,6]- and the nodes at the [5,6]-sites (Fig. 8).

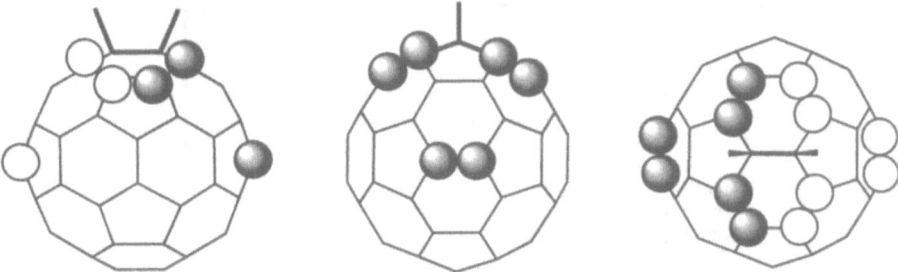

Fig. 8. PM3 calculated HOMO of $C_{60}H_2$ shown from three perspectives

This implies a useful approximation, namely to look at a 1,2-dihydro[60]-fullerene as a stereoelectronically slightly perturbated C_{60}. The introduction of a double bond into a five-membered ring costs about 8.5 kcal/mol [3, 103] (Figs. 9 and 10). In a 1,4-adduct (1,4-dihydro[60]fullerene) one and in a 1,6-adduct (1,16-dihydro[60]fullerene or 1,6-dihydro[60]fullerene) two bonds in five-membered rings are required for the corresponding lowest-energy Kekulé structure. This VB consideration is also confirmed experimentally and by computations (Fig. 7).

The 1,4 and 1,16 addition patterns are only preferred for sterically demanding addends, since in the corresponding 1,2-adducts or even in some 1,4-adducts the eclipsing interactions become predominant (Table 3). A 1,6-addition pattern with a cluster closed structure is unfavorable and has never been isolated due to both introduction of two [5,6]-double bonds and eclipsing interactions. In $1,2-C_{60}H_2$, the eclipsing interaction was estimated to be about 3–5 kcal/mol. This value increases with an increasing size of the addend. The

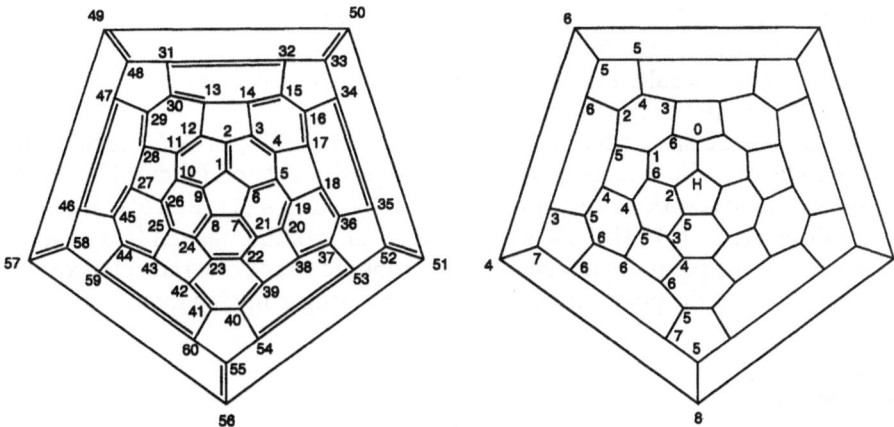

Fig. 9. Numbering system within the lowest energy VB structure of C_{60} and dependence of the topologically counted minimum number of [5,6]-double bonds which have to be introduced by the addition of two hydrogens on the positions of the hydrogens

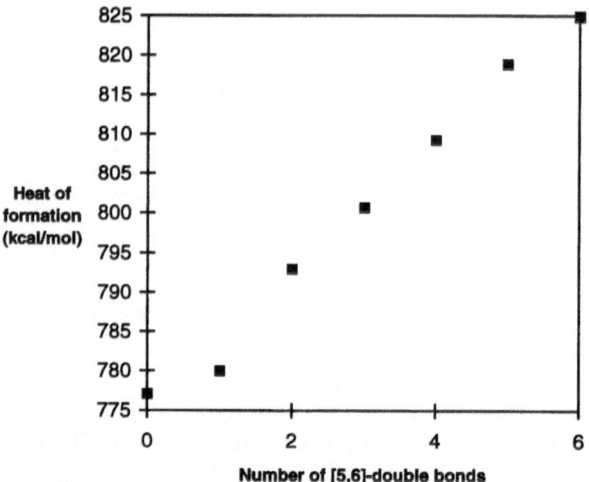

Fig. 10. Dependence of the MNDO heats of formations of the $C_{60}H_2$ isomers on the number of [5,6]-double bonds introduced by hydrogen addition

Table 3. Calculated relative PM3 energies (kcal/mol) of dihydro [60] fullerene derivatives with addends of different sterical requirement

	1,2-Adduct	1,4-Adduct	1,6-Adduct	1,16-Adduct
$C_{60}H_2$	0	3.8	18.3	15.5
$C_{60}Ht$-Bu	0	3.1	17.8	15.0
$C_{60}(t$-Bu$)_2$	31.3	0	40.6	9.2
$C_{60}[Si(t$-Bu$)_3]_2$	–	13.3	–	0

relative product distribution is balanced by minimizing [5,6]-double bonds on the one hand an eclipsing interactions on the other hand.

Dihydro[60]fullerenes having the addends attached further apart as those described above would require the introduction of even more [5,6]-double bonds and have not been isolated.

The charge or spin density of primary adducts RC_{60}^- or $RC_{60}^{\cdot}$ formed by nucleophilic or radical additions is highest at position 2 and followed by 4 (11) for nucleophilic additions and at position 2 followed by 4 (11) and 6(9) for radical additions (Fig. 11) [3]. This is also a result of avoiding formal [5,6]-double bonds. From this point of view also, an electrophilic attack or a radical recombination process at the position 2 is favored, neglecting eclipsing interactions.

3.1.2
Cycloadditions

In cycloadditions exclusively the [6,6]-double bonds of C_{60} act as dienes or dienophiles [3, 5, 6]. A large variety of cycloadditions were carried out with C_{60} and the complete characterization of both monoadducts and multiple adducts

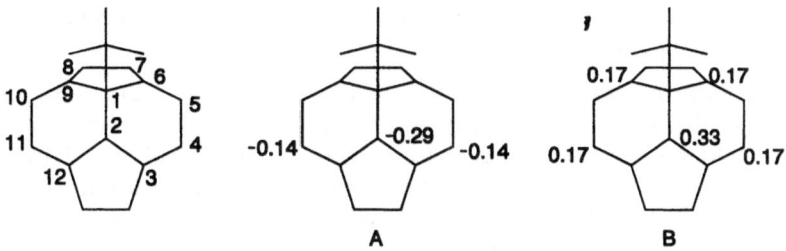

Fig. 11. Charge distribution (A) (Mulliken charges) and spin distribution (B) in the intermediates t-BuC$_{60}^-$ and t-BuC$_{60}^\cdot$

greatly increased the knowledge of fullerene chemistry. These chemical transformations also provide a powerful tool for fullerene functionalization. Almost any functional group can be covalently linked to C_{60} by cycloadditions of suitable addends. Many cycloadducts are very stable. This is an important requirement for further side chain chemistry as well as for possible applications of fullerene derivatives with unprecedented materials and biological properties. Among the most important cycloadditions are [4 + 2] cycloadditions like Diels-Alder and Hetero-Diels-Alder reactions, where C_{60} reacts always as dienophile, [3 + 2] cycloadditions with 1,3 dipoles, thermal or photochemical [2 + 2] cycloadditions, [2 + 1] cycloadditions (Scheme 2) and others for example [8 + 2] cycloadditions.

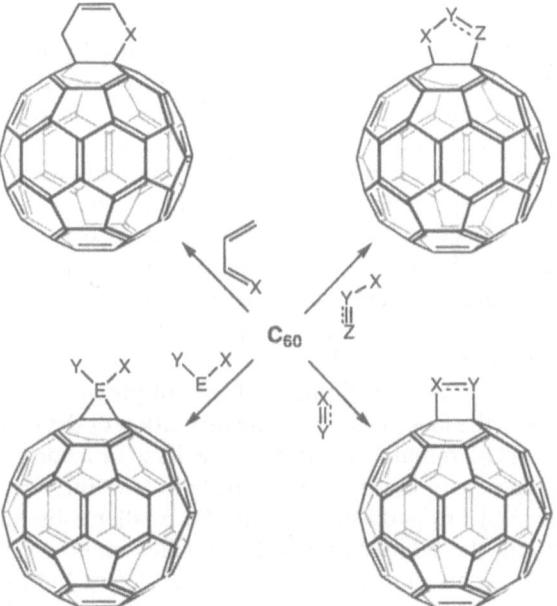

Scheme 2. General examples for cycloadditions to C_{60}

Among these general reactions several examples deserve special attention since they reflect characteristic chemical properties of C_{60}.

1. Reversibility of Diels-Alder-reactions: some of [4+2] cycloadditions, for example, the reaction of C_{60} with anthracenes [104–108] or cyclopentadiene [104, 109] are reversible. The parent components are obtained from the adducts upon heating. In the case of 9,10-dimethylanthracene (DMA) as diene the Diels-Alder reaction is already reversible at room temperature, making it difficult to isolate the adduct. Such facile retro-reactions have been used for regioselective formations of stereochemically defined multiple adducts like the template mediated syntheses using DMA as reversibly binding diene (Scheme 3) [107] and the topochemically controlled solid state reaction of C_{60} with C_{60}(anthracene) to give trans-1-C_{60}(anthracene)$_2$ (Scheme 4) [108].

2. [2+2] Reactivity with benzyne: even with the good dienophile benzyne C_{60} does not react as diene but rather forms a [2+2] cycloadduct (Scheme 5) [110–112]. One reason for this type of reactivity is certainly due to the fact that in a hypothetical [4+2]-adduct one unfavorable [5,6]-double bond in the lowest energy VB structure is required.

3. Formation of cluster opened methano- and imino[60]fullerenes (fulleroids and azafulleroids): thermal [3+2]-cycloadditions of diazo compounds or azides lead to the formation of fulleropyrazolines or fullerotriazolines. The thermolysis of such adducts after extrusion of N_2 affords as kinetic products the corresponding [5,6]-bridged methano and iminofullerenes with an intact 60 π-electron system and an open transannular bond (Scheme 6) [113–128]. The corresponding [6,6]-bridged structures with 58 π-electrons and a closed transannular bond are formed only in traces.

However, if the substituents on the methano or imino bridges contain at least one substituent such as a phenyl or alkoxycarbonyl group stabilizing radical intermediates, facile thermal rearrangements from [5,6]-bridged to the thermo-dynamically more stable [6,6]-bridged isomers can take place, allowing in some cases for a complete conversion [129–131]. The photochemical extrusion of N_2 leads preferably to [6,6]-adducts [132]. The exclusive formation of cluster open [5,6]-bridged and closed [6,6]-bridged isomers is also a consequence of the principle of the minimization of [5,6]-double bonds. Hypothetical open [6,6]- and closed [5,6]-bridged isomers would require the formation of three and two unfavorable [5,6]-bonds, respectively (Fig. 12). No monoadduct with such a structure has so far been observed.

The formation of [5,6]-bridged fulleroids out of precursor pyrazolines or triazolines is driven by the fact that during the formation of the methylene or imino bridge the position C-2 remains blocked by the diazo or azide site of the addend. Significantly, in the case of methano addends with two different substituents the formation of the [5,6]-bridged isomers with the smaller substituent above a six-membered ring and the bulkier group above a five-membered ring is the preferred process [126]. This can be understood considering both mechanisms that have been suggested for the fulleroid formation out of pyrazoline or triazoline precursors, namely a concerted orbital symmetry controlled [π^2s + π^2s + σ^2s + σ^2a]

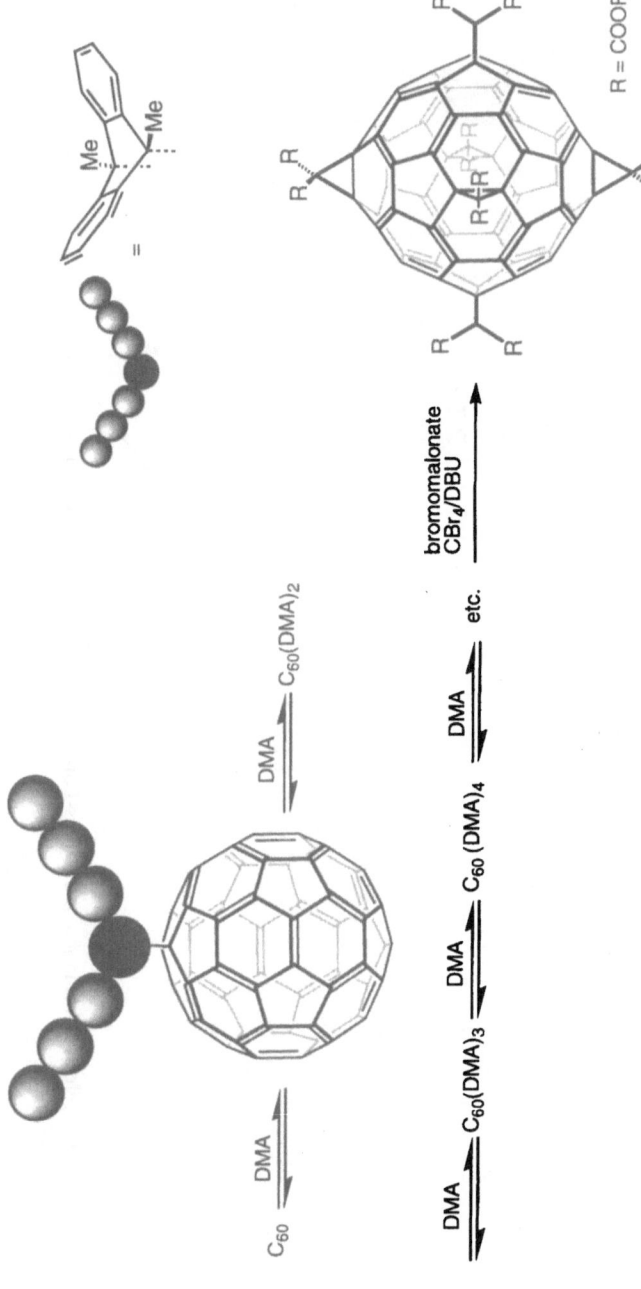

Scheme 3. Example for a template-mediated formation of a hexaadduct with an octahedral addition pattern

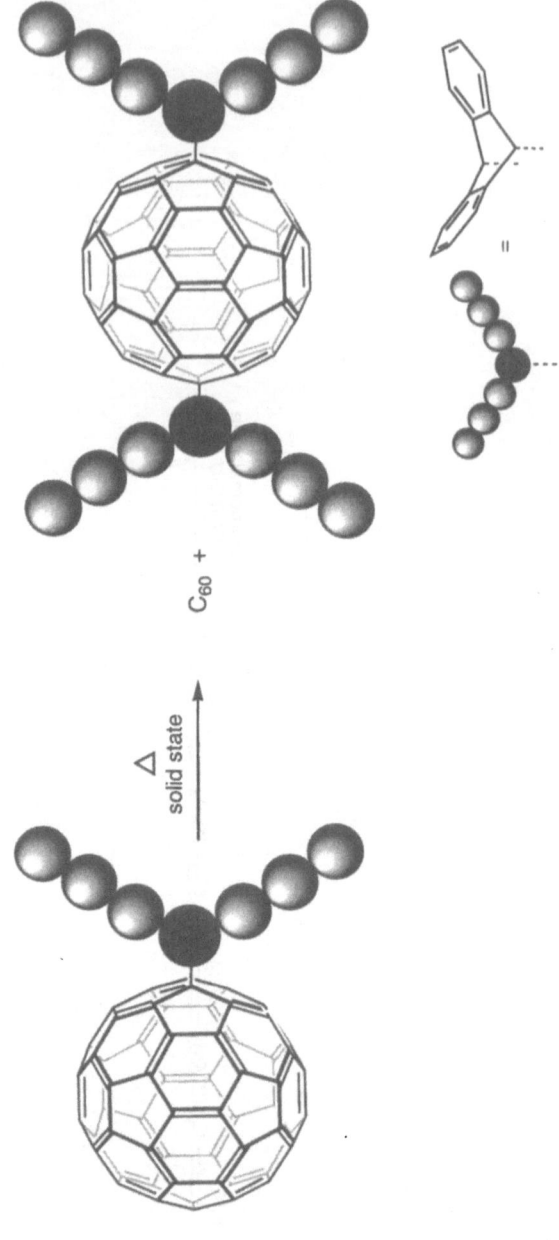

Scheme 4. Selective formation of the *trans*-1 bisadduct of C_{60} with anthracene in the solid state

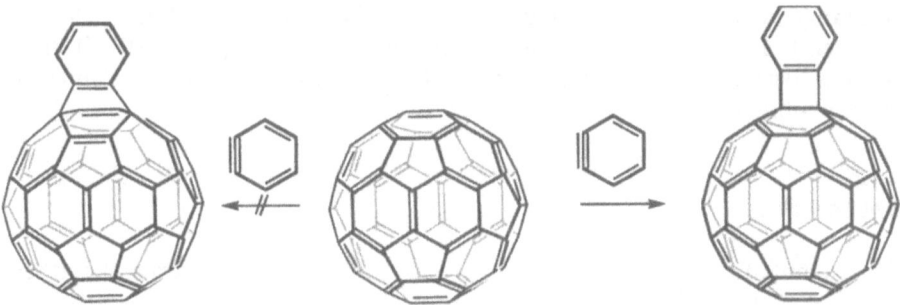

Scheme 5. Reaction of C$_{60}$ with benzyne

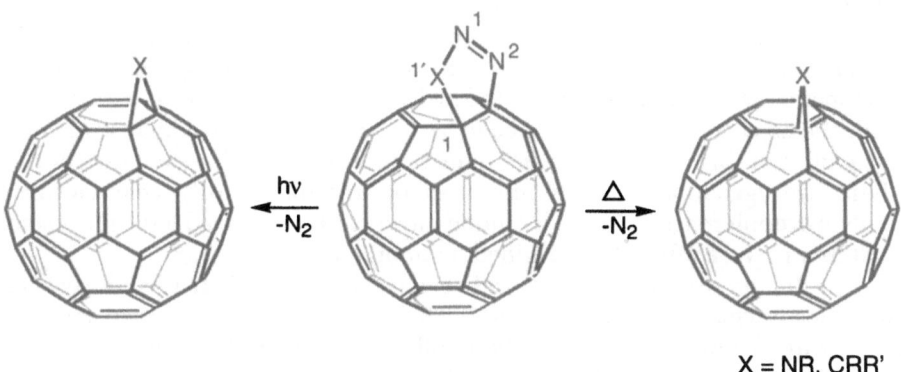

X = NR, CRR'

Scheme 6. Formation of methano- and imino[60]fullerenes out of pyrazolines or triazolines

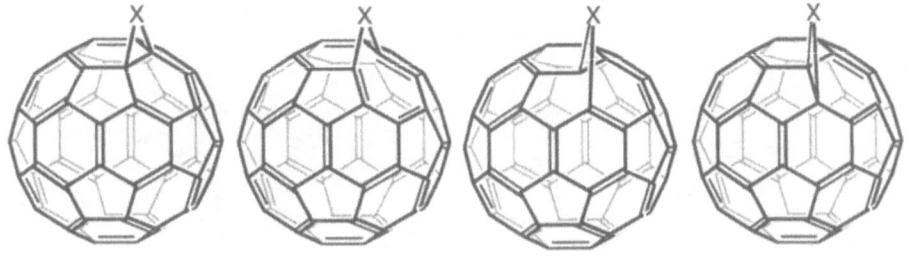

X = NR, CRR'

Fig. 12. Valence and constitutional isomers of [5,6]- and [6,6]-bridged imino and methano[60]fullerenes

rearrangement [127] or a stepwise process involving a homolytical cleavage between C-1' (N-1') and N-1 followed by rotation about C-1/C-1'(N-1') in the sense of lowest sterical hindrance and a subsequent nitrogen extrusion and radical recombination [126, 128].

3.1.3
Hydrogenations

All attempts to synthesize and characterize the completely hydrogenated icosahedral $C_{60}H_{60}$ (all-outside isomer) failed. This compound is expected to be very unstable, since it contains an enormous amount of strain energy due to 20 planar cyclohexane rings and 90 eclipsing H-H interactions [3]. Molecular mechanics (MM3) [133] and semiempirical PM3 calculations on $C_{60}H_{60}$ show that moving just one hydrogen inside the cage and forming a new conformer is an exothermic process (Fig. 13).

A conformer of $C_{60}H_{60}$ with ten H-atoms added inside the cage was predicted to be the lowest energy isomer with an MM3-heat of formation being 400 kcal/mol lower than for the all-outside isomer [133]. However, a formation of such systems via outside hydrogenation and subsequent isomerization is hampered by the high barrier for the hydrogen penetration from outside to inside the C_{60} cage which is calculated to be at least 2.7 eV/atom. Exhaustive hydrogenations of C_{60}, via Birch-Hückel reductions, transfer hydrogenations, or catalytic hydrogenations lead to mixtures of unstable, hardly characterizable polyhydrofullerenes $C_{60}H_{2n}$ [3]. Among those $C_{60}H_{36}$ seems to be the most stable but its structure was not determined unambiguously [134–138].

The di- and some of the tetra- and hexahydrides on the other hand have been completely characterized [139–143]. Their synthesis was achieved either by hydrozirconation, by hydroboration followed by hydrolysis or by zinc/acid reduction. The addition of one pair of H-atoms is always 1,2 and occurs at [6,6]-sites as shown for the synthesis of the parent 1,2-dihydro[60]fullerene (Scheme 7).

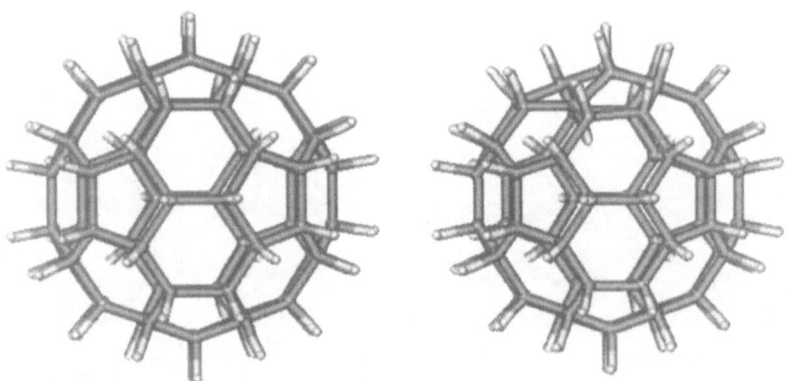

Fig. 13. PM3 calculated structure of the all outside isomer (HOF = 396.69 kcal/mol) and the one inside isomers (HOF = 329.49 kcal/mol) of $C_{60}H_{60}$

The regiochemistry of the formation of the tetra-[142] and hexahydrides [143] is discussed in Sect. 3.2.2.

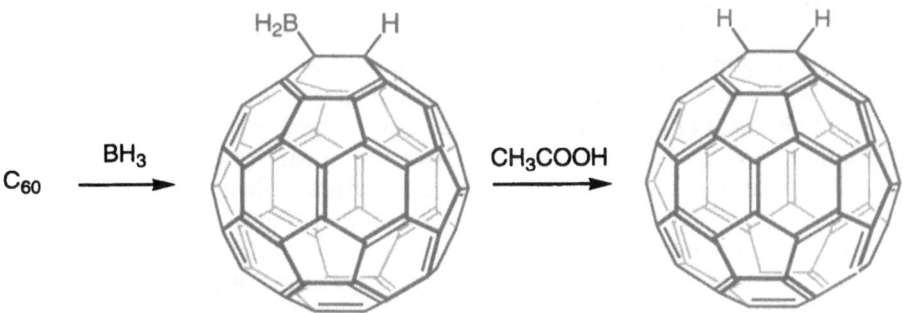

Scheme 7. Formation of 1,2-dihydro[60]fullerene via hydroboration

3.1.4
Transition Metal Complex Formation

The reactivity of C_{60} comparable to that of electron deficient conjugated olefins is nicely reflected by reactions with transition metal complexes. A variety of single crystal structures and spectroscopic studies show that the complexation of transition metals to the fullerene core proceeds in a dihapto manner or as hydrometalation reactions rather than in η^6- or η^5-binding mode. This was elegantly demonstrated by the reaction of C_{60} with ruthenium complexes (Scheme 8) [144]. A variety of iridium complexes $(\eta^2$-$C_{60})Ir(CO)Cl(PR^1R^2R^3)_2$ were synthesized by allowing C_{60} to react with different Vaska-type complexes $Ir(CO)Cl(PR^1R^2R^3)_2$ [145]. η^2-Complex formation was also observed upon reaction of C_{60} with other Ir [146] as well as Rh [147] complexes. Hydrometallation was obtained with $Cp_2Zr(H)Cl$ [140].

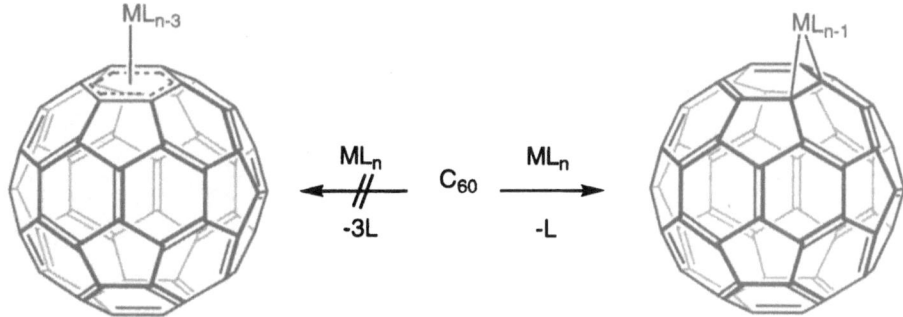

Scheme 8. Typical reactivity of C_{60} towards transition metal complex reagents

3.1.5
Oxidation and Reactions with Electrophiles

Although the reduction of the fullerenes is far more facile than their oxidation, a variety of oxidative functionalization as well as electrophilic additions to C_{60} have been reported [3]. Oxygenated fullerenes $C_{60}O_n$ were found in the fullerene mixture generated by graphite vaporization [148]. The formation of these oxides is due to a small amount of molecular oxygen present in the fullerene reactor. Mixtures of $C_{60}O_n$ (n up to 5) can also be generated by electrochemical oxidation of C_{60} [149] or by photolysis of the crude fullerene extract. More drastic conditions, such as UV irradiation in hexane or heating in the presence of oxygen, [148] lead to an extensive oxygenation or fragmentation of C_{60} [150]. Upon photooxygenation of C_{60} by irradiating an oxygenated benzene solution for 18 h at room temperature and subsequent purification by flash chromatography followed by semipreparative HPLC, $C_{60}O$ can be obtained in 7% yield [150b]. Chromatography of $C_{60}O$ on neutral alumina leads to an efficient conversion to C_{60} in 91% yield. $C_{60}O$ can also be prepared by allowing toluene solutions of C_{60} to react with dimethyldioxirane, whereupon a 1,3-dioxolane adduct of C_{60} was formed simultaneously [151]. Epoxides of C_{60} can also be obtained by the reaction with ozone. Presumably, epoxides are formed after O_2 extrusion of the primary ozonides [152].

The reaction of C_{60} with osmium tetroxide leads to osmylated fullerenes [153]. The monoadduct 6 was obtained in high yields by osmylation of C_{60} followed by the addition of pyridine or by using a stoichiometric amount of OsO_4 in the presence of pyridine. It was observed that the osmylation product fully reverted back to C_{60} when heated under vacuum. The pyridine molecules in 6 can be exchanged for other co-ordinating ligands, such as 4-*tert*-butylpyridine. This ligand exchange reaction has been used to increase the solubility of osmylated fullerenes, which allows the growth of high quality single crystals.

The fluorination of C_{60} was achieved either by the treatment of dichloromethane solutions with XeF_2 [154] or by allowing fluorine gas at low pressure to

6

react with the solid fullerene [155, 156]. In this way mixtures of polyfluorofuller-enes $C_{60}F_{2n}$ with n = e.g., 15–22 were obtained. A defined regioisomer of $C_{60}F_{48}$ was synthesized by the reaction of solid C_{60} with fluorine gas in the presence of NaF [157]. As another example for structurally characterized fluoride, $C_{60}F_{18}$ was obtained after fluorination with K_2PtF_6 [158, 159]. Higher degrees of fluorina-tion were obtained by the reaction of prefluorinated fullerenes $C_{60}F_{2n}$ with fluo-rine under simultaneous UV radiation [160]. Significantly, fluorinated fuller-enes $C_{60}F_{2n}$ beyond n = 30 were also formed by this preparation method. This hyperfluorination requires the rupture of σ-bonds of the fullerene framework.

A polychlorination of C_{60} was carried out by allowing a slow stream of chlor-ine gas to react with C_{60} in a hot class tube at temperatures between 250 and 400 °C [161] or by the treatment of solid C_{60} with liquid chlorine at –35 °C [162]. The chlorofullerene $C_{60}Cl_6$ (see Sect. 3.2.4) as an isomerically pure single pro-duct was synthesized by the reaction of C_{60} with an excess of IC1 in benzene or toluene at RT [163].

The treatment of C_{60} with liquid bromine afforded a T_h-symmetric $C_{60}Br_{24}$ (7) as yellow orange crystalline compound (Fig. 14) [164]. Bromination in CS_2 gave $C_{60}Br_8$ (8) as dark brown crystals in 80% yield [165]. The bromide $C_{60}Br_6$, iso-structural to $C_{60}Cl_6$, was formed by bromination in benzene or CCl_4 [165, 166].

A review on halogenated fullerenes was provided by Taylor [167]. Here their properties are also compared and summarized. The stability order of the halo-genofullerenes is comparable to that generally of other organic halides, since it increases along the series: iodo-<bromo-<chlorofullerene(<fluorofullerene).

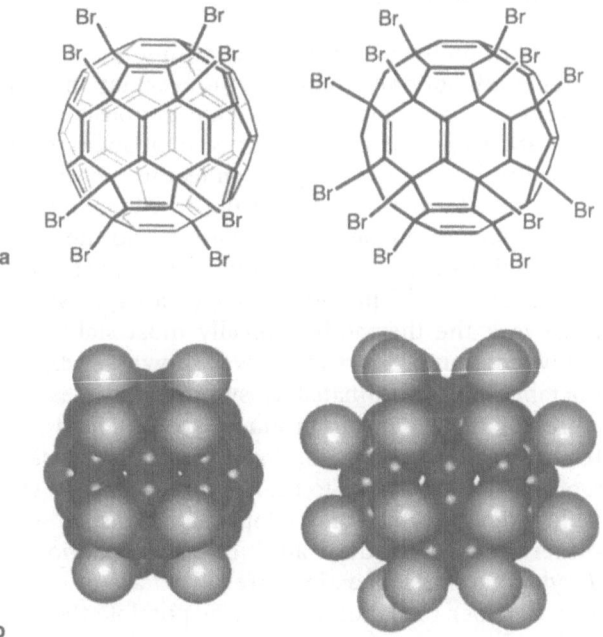

Fig. 14. a VB structures of $C_{60}Br_8$ and $C_{60}Br_{24}$. **b** Space filling models of $C_{60}Br_8$ and $C_{60}Br_{24}$

However, the halogenated fullerenes are significantly less stable than other alkyl halides (and also aryl halides). The maximum number of halogens (excluding fluorine here) that can be attached to the cage varies according to the size of the halogen. For bromination it is 24, and in these perbrominated derivatives, no halogen atoms are adjacent; they are never closer than 1,3- to each other. For chlorination the maximum number is approximately 40. What is not known is the mechanism of addition, though it is a reasonable assumption that radicals are involved.

Persistent radical cations of C_{60} have been observed in super acidic media, for example in magic acid (FSO_3H, SbF_5) [168]. Multiply charged fullerene radical cations were also stabilized in a mixture of fuming sulfuric acid and SO_2ClF at low temperatures [169]. The reaction of C_{60} with sulfuric acid and nitric acid and the subsequent hydrolysis of the intermediates with aqueous base resulted in the formation of fullerenols [170, 171].

The reaction of fullerenes with the Lewis acid BH_3 were used to synthesize hydrogenated fullerenes (Scheme 7). Also, other Lewis acids, for example $AlBr_3$, $TiCl_4$, $SnCl_4$, and $FeCl_3$, have been allowed to react with C_{60} [172–174]. If these reactions are carried out in CS_2, fullerene-Lewis acid complexes precipitate.

The treatment of C_{60} in aromatic hydrocarbons with Lewis acids, such as $AlCl_3$, $AlBr_3$, $FeBr_3$, $FeCl_3$, $GaCl_3$, or $SbCl_5$ in hydrocarbons leads to a fullerylation of aromatics. In this case, the Lewis acid serves as a catalyst and increases the electrophilicity of the fullerene. Mixtures of polyarylated fullerenes were obtained.

3.2
Regiochemistry of Multiple Additions

Systematic investigations on the regioselectivity of multiple 1,2-additions [102, 175–186] to fullerenes are not only important to reveal intrinsic chemical properties of these carbon cages – they also provide the basis for the design of highly symmetrical and stereochemically defined oligoadducts. The multitude of possible aesthetically pleasing and unique architectures in which the fullerene core serves as structure determining tecton is unprecedented in organic chemistry. Controlled variations of the degree of addition and the addition pattern allow characteristic tuning of the physical properties, in particular the electronic and chemical properties, of fullerenes. Whereas for reversible additions to the fullerene framework the thermodynamically most stable isomer can be obtained as the only reaction product which was shown, for example, with the synthesis of an octahedrally coordinated T_h-symmetrical hexaplatinum complex [184], mixtures of regioisomeric products are obtained for irreversible kinetically controlled addition reactions. This paragraph focuses on the most important types of multiple additions to the C_{60} core leading to stable, isolable, and characterizable products, which are cyclopropanations, iminofullerene formations, Diels-Alder reactions, radical additions, halogenations, aminations, and alkylations/arylations. These reactions are sorted according to prototypes of addition modes. In order to elaborate as far as possible the inherent regioselectivity principles of subsequent additions the consideration of elegant synthesis strategies like tether controlled additions are neglected.

3.2.1
Bond and Site Labeling in Fullerenes and Fullerene Adducts

Most of the exohedral fullerene derivatives of preparative importance are formed by one or several formal 1,2-additions to [6,6]- or [5,6]-bonds. For a suitable discussion of the regiochemistry of such fullerene derivatives it is very valuable to introduce a simple and clear site labeling system, which allows facile description of the constitution of a given fullerene derivative. We first introduced a very descriptive nomenclature [177, 188] for the assignment of the relative positional relationships (like *ortho, meta* and *para* in benzene chemistry) of addend carrying bonds in C_{60} derivatives with labeling the corresponding bonds as *cis*-n (n = 1–3), e', e'', *trans*-n (n = 1–4) (Fig. 15).

This nomenclature is increasingly being used in fullerene literature. However, it can only be applied to [6,6]-bonds within C_{60}-derivatives. But, since, for example, the regiochemistry of multiple [5,6]-adducts of C_{60} and multiple adducts of higher fullerenes already started to develop, we also introduced a general site labeling algorithm, which can be applied to [5,6]- and [6,6]-positions of all fullerenes and their derivatives [189].

The basis for the algorithm is the contiguous numbering scheme of the C-atoms of a given fullerene in a spiral fashion, which is used for the logic structure assignment of any fullerene adduct [3]. Suitable two-dimensional representations of the C-atom numbering are provided in the Schlegel diagram of the fullerene. The spiral numbering guarantees to define priority orders of C-atoms and C-C-sites within the fullerene framework, which is a necessary requirement for an unambiguous structure assignment of a fullerene derivative. The algorithm can be used for bond labeling in parent fullerenes and bond or site labeling in fullerene derivatives and labeling sets of C-atoms in fullerenes and their derivatives.

In parent fullerenes not every individual bond but, rather, sets of bonds are labeled. Sets of bonds are bonds with identical geometrical environment. Each bond of a given set can be transferred into any other bond of this set using suitable symmetry operations. In fullerene adducts, in addition, reference sites as well as further priority orders need to be defined. For details of the bond labeling algorithm see [189].

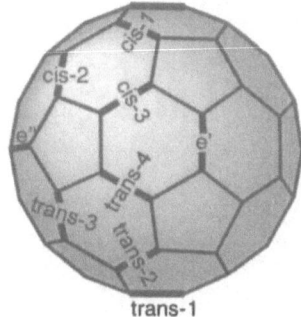

Fig. 15. Relative positional relationships of [6,6]-bonds in a C_{60} derivative

3.2.1.1
Parent Fullerenes

The various sets of [6,6]-bonds are labeled in decreasing order of priority as I, II, III ... and the [5,6]-bonds as A, B, C etc. The definition of a priority order is based on the nomenclature rule of the lowest set of locants.

3.2.1.2
Fullerene Derivatives

In fullerene derivatives reference sites have to be defined in order to assign unambiguously each site of the fullerene framework of adducts, which in turn allows one to specify addition patterns, positional relationships of addends, as well as the absolute configuration of an adduct. All adducts of a given fullerene can be described with the same labeling scheme. The first step in the site labeling process is the definition of the reference site, which is the site of highest priority. The highest priority site has the locant set 1,2. This site is labeled as I ([6,6]-bond) or as A ([5,6]-bond) and distinguished from the other sites of this set of the parent fullerene, which breaks the symmetry at least to C_{2v}. The remaining sets of [5,6]- and [6,6]-sites are then labeled as described for the parent fullerenes. The result can be "stored" for example in a Schlegel diagram and used for the assignment of sites in any derivative of a given fullerene. The labeling scheme for C_{60} derivatives is represented in Figs. 16 and 17. The priority

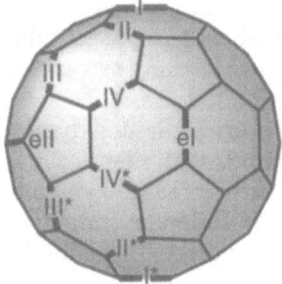

Fig. 16. Labeling of [6,6]-bonds within C_{60} derivatives

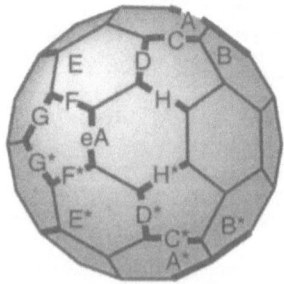

Fig. 17. Labeling of [5,6]-bonds within C_{60} derivatives

of sites decreases in C_{60}-derivatives in the order I, II, III ... eI, eII, ... III*, II* I* and A, B, C .. eA, H, H*.. C*, B*, A*.

3.2.1.3
Assignment of the Constitution and Configuration of a Fullerene Derivative

In order to take into account fullerene derivatives with an inherent chiral addition pattern, two mirror images of each labeling diagram with clockwise and anti-clockwise numbering of the C-atoms have to be considered. This guarantees the absolute compatibility of the general configurational description of fullerene derivatives with a chiral addition pattern introduced by Diederich [187] using the descriptors ^{f}C and ^{f}A. These descriptors indicate the mode of numbering as fullerene clockwise or fullerene anti-clockwise. As examples for the site labeling of fullerene derivatives with an inherent chiral addition pattern chiral trisadducts are shown in Fig. 18.

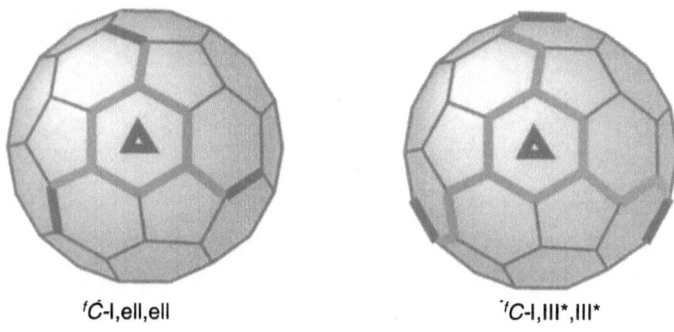

$^f\acute{C}$-I,eII,eII $^{'f}\acute{C}$-I,III*,III*

Fig. 18. Bond labeling of trisadducts with an inherent chiral addition pattern

3.2.2
1,2 Additions with Preferred e- and cis-1 Modes: The trans-1 Effect

Cycloadditions to [6,6]-double bonds of the fullerene framework are the most versatile and suitable reactions in fullerene chemistry. For a second attack to a [6,6]-bond of a C_{60} monoadduct nine different sites are available (Fig. 15). Hence, for two different symmetrical addends nine regioisomeric bis-adducts are in principle possible, whereas for identical addends only eight regioisomers can be considered, since attack to both e'- and e''-positions leads to the same products.

Bisaddition reactions represent the simplest and most instructive model cases for regioselectivity investigations and form the basis for more complex situations. In a comprehensive study of bisadduct formations with two identical as well as with two different addends, nucleophilic cyclopropanations, Bamford-Stevens-reactions with dimethoxybenzophenone-tosylhydrazone and nitrene additions were used (Fig. 19, Scheme 9) [102].

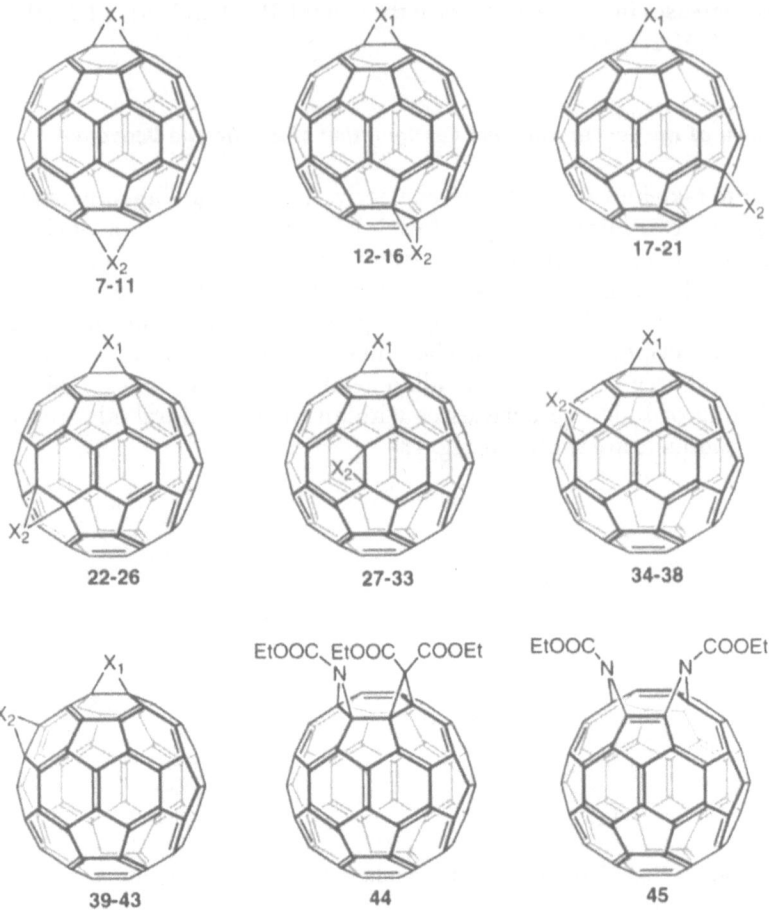

Fig. 19. Regioisomers **7–45** of the bis-adducts $C_{62}(COOEt)_4$, $C_{60}(NCOOEt)_2$, $C_{62}(anisyl)_4$, $C_{62}(anisyl)_2(COOEt)_2$, and $C_{61}(COOEt)(NCOOEt)_2$ **7,12,17,22,27,34,39**:$X^1 = X^2 = C(COOEt)_2$; **8,13,18,23,28,35,40**:$X^1 = X^2 = NCOOEt$; **9,14,19,24,29,36,41**:$X^1 = X^2 = C(anisyl)_2$; **10,15,20,25, 30,37,2**:$X^1 = C(anisyl)_2$, $X^2 = C(COOEt)_2$; **11,16,21,26,32,38,43**:$X^1 = NCOOEt$, $X^2 = C(COOEt)^2$; **31**:$X^1 = C(COOEt)_2$; $X^2 = C(anisyl)_2$; **33**:$X^1 = C(COOEt)_2$; $X^2 = NCOOEt$

For a comparative analysis of the regioselectivities of these two-fold additions to C_{60} (Scheme 9) (Fig. 20) the product distributions show the following characteristics: (1) the product distributions are not statistical (one possibility for a *trans*-1 attack, two possibilities for e'- or e''-attacks and each four possibilities for attack to the other *trans*- or *cis*-positions), (2) in most cases *e*-isomers followed by the *trans*-3-isomers are the preferred reaction products, (3) *cis*-1-isomers are formed only if the steric requirement of the addends allows their suitable arrangement in such a close proximity (e.g., at least one imino bridge chain is required, which unlike methano bridges contains only one flexible side chain), (4) together with the *e*-isomers the *cis*-1-adducts are the major products

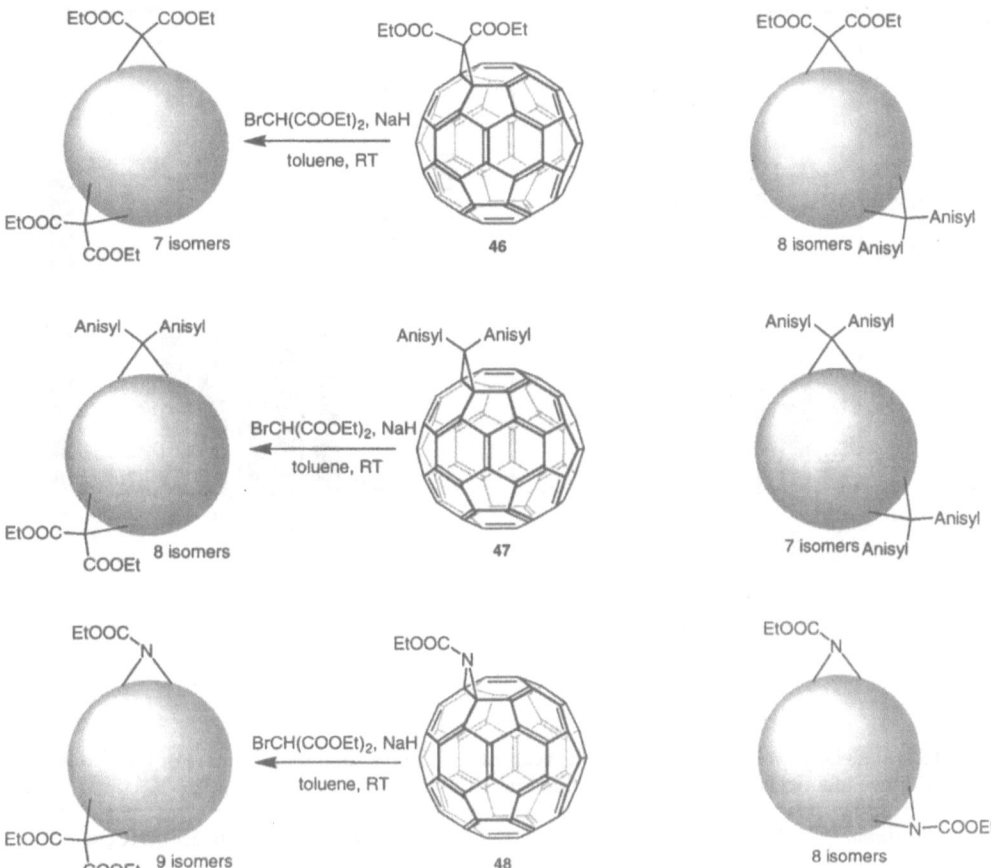

Scheme 9. Synthesis of regioisomeric bisadducts of C$_{60}$

if their formation is possible at all, (5) an attack to an e″-position is slightly preferred over an attack to an e′-position, and (6) the regioselectivity of bis-adduct formation is less pronounced if more drastic reaction conditions are used (e. g., less regioselectivity for nitrene additions in refluxing 1,1,2,2-tetrachloroethane compared to reactions with diethylbromomalonate at room temperature).

Similar product distributions were observed for two-fold additions of diamines and for the formation of the tetrahydro[60]fullerenes [142].

For an interpretation of these experimental findings it is useful to distinguish between the properties of the reaction products **7–45** (bis-adducts) like their relative stabilities and the geometric and electronic structure of the precursor molecules **46–48** themselves. The AM1-calculated stabilities of various series of bis-adducts are listed in Table 4.

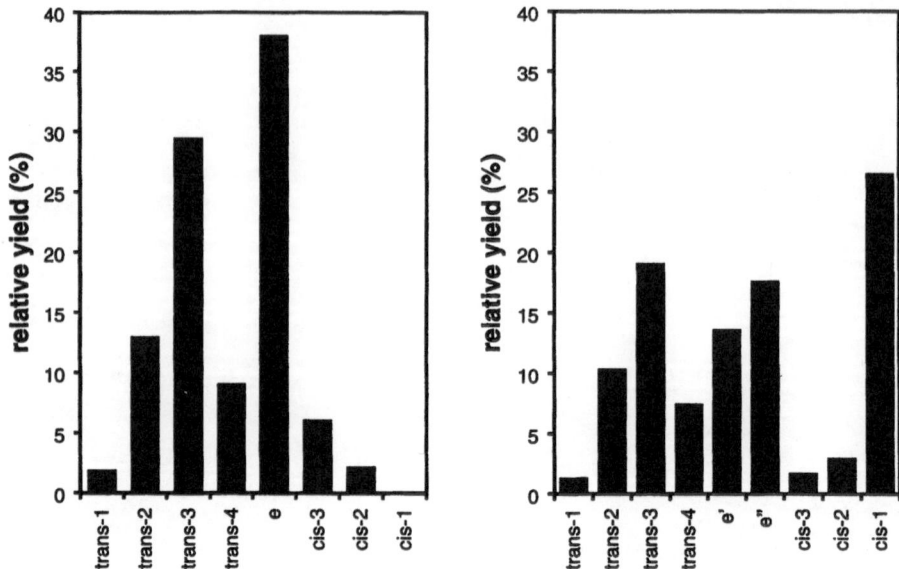

Fig. 20 a, b. Relative yields of the isolated regioisomeric bisadducts of: **a** $C_{62}(COOEt)_4$; **b** $C_{61}(COOEt)_2(NCOOEt)$

Table 4. Relative stabilities ($AM1_{HOF}$) in kcal/mol of the possible regioisomers of bis-adducts with two identical and two different addends [102]

Positional relationship	$C_{62}(COOMe)_4$	$C_{62}(phenyl)_2$	$C_{60}(NCOOMe)_2$	$C_{61}(COOMe)_2(NCOOMe)$
trans-1	0.2	0.2	4.4	1.3
trans-2	0.2	0.3	4.5	1.2
trans-3	0.1	0.2	4.3	1.1
trans-4	0.0	0.1	4.4	1.1
e′	0.0[a]	0.0[a]	4.2[a]	0.8[b]
e″	0.0[a]	0.0[a]	4.2[a]	0.8[c]
cis-3	1.3	2.3	5.9	3.7
cis-2	1.8	3.3	6.7	3.8
cis-1	17.7	24.9	0.0	0.0

[a] e′- and e″-isomers are identical.
[b] e′-isomer referred to $C_{61}(COOEt)_2$ as precursor molecule.
[c] e″-isomer referred to $C_{61}(COOEt)_2$ as precursor molecule.

Two important results arise from these calculations: (1) in bis-adducts with two dialkoxycarbonylmethylene or diarylmethylene groups the cis-1-adducts are considerably less stable than the other isomers, which exhibit very similar AM1 heats of formation, and (2) in bis-adducts with at least one imino addend the cis-1-isomers are not destabilized compared to the others; the cis-1- and to a minor extent the *e*-isomers are slightly more and the *cis*-2- and *cis*-3-isomers slightly less stable than the average. In all cases the *trans*-isomers always have about the same calculated heat of formation. As can be seen from space filling models, the instability of a *cis*-1-isomer like *cis*-1-C_{62}(anisyl)$_4$ (Scheme 9) is due to the pronounced sterical repulsion of the addends leading to considerable deformations of typical bond angles. On the other hand a strain free situation is provided if at least one imino addend is present, since in low energy invertomers unfavorable interactions between the addends are avoided.

Analogous behavior was observed for the various regioisomers of $C_{60}H_4$ [142], where, except for eclipsing H-interactions, no additional strain due to the addends is present. The corresponding *cis*-1-adduct is the major product formed by a two-fold hydroboration followed by hydrolysis. AM1-calculations predict the *cis*-isomers to be somewhat less stable than the *trans*- and e-isomers. According to ab initio calculations (HF/3–21G) the *cis*-1- followed by the e-isomer is the most stable.

In order to evaluate the influence of the geometric and electronic properties of the educts on the observed product distributions a variety of experimental and calculated monoadduct structures were analyzed (Fig. 21, Table 5). For example, as can be seen from the X-ray crystal structure of 47 the average values of the [5,6]-bonds are 1.451 Å and those of the [6,6]-bonds excluding the cyclopropanated [6,6]-bond between C-1 and C-2 1.388 Å [102]. This clearly shows that the [5]radialene-type structure of the fullerene cage is preserved.

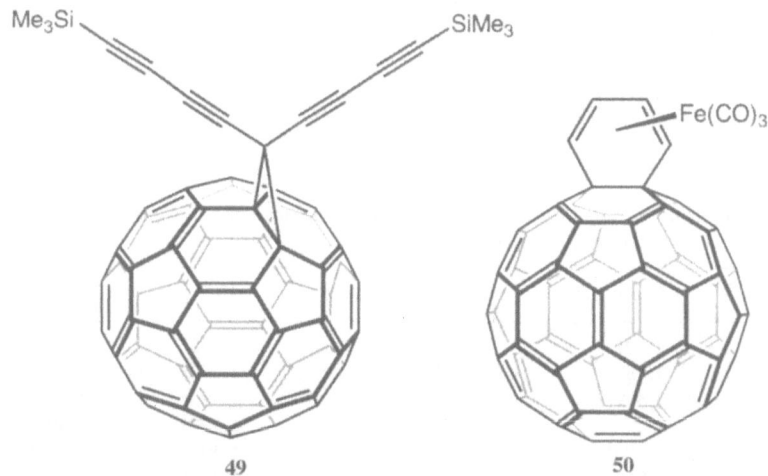

49   50

As can be seen from Table 5 the *cis*-1 and the e″-bonds are the shortest, almost independent of the nature of the addend. As a consequence, the diameters of the carbon cage perpendicular and parallel to the cyclopropane ring differ by 0.17 Å and the sphere is flattened along the axis a⊥ perpendicular to the cyclopropanated double bond. The most pronounced elongation compared to free C_{60} is observed along the axis leading through the poles (a_{pol}). This slight but signifi-

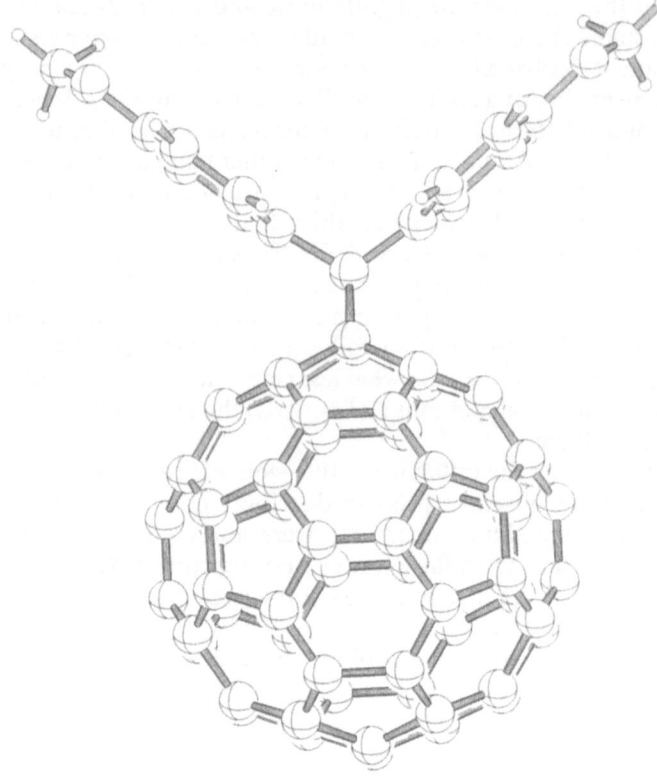

Fig. 21. X-ray structure of **47**

Table 5. Comparative [6,6]-bond length in Å (mean values) of the C_{60} monoadducts determined by X-ray crystallography [102]

Bond type	C(anisyl)$_2$	C$_{61}$(COOEt)$_2$ [25]	49	50
trans-1	1.401	1.384	1.385	1.381
trans-2	1.388[a]	1.391[a]	1.392[a]	1.389[a]
trans-3	1.399[a]	1.395[a]	1.393[a]	1.399[a]
trans-4	1.388[a]	1.386[a]	1.391[a]	1.391[a]
e'	1.396[b]	1.394[b]	1.394[b]	1.385[b]
e"	1.362[b]	1.384[b]	1.379[b]	1.379[b]
cis-3	1.394[a]	1.389[a]	1.391[a]	1.398[a]
cis-2	1.390[a]	1.406[a]	1.393[a]	1.390[a]
cis-1	1.377[a]	1.379[a]	1.378[a]	1.367[a]
C(1)–C(2)	1.625	1.606	1.574	1.586

[a] Mean value out of four bond lengths.
[b] Mean value out of two bond lengths.

cant deviation from the ideal C_{60} symmetry leads to a characteristic droplet-like distortion (Fig. 22).

Qualitatively the same trend is predicted by calculations (AM1). In Fig. 23 the deviations of [6,6]-bond lengths from that of C_{60} are represented for a variety of adducts including **46** and **47**. Another trend that comes out from the analysis of these data is that the *cis*-2- and *cis*-3-bonds are somewhat elongated and that the opposite hemisphere is less disturbed. Similar to the geometrical distortions the polarizations (AM1-Mulliken-charges) of the C-framework in these mono-adducts is somewhat enhanced in the neighborhood of the first addend but essentially zero in the opposite hemisphere.

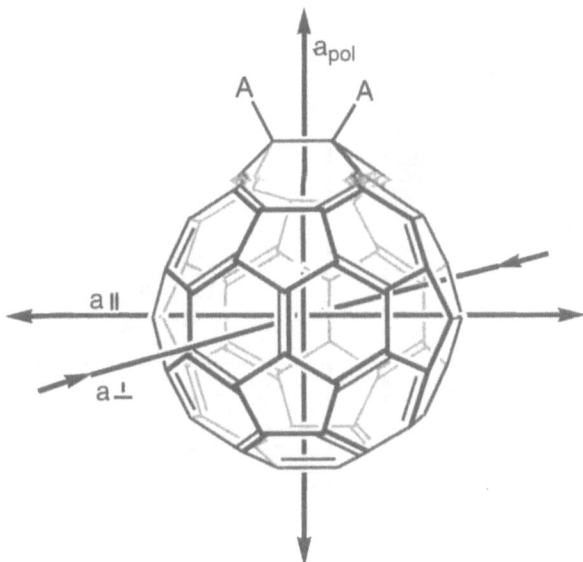

Fig. 22. Schematic representation of the cage distortion in a 1,2-monoadduct of C_{60}

Coefficients of frontier orbitals in monoadducts like $C_{61}(COOEt)_2$ (**46**) at a given [6,6]-bond are presented in Figs. 24 and 25. LUMO coefficients in a precursor molecule are expected to influence the product distribution of nucleophilic additions and the HOMO coefficients those of additions of electrophiles, e. g., nitrenes or carbenes. Such MO-structures exhibit the following characteristics. (1) In a first approximation the distribution of the MO coefficients to specific sites within monoadducts is totally independent of the nature of the addend and of the experimentally observed and calculated deviation of the lengths of [6,6]-double bonds (Table 5, Fig. 23) compared to the average value in free C_{60}. The latter can be demonstrated by quantum mechanical single point calculations on symmetries with equal lengths of the [6,6]-bonds and even with those structures having essentially no bond length alternations between [6,6]- and [5,6]-bonds. (2) The distribution of the MO coefficients to specific sites

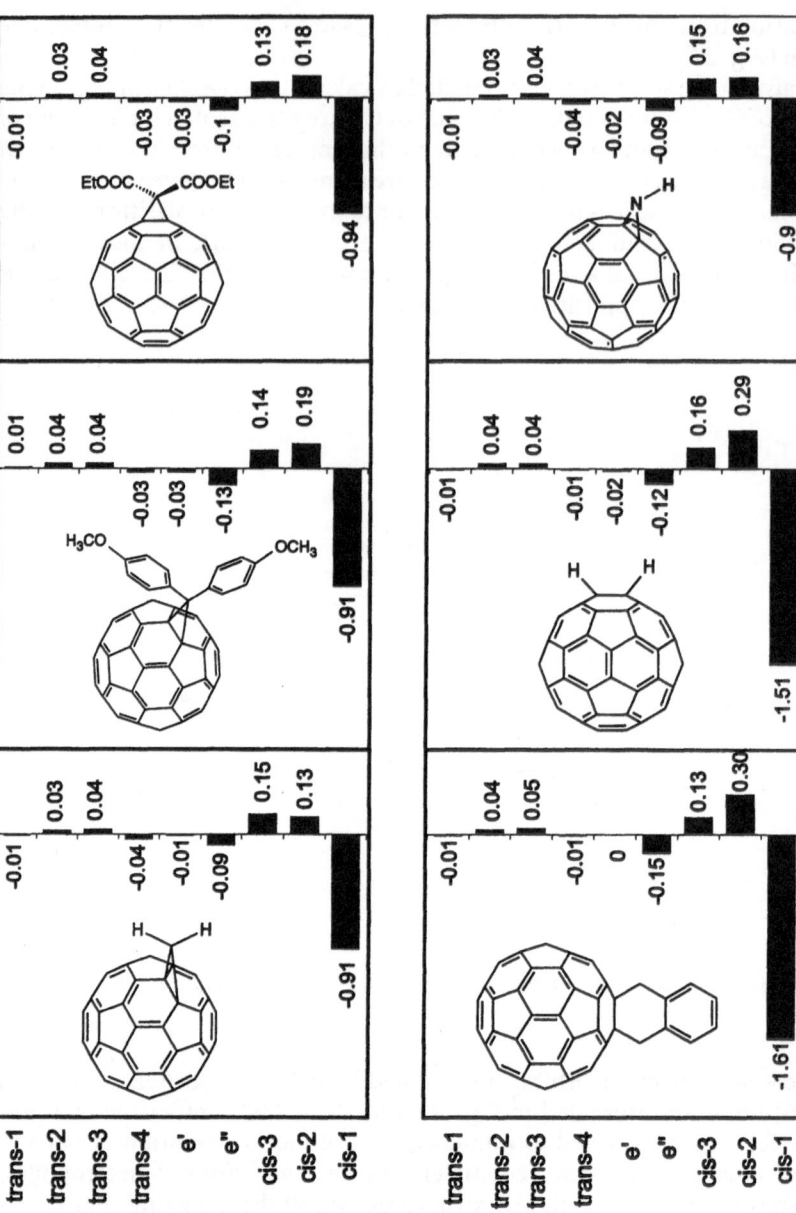

Fig. 23. AM1 calculated [6,6]-bond lengths distortions in pm in 1,2-monoadducts. The distortions are relative to the calculated lengths of the [6,6]-bonds of free C_{60}

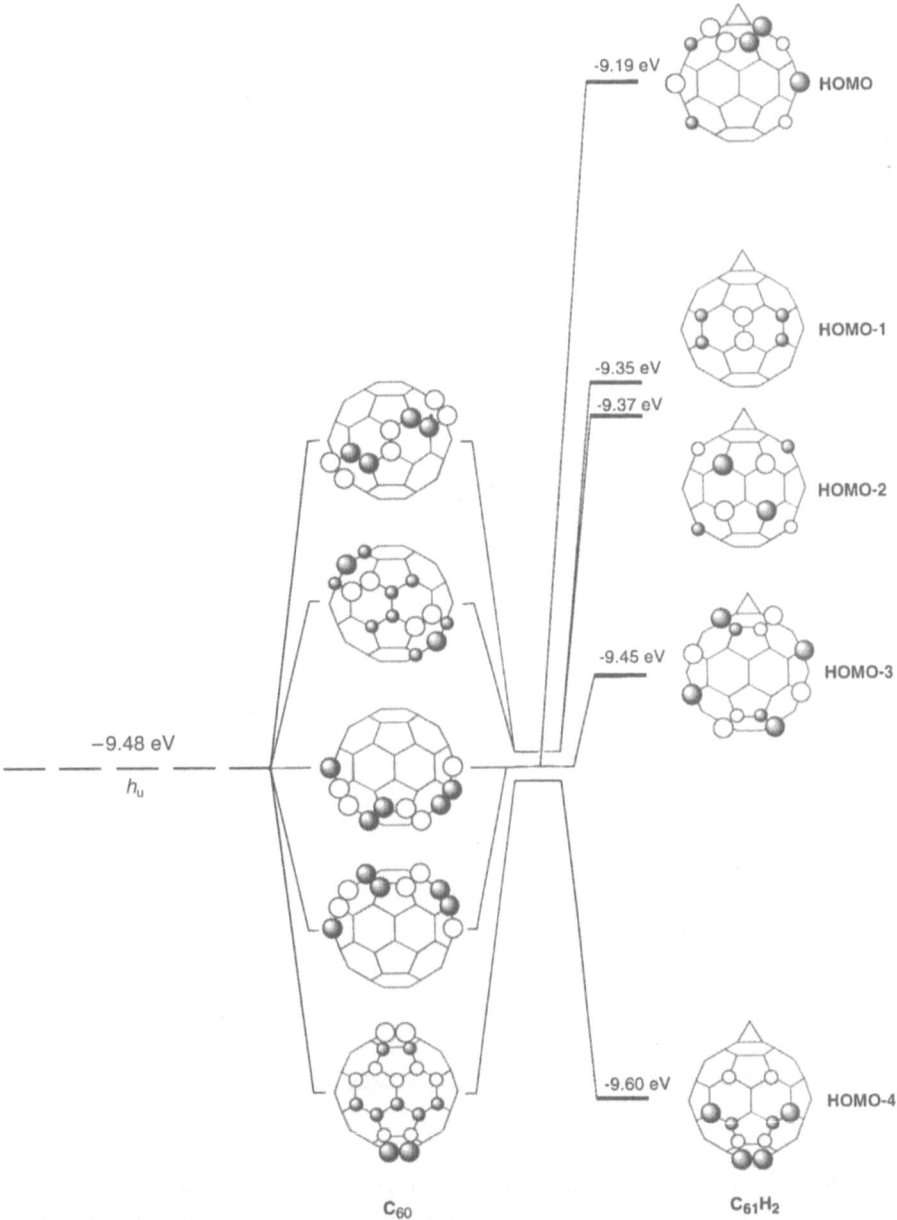

-9.19 eV   HOMO

HOMO-1

-9.35 eV
-9.37 eV

HOMO-2

-9.45 eV   HOMO-3

−9.48 eV
h_u

-9.60 eV   HOMO-4

C_{60} $C_{61}H_2$

Fig. 24. Correlation of the HOMOs of C_{60} with the HOMOs of $C_{61}H_2$

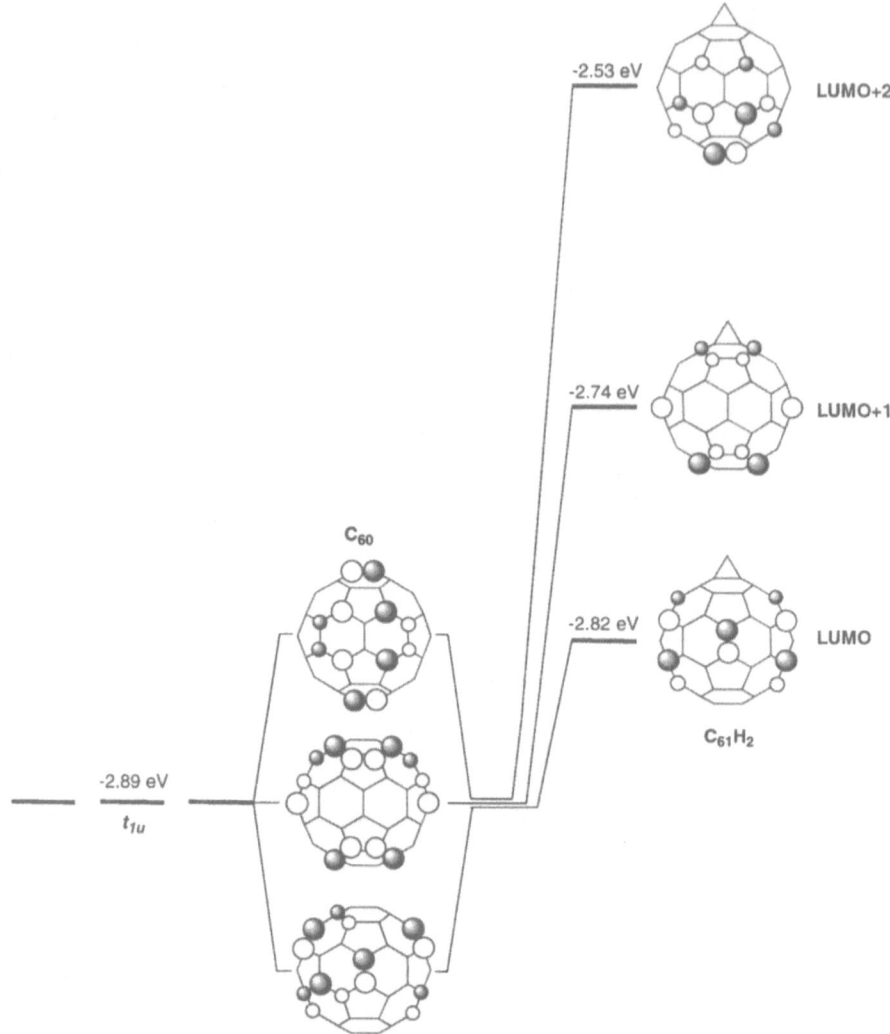

Fig. 25. Correlation of the LUMOs of C_{60} with the LUMOs of $C_{61}H_2$

within monoadducts resembles that within free C_{60} as can be seen from the diagrams correlating the HOMOs and LUMOs of C_{60} with those of a monoadduct (Figs. 24 and 25). The lowest lying HOMO–4 of the monoadduct correlates with that h_u orbital having highest coefficients in two opposing [6,6]-bonds. Relative to this h_u orbital three of the others have high coefficients in the equatorial sites and none in the opposing I and I* (*trans*-1) sites (*trans*-1 effect). As a consequence, since only HOMO–4 correlates with h_u^I the higher lying HOMOs have no coefficients in the *trans*-1 site. In the HOMO highest coefficients are located in the *cis*-1 and e″ site and in the HOMO–1 in the e′ sites. One of the three

t_{1u} orbitals, which are oriented perpendicular to each other, correlates with the LUMO + 2. Due to the perpendicular orientation of the t_{1u} orbitals in the LUMO and LUMO+1 there are no coefficients in *trans*-1 but preferably in the *e*-positions (*trans*-1 effect). In the LUMO pronounced coefficients are also found in *trans*-3 and *cis*-2 sites. Only in the LUMO+2 also the *trans*-1-, *trans*-4- and *cis*-3-bonds exhibit enhanced coefficients.

Although the distribution of the orbital coefficients in the HOMOs is independent of the characteristic distortion of a monoadduct, the differences in [6,6]-bond lengths are influenced by the HOMO coefficients since the shortest [6,6]-bonds are those with the highest coefficients in the HOMO. Pronounced bonding interactions at a binding site in general cause a contraction of the bond length. The fact the *cis*-1 bond is shorter than the *e″* bond although the coefficients in the HOMO are comparable could be due to the removal of cyclic conjugation within the six-membered ring involving the *cis*-1 bonds.

In conclusion, the typical product distributions of two-fold addition to [6,6]-bonds, especially the preferred attacks in *e*- for sterically demanding and *e*- and *cis*-1 sites for sterically less demanding addends, correlate with enhanced frontier orbital coefficients. The preference of *cis*-1 attacks within the series $C_{61}(COOEt)_2(NCOOEt)$ indicate that not only the nitrene additions but also the cyclopropanations should be mainly HOMO-C_{60} controlled. In this case, the cyclopropanation via malonates would be due to addition or carbenes. If it would be due to an initial nucleophilic attack, then the preferred formation of *e*-adducts would also be reflected by the coefficients of the LUMO and LUMO + 1. For the explanation of the pronounced formation of *cis*-1 adducts however additional factors have then to be considered which could be thermodynamic arguments since in the case of sterically non-demanding addends the *cis*-1 adducts followed by the *e*-adducts are the most stable. Another driving force could be simply due to the fact that the *cis*-1 bonds are the shortest having the most double bond character. The extend of pyramidalization (curvature) at a given site, which is a predominant factor governing the preferred attack, for example to C_{70} ([6,6]-bonds at the poles), does not play a role for these successive additions to C_{60}. This can be clearly seen from the fact that the reactive *cis*-1 sites are the least pyramidalized in a mono-adduct.

Among the bisadducts series represented in Fig. 19 the *cis*-1 adducts of $C_{60}(NCOOR)_2$ stand out since they represent the first examples of [6,6]-adducts with open transannular [6,6]-bonds (Fig. 26) [124]. Characteristic features within the fullerene framework of these valence isomers are the presence of (i) a doubly bridged 14-membered ring with a phenanthrene perimeter and (ii) an eight-membered 1,4-diazocine heterocycle. Upon changing the addition pattern, as demonstrated by the investigation of the other possible regioisomers of $C_{60}(NCOOR)_2$, "regular" behavior with closed transannular [6,6]-bonds is observed. The fullerene cage can be reclosed again via an intra-ring Diels-Alder reaction by transferring *cis*-1-$C_{60}(NCOOt-Bu)_2$ into *cis*-1-$C_{60}(NH)_2$ **52** as cluster closed valence isomer. The latter phenomenon clearly demonstrates the role of the addend. An extensive AM1 and DFT study revealed that only in *cis*-1 adducts that prefer planar imino bridges (e. g., carbamates or amides) are the open forms more stable than the closed forms. These are the first chemical modifications

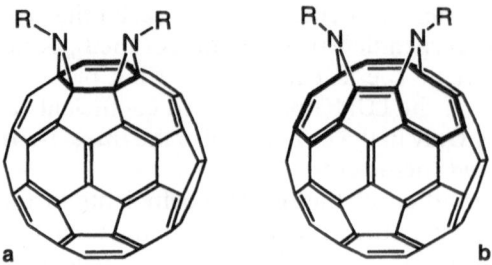

Fig. 26 a, b. Valence isomers of *cis*-1 bisimino[60]fullerenes with: **a** closed; **b** open transannular [6,6]-bonds

of the fullerene core, which allow the synthesis of open as well as closed valence isomers with the same addition pattern. The following conclusions can be deduced.

1. With the exception of a *cis*-1 adduct, whereupon the two-fold ring opening due to the location of the imino bridges in the same six-membered ring require only three [5,6]-double bonds to be introduced, six of these energetically unfavorable bonds would be required for the hypothetical open structures of the other seven regioisomers (*trans*-1 to *cis*-2). In the latter regioisomers the transannular [6,6]-bonds are always closed.
2. In a *cis*-1 adduct a closed valence isomer bears a strained planar cyclohexene ring but the introduction of an unfavorable [5,6]-double bonds is avoided, whereas in an open valence isomer no strained planar cyclohexene but three [5,6]-double bonds are present.
3. For the *cis*-1 adducts open valence isomers are favored for imino addends with planar imino bridges like carbamates and the closed isomers are favored for imino addends with pyramidalized imino bridges like alkylimines or NH.
4. Carbamates or amides prefer planar arrangements of the nitrogen due to resulting favorable conjugation of the free electron pair with the carbonyl group. This has consequences for imino[60]fullerenes, since the planar arrangement of the carbamate N-atoms and the required enlargement of the bond angles between C-1,N,C-2 or C-3,N,C-4 are most favorably realized if the transannular [6,6]-bonds are open.

As an example for regioselectivity investigations on multiple additions to C_{60}, three- to six-fold cyclopropanations with diethyl bromomalonate are now discussed as model cases [178,189]. Using this cyclopropanation next to the seven bisadducts a large variety of tris- up to hexakisadducts of C_{60} were isolated and completely characterized. Among those adducts are the examples shown in Fig. 27. Each compound was synthesized from its direct precursor adduct.

The cyclopropanation of the I,eI- and I,III*-bisadducts 25 and 15 proceeds with an even more pronounced regioselectivity than that of 46. Cyclopropanation of 25 yields 40% of the C_3-symmetrical I,eI,eII-trisadduct 52 with all

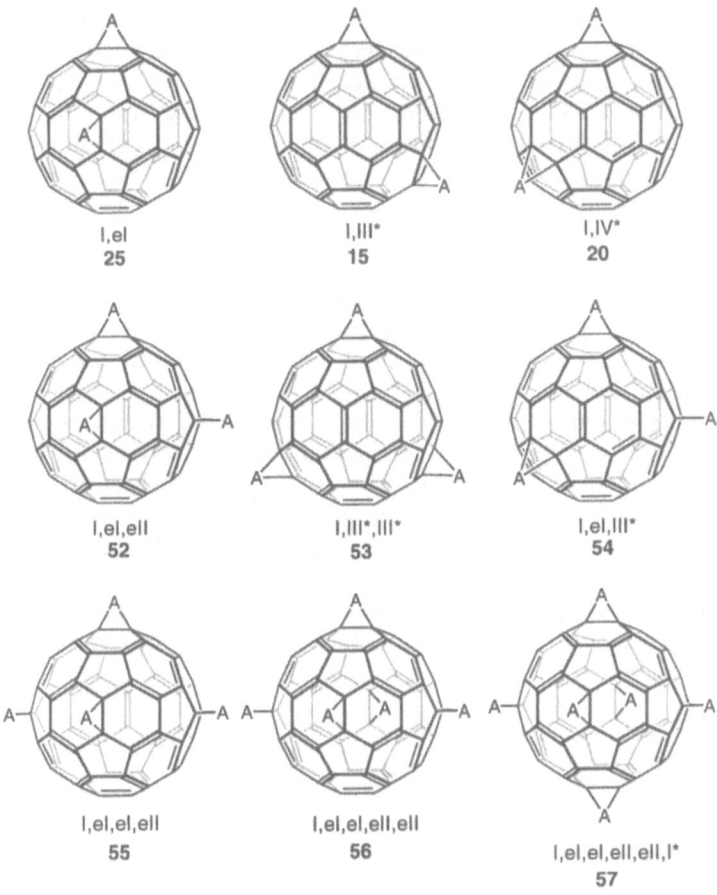

I,el **25**	I,III* **15**	I,IV* **20**
I,el,ell **52**	I,III*,III* **53**	I,el,III* **54**
I,el,el,ell **55**	I,el,el,ell,ell **56**	I,el,el,ell,ell,I* **57**

A = C(COOEt)₂

Fig. 27. Examples of isolated bis- to hexakisadducts of C_{60}

addends in *e*-positions, and the major product of the cyclopropanation of **15** is the D_3-symmetrical I,III*,III*-trisadduct **53** with all addends in *trans*-3 positions.

In addition to these two trisadducts the another characterizable C_1-symmetrical trisadduct **54** was isolated in fairly high yields. Due to the C_1 symmetry it is not possible to deduce its structure with spectroscopic methods. However, the structure of **54** can be assigned unambiguously considering its possible routes of formation. Since it can be synthesized by starting from **25**, **15** and the I,IV*-isomer **20**, the three relative positional relationships *e*- (*e'* or *e"*), *trans*-4 and *trans*-3 must be involved in its structure. This leaves only one possibility for its structure, namely the pair of enantiomers (^fC)- and (^fA)-I,eII,III*-tris[di(ethoxycarbonyl)methano]hexahydro[60]fullerene. The relative yields of **54** are 9% (based on **25**), 15% (based on **15**), and 39% (based on **20**).

In addition to *cis*-1 positions, which are forbidden for sterically demanding addends, the I,eI-bisadduct has enhanced HOMO and LUMO coefficients in e- and *trans*-3 positions, especially in the I,eI,eII-sites, which corroborates the preferred formation of **25** from **46** (Fig. 28). Enhanced coefficients in the I,eI,III*-site are present only in the HOMO–1 and the LUMO + 1. This accounts for the formation of **54** out of **25** in only 9% relative yield. As for the monoadduct **46** the AM1 optimized molecular structure of I,eI-$C_{62}H_4$ exhibits the significantly shortest [6,6]-double bonds, with an average length of 1.377 Å in *cis*-1-positions relative to the two addends. Next to these the shortest double bonds are those in the I,eI,eII-sites, with an average length of 1.384 Å.

MO calculations of the model compounds I,III*-$C_{62}H_4$ and I,IV*-$C_{62}H_4$ corroborate the facile formation of **53** from **15** and of **54** from **15** and **20** as orbital-allowed processes, respectively. A $C_{60}H_6$ isomer with the same all *trans*-3 I,III*,III* addition pattern was found as the preferred adduct upon the hexahydrogenation of C_{60} via Zn-Cu acid reduction [142].

Further cyclopropanation of the trisadduct **52** proceeds with high regioselectivity with formation of only two tetrakisadducts $C_{64}(COOEt)_8$ in the ratio of 1:2. The first product is C_1-symmetrical and cannot be characterized unambiguously; the second product is the C_s-symmetrical I,eI,eI,eII-tetrakisadduct **55**, which is formed predominantly in a relative yield of 64%. The precursor model I,eI,eII-$C_{63}H_6$ of **55** again has pronounced orbital coefficients in the remaining e- positions, considering that most of the other sites with enhanced coefficients are unproductive *cis* positions (Fig. 29). Attack at either of the remaining e-positions of **52** leads to **55**.

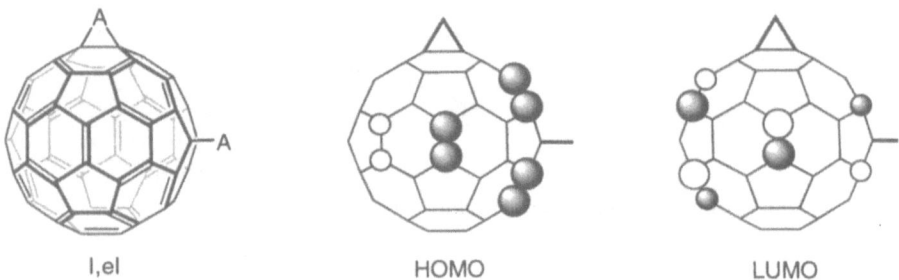

Fig. 28. Frontier orbitals of I,eI bisadducts

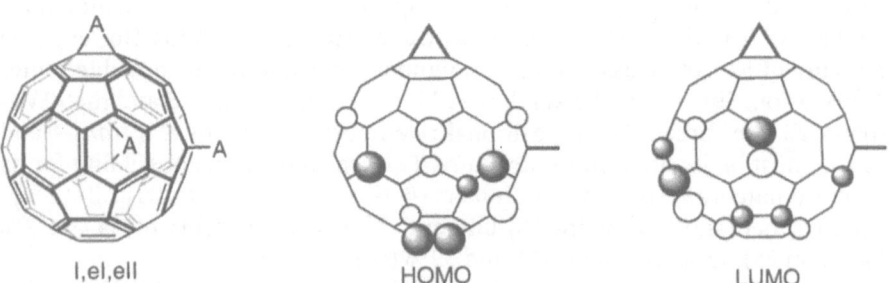

Fig. 29. Frontier orbitals of I,eI,eII trisadducts

The frontier orbitals of **55** again exhibit highest coefficients in the two remaining equatorial sites, especially if unproductive *cis*-1 bonds are not considered (Fig. 30). Consequently, the cyclopropanation of **55** proceeds with very high regioselectivity with formation of the C_{2v} symmetrical I,eI,eI,eII,eII-adduct **56** as the only regioisomeric pentakisadduct.

The T_h-symmetrical I,eI,eI,eII,eII,I*-hexakisadduct **57** was formed as the only regioisomer out of **56** and no higher adducts were observed. In this case the structures of the HOMO and the LUMO of **56** clearly show that I* is essentially the only accessible bond of the attack of both sterically demanding electrophiles and nucleophiles (Fig. 31). For both types of addition to **56** a hexaadduct with a I,eI,eI,eII,eII,I*-addition pattern was always formed as the only, or by far the predominant, product.

Hexakisadducts with an I,eI,eI,eII,eII,I*-addition patterns having identical or different addends can also be synthesized in high yield in one step starting from C_{60} or monoadducts via a template mediated reaction using 9,10-dimethylanthracene as reversibly binding addend. Since the regioselectivity for subsequent attacks into equatorial sites increases with the number of bound addends, mixed hexakisadducts starting from precursor with addends attached to octahedral sites are also easily accessible on a preparative scale. T_h-symmetrical hexaadducts were also synthesized via six-fold Diels-Alder reactions in comparatively high yields by allowing an access of dienes to react with C_{60}. However, the latter results imply that reversible reactions are involved, allowing for thermodynamic control.

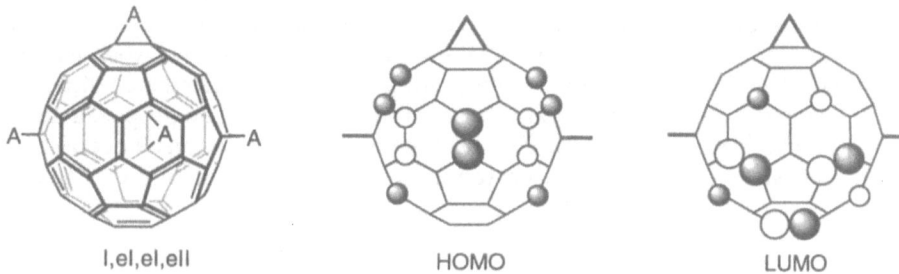

I,eI,eI,eII   HOMO   LUMO

Fig. 30. Frontier orbitals of I,eI,eI,eII tetrakisadducts

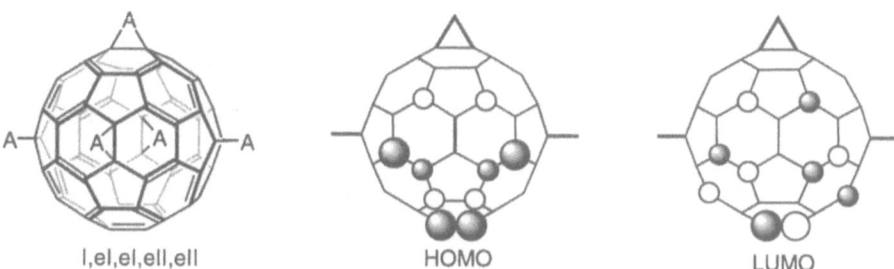

I,eI,eI,eII,eII   HOMO   LUMO

Fig. 31. Frontier orbitals of I,eI,eI,eII,eII pentakisadducts

Thermodynamic arguments increasingly gain importance for the regioselectivity of subsequent additions as the number of the addends already bound increases. Hexaadduct **57** is stabilized by at least 5 kcal/mol compared to all the other 43 regioisomeric hexakisadducts which can be formed from **52** without introducing unfavorable *cis*-1 additions [189]. This corresponds with the reduced reactivity of the higher adducts towards diethyl bromomalonate. The cyclopropanation of **52** requires a reaction time of four days to obtain sufficient amounts of products, compared to the formation of **25** and **52** requiring only a few hours. As a consequence of the decrease of the reaction rate with an increasing number of addends bound to the fullerene core, the formation of the T_h symmetrical I,eI,eI,eII,eII,I*-hexakisadduct **57** with octahedral addition pattern requires an excess of 12 equivalents of diethyl bromomalonate and a reaction time of seven days in order to obtain acceptable yields. Due to the low reaction rate it can be assumed that the reaction **56** affording **57** is controlled thermodynamically as well.

The single crystal structure of **57** confirms computational results. Very remarkable is the bonding in the remaining π-electron system which is a new type of an oligocyclophane (Fig. 32). The alternation of bond lengths between [6,6]- and [5,6]-bonds within the benzenoid rings is reduced by half to approximately 0.03 Å compared to the parent C_{60}. Relative to solutions of C_{60} (purple) or its adducts $C_{61}(COOEt)_2$ to $C_{65}(COOEt)_{10}$ (red to orange) the light yellow solutions of **57** show only weak absorptions in the visible region of the spectrum. Hexaadducts with other addends and mixed adducts have similar optical properties.

In all calculated structures of the model compounds depicted in Fig. 27 the [6,6]-double bonds in *cis*-1 positions relative to the addends already bound are the significantly shortest and have a length of about 1.375 Å. The average bond length of the other [6,6]-double bonds is about 1.386 Å. The calculated difference between the latter is comparatively small in most cases. The proposed additional shortening of the *cis*-1 bonds due to the removal of electron delocalisation within the six-membered ring is corroborated by an elongation of the AM1-cal-

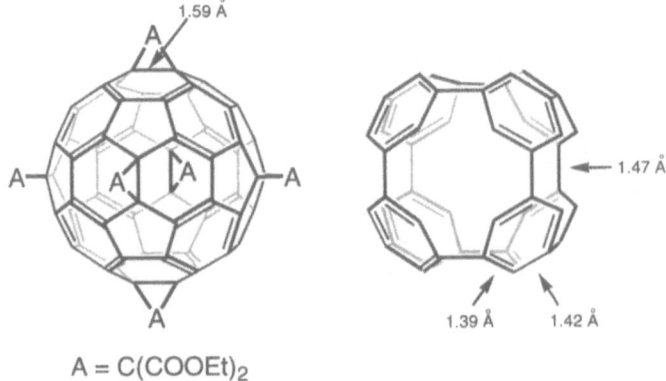

A = C(COOEt)₂

Fig. 32. Schematic of characteristic bond lengths and of the remaining benzenoid π-electron system of **57**

culated [5,6]-bonds within the six-membered ring. In **46** the A-bonds have a length of 1.50 Å and the C-bonds a length of 1.48 Å; the corresponding values for C_{60} are 1.46 Å. Hence, the bond length alternation between [6,6]-"double bonds" and [5,6]-"single bonds", which is already present in C_{60}, is considerably enhanced in addend carrying six-membered rings of [6,6]-adducts.

3.2.3
The Vinylamine Mode

The introduction of two [5,6]-aza bridges shows a remarkable regioselectivity even if segregated alkylazides are used [125]. The bisiminofullerenes **60** (Scheme 10) are by far the major products and only traces of one other bisadduct with unidentified structure are found, if, for example a two-fold excess of azide is allowed to react with C_{60} at elevated temperatures [125]. To obtain clues on the mechanism of this most regioselective bisadduct formation process in fullerene chemistry a concentrated solution of an azafulleroid precursor **58** was treated with an alkyl azide at room temperature.

Under these conditions only one mixed [6,6]-triazoline-[5,6]-iminofullerene isomer **59** was formed which is explained by the fact that **58** behaves as a strained electron-poor vinylamine. The significantly highest Mulliken charge of 0.06 (AM1) is located at C-1 and C-6, and the lowest of –0.07 at C-2 and C-5 (Fig. 33). The most negatively polarized N-atom of the azide (AM1) is that bearing R. A kinetically controlled attack of the azide, therefore, leads predominantly to **59**.

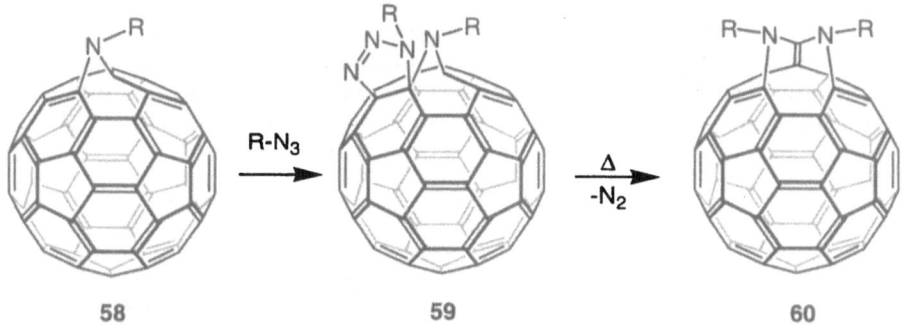

Scheme 10. Regioselective formation of bisazafulleroid **60**

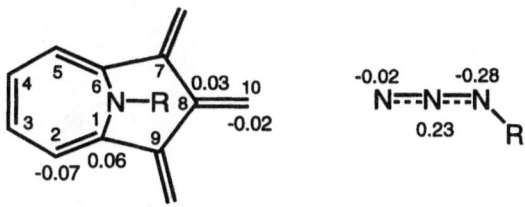

Fig. 33. AM1 Mulliken charges near the imino bridge of **58** and in N_3R (R = CH_2COOMe)

The further ring expanded doubly bridged bisiminofullerenes **60** exhibit three seven-membered rings and one eleven-membered ring within the fullerene cage. One C-atom is already halfway decoupled from the spherical carbon core. The pronounced relative reactivity of the vinylamine type double bonds within [5,6]-bridged iminofullerenes was also demonstrated by the facile addition of mild nucleophile such as water and amines. For example, the first stable fullerenol was synthesized by treatment of a toluene solution of **58** in the presence of water and neutral alumina and an almost quantitative reaction. This finding that the reactivity of the vinylamine type [6,6]-double bonds in **58** is dramatically enhanced over the remaining [6,6]-bonds turned out to be a key for further cluster opening reactions and the formation of nitrogen heterofullerenes. Subsequent attacks to [6,6]-bonds within [6,6]-adducts and also to [6,6]-bonds within the [5,6]-bridged methanofullerenes are by far less regioselective [3,4].

Bis-[5,6]-bridged iminofullerenes with another addition pattern can only be obtained in good yields if a tether directed synthesis is applied. If the tether between two azide groups is rigid enough than the second addition is forced to occur at specific regions of the fullerene cage [128].

3.2.4
The Cyclopentadiene Mode

Upon the reaction of C_{60} with a given excess of free radicals [97,98], amines [63], iodine chloride [166] or bromine [165], specific organocopper [74] and organolithium reagents [73] the formation of substituted 1,4,11,14,15,30-hexahydro-[60]fullerenes is the predominant or almost exclusive process (Fig. 34). In these hexa- or pentaadducts of C_{60} the addends are bound in five successive 1,4- and one 1,2-positions. The corresponding corannulene substructure of the fullerene core contains an integral cyclopentadiene moiety whose two [5,6]-double bonds are decoupled from the remaining conjugated π-electron system of the C_{60} core.

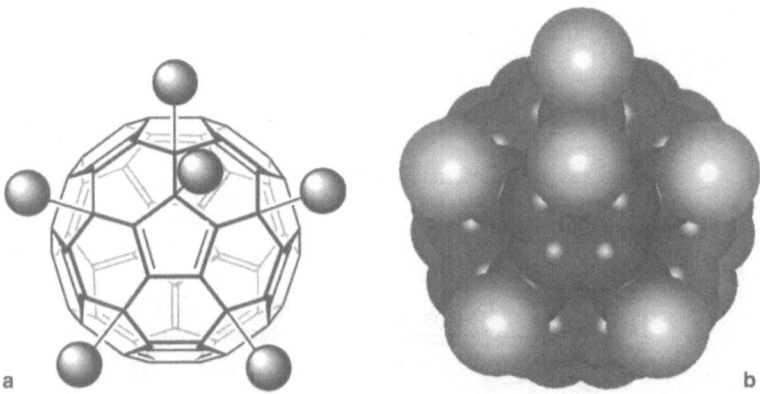

a   b

Fig. 34a, b. Schematic representation of: **a** the VB structure of 1,4,11,14,15,30-hexahydro-[60]fullerenes; **b** space filling model of the hexachloro derivative

This characteristic structure type is completely different from those polyadducts of C_{60} described in the preceding paragraphs. In many cases the detailed mechanisms of these polyadduct formations remain unclear, although it can be assumed that sequential additions of radicals or nucleophiles like the cuprate Ph_2Cu^- are involved.

Valuable insight of multiple additions of free radicals came from the ESR spectroscopic investigations of benzyl radicals, ^{13}C-labeled at the benzylic positions [97, 98]. These radicals can be prepared in situ by photolysis of saturated solutions of C_{60} in labeled toluene containing about 5% di-*tert*-butyl peroxide. Thereby, the photochemically generated *tert*-butoxy radicals readily abstract a benzylic hydrogen atom from the toluene. Two radical species with a different microwave power saturation behavior can be observed. One radical species can be attributed to an allylic radical **63** and the other to a cyclopentadienyl radical **65** formed by the addition to three and five adjacent [5]radialene double bonds, respectively (Scheme 11). In these experiments no evidence for the radical **61** is found, which is very likely a short-lived species.

The ESR spectra of **63** and **65** do not provide information on whether or not the corresponding radical species carry an even number of benzyl groups attached elsewhere on the surface of the C_{60} molecule. Looking at the high reactivity of radicals towards C_{60}, it is very unlikely that only single radical adducts, such as R_3C_{60} or R_5C_{60}, are responsible for the observed ESR spectra. But very important information about the electronic and chemical properties of a corannulene substructure are extractable from these experiments. The orbitals with a

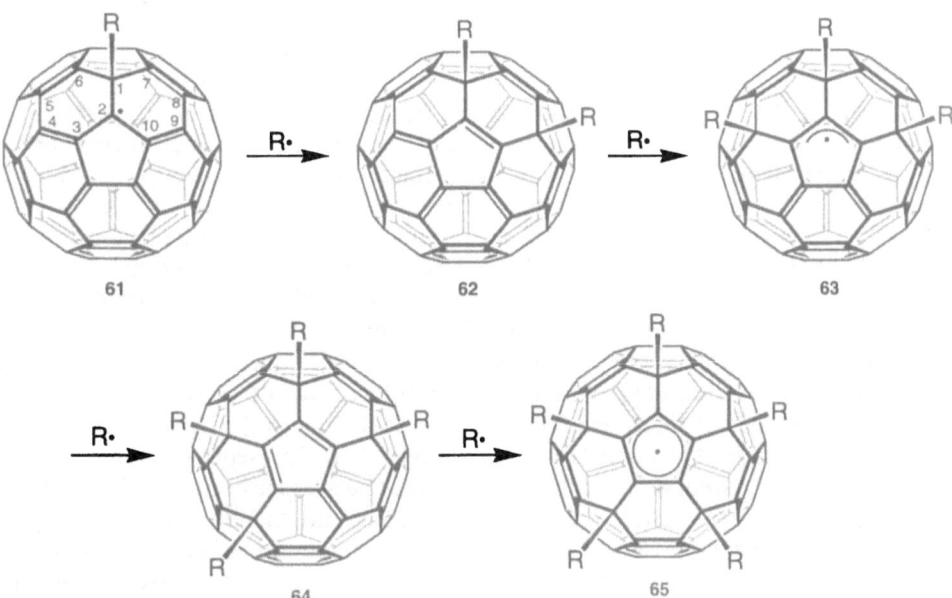

Scheme 11. Successive addition of sterically demanding radicals to C_{60} – the cyclopentadiene mode

spin density in radical species containing a moiety **65** are highly degenerate due to the local five-fold symmetry. The formation of species **63** and **65** from C_{60} proceeds by initial addition of one benzyl radical leading to the monoadduct **61**. The unpaired spin in this radical is mostly localized on carbon 2, (4,11) and (6,9) (Fig. 11). This electronic localization as well as the steric requirement of the benzyl group will direct a second attack accompanied with a radical recombination to position 4 or 11. A third attack to the diamagnetic **62** can occur anywhere on the fullerene surface. However, upon the formation of **63** the unfavorable [5,6]-double bond in **62** disappears and a resonance stabilized allyl radical is formed. Significantly, the formation of the intermediate is also an orbital controlled process, since high HOMO coefficients are located in the γ positions whose attack leads to adducts like **63** and **64**, respectively. Moreover, semiempirical calculations on two different adducts $H_3C_{60}^{\cdot}$, **66** and **67**, show that **66**, in which the three groups are all added in adjacent 1,4-positions, is about 8.5 kcal/mol lower in energy than **67**, where the groups are added in non-adjacent positions.

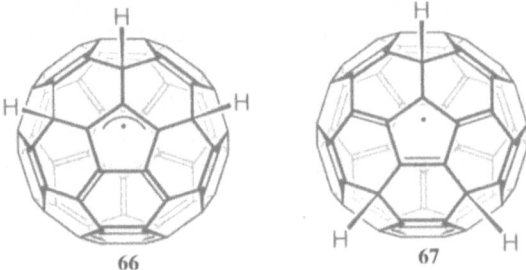

A radical $R_3C_{60}^{\cdot}$, in which a third attack occurred far away from the groups already attached in $C_{60}R_2$ (**62**), would be very unstable due to a facile formation of a diamagnetic $C_{60}R_4$, which is not detectable in the ESR spectra. Another reason for the stability of **63** is the steric protection provided by the three attached benzyl groups. The radical recombination of **63** with a benzyl radical leads to **64** for the same reasons for which **62** is formed from **61**. The stability of **64** is restricted due to the two unfavorable [5,6]-double bonds, which directs another attack to form the resonance stabilized **65**.

A stable representative of an intermediate **64** was obtained by Komatsu et al. [73] upon the reaction of C_{60} with potassium fluorenide in the presence of neutral fluorene without rigorous exclusion of air in 40% yield (Scheme 12).

The single-crystal structure of **68** clearly reflects the presence of the integral fulvene-type π-system on the spherical surface. The average bond length for the [5,6]-bond between C1-C2 and C3-C4 is 1.375 Å and therefore considerably shorter as a typical [5,6]-bond in C_{60}. In contrast, the bond between C2 and C3 (1.488 Å) exhibits a prominent elongation. The bond length of the fulvene C5-C6 bond (1.350 Å) is the shortest reported for C_{60} derivatives. This tendency for the bonds to alternate in length is comparable to that observed in fulvene derivatives. The typical fulvene reactivity towards nucleophiles was demonstrated

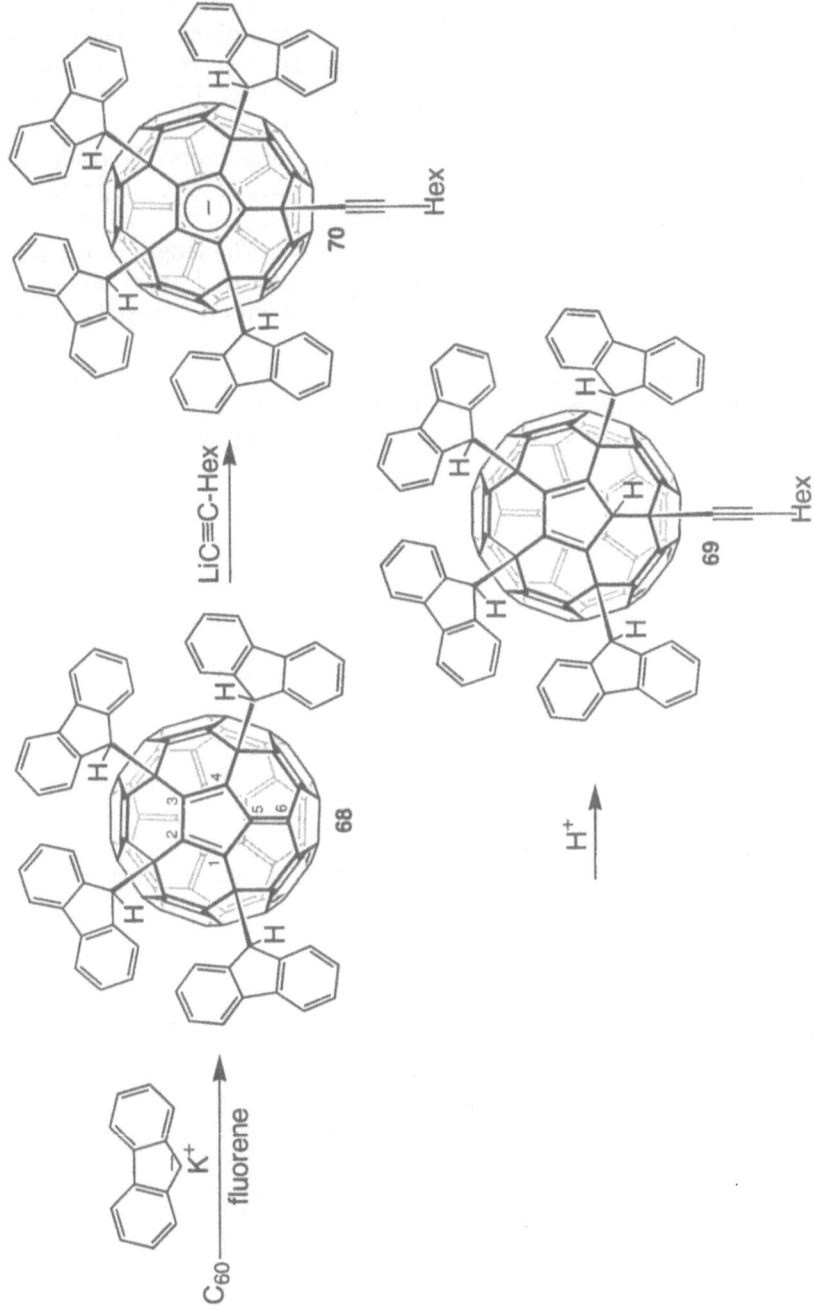

Scheme 12. Selective formation of C$_{60}$ adducts with integral fulvene, cyclopentadienide and cyclopentadiene moieties

with the reaction of **68** with 1-octynyllithium. After quenching with acid hexaadduct **69** the cyclopentadiene addition pattern was obtained. The reaction intermediate is the cyclopentadienide **70** (Scheme 12). The results of this beautiful preparative fullerene chemistry support the earlier ESR investigations (Scheme 11).

A pentahaptofullerene metal complex **71** was obtained upon the reaction of the hexaadduct **72** with TlOEt demonstrating the acidity of the cyclopentadiene (Scheme 13) [74]. A single crystal investigation confirmed the cyclopentadienide character of the complexed five-membered ring. The length of the [6,6]- and [5,6]-bonds in the remaining C_{50} moiety are similar to those of monofunctionalized C_{60} derivatives which is also demonstrated with the crystal structure of **68**.

The analogous heterocyclopentadiene addition pattern was obtained by the reaction of the heterofullerene monoadducts **73** with ICl. In this case a pyrrole substructure within the adduct **74** is decoupled from the remaining 50 π-electron system (Scheme 14) [192].

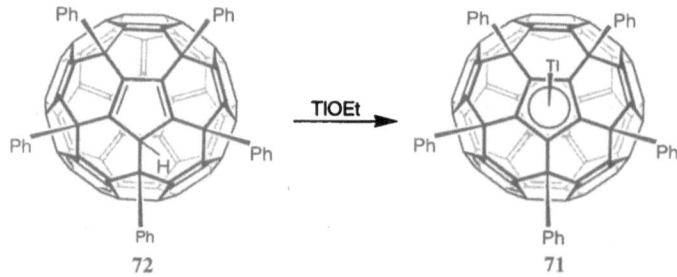

Scheme 13. Formation of a cyclopentadienide complex of a C_{60} derivative with Tl

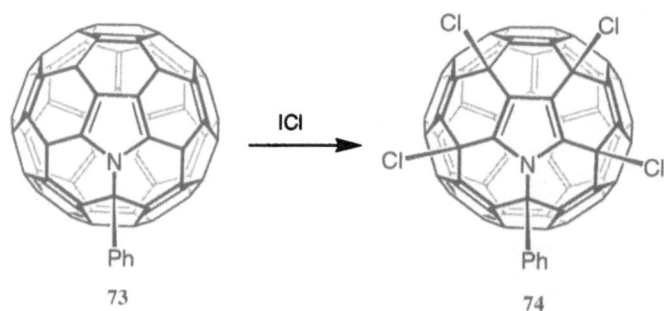

Scheme 14. Formation of a heterofullerene derivative with an integral pyrrole substructure

4
Inside Reactivity

The spherical structure of fullerenes allows exohedral and endohedral chemistries to be distinguished. As demonstrated in the preceding paragraphs the covalent exohedral chemistry of C_{60} has been well established over the last few years. For this type of chemistry, which is mainly addition chemistry, the fullerenes are certainly more reactive than planar aromatics because an important driving force for such addition reactions is the reduction of strain, which results from pyramidalization in the sp^2-carbon network. For the investigation on the potential covalent chemistry taking place in the interior of a fullerene access to suitable endohedral model systems is required. Most of the endohedral fullerenes known so far are either metallofullerenes [193, 194] where electropositive metals are encapsulated or endohedral complexes involving a noble gas guest [54, 55]. However, these prototypes do not serve as model systems for investigations on endohedral covalent bond formation since the electropositive metals transfer electrons to the fullerene shell, leading to components with ionic character, and due to their inert nature the noble gas hosts do not undergo any reaction with the fullerene guest at all. In contrast to these guests, reactive non-metal atoms or a reactive molecular species like a methyl radical would be more instructive probes for screening the chemical properties of the inner concave surface of C_{60}.

An ideal example for this scenario was recently provided by Weidinger et al. [11] with the endohedral fullerene N@C_{60} (2). For the production of N@C_{60} the empty fullerenes, during deposition on a substrate, were bombarded with low energy nitrogen ions. (Fig. 35) [11, 195, 196c]. The simultaneous deposition and bombardment makes it possible that freshly deposited fullerenes are always exposed to an ion beam. After bombardment for several hours the fullerene film can be scratched from the target, dissolved in toluene and filtered. The soluble material consists of a mixture of N@C_{60} and empty fullerenes in typical ratios of 10^{-4} to 10^{-5}.

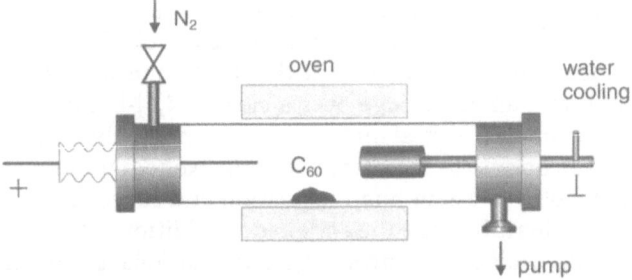

Fig. 35. Schematic view of the glow discharge reactor for the production of N@C_{60}A quartz tube with two electrodes is continuously flushed with nitrogen gas. The middle part of the tube in which C_{60} powder is deposited is surrounded by an oven. Typical working parameters for the glow discharge are: $U = 500$ V, $l = 1$ mA and $p_{N2} = 0{,}1$ mbar. The C_{60} fullerenes are sublimed at 450 °C and condense on the water cooled cathode where they are bombarded by nitrogen ions from the glow discharge

The analysis of the ESR and ENDOR spectra provided convincing evidence that nitrogen is in its atomic $^4S_{3/2}$ state and that no charge transfer occurs [11, 195]. In addition the absence of an electric field gradient (EFG) indicates that the N-atom is located in the center of C_{60}. Both the fine structure and the quadrupole interaction are zero. The only configuration consistent with this high symmetry is atomic. These results reveal an astonishing and unprecedented situation: The N-atoms do not form covalent bonds with C-atoms of the fullerene cage, showing that in contrast to the outer, the inner concave surface is extremely inert. In other words the reactivity of a carbon network depends on its shape.

This conclusion was corroborated by a variety of semiempirical (PM3-UHF/RHF) and density functional calculations(UB3LYP/D95//PM3) on the system N/C_{60} [196]. The structure of N@C_{60} with nitrogen in the center of the cage is the global minimum of the endohedral complexation (Table 6). The formation of N@C_{60} from the free compounds is more or less thermoneutral. The Coulson and Mulliken charges (PM3 and DFT) of the nitrogen are both zero. The spin density is exclusively localized at the nitrogen. The fact that no charge transfer from the nitrogen to the fullerene takes place is due to the low, even slightly negative electron affinity of N ($E_A = -0.32$ eV) which is comparable to that of He ($E_A = -0.59$ eV). The transition from the quartet to the doublet state of the encapsulated N-atom needs 14 kcal mol^{-1} (PM3-UHF).

With a fixed cage-geometry the energy increases continuously when the N-atom moves from the center to the cage, independent of the way of approach (Fig. 36). In contrast, there are strong attractive interactions if the N-atom approaches a C-atom, a [6,6]-bond or a [5,6]-bond from the outside.

This result can be explained with the pyramidalization of the C-atoms of C_{60} and the reduced p-character of the π-orbitals. Due to electron pair repulsion the charge density on the outside of the fullerenes is higher than on the inside. This implies that, in contrast to the reactive exterior, the orbital overlap with an N-atom is essentially unfavorable inside C_{60} and at the same time there is a repulsion of the valence electron pairs (Fig. 37).

If the fullerene cage is allowed to relax during the various approaches of the N-atom to the cage, there are local minima in the endohedral case corresponding to the covalently bound structures with a C-N bond, an aza-bridge over a closed [6,6]-bond and an aza-bridge over a closed [5,6]-bond (Fig. 38). These local minima are about 20–50 kcal mol^{-1} higher in energy than the global minimum of N@C_{60}. In contrast to the corresponding exohedral binding no energy gain is accompanied by covalent bond formation and subsequent relaxation of the cage geometry. This is because the endohedral addition at the relatively rigid cage structure of C_{60} leads to the introduction of additional strain. In the case of the corresponding exohedral adducts the opposite effect is operative. Here, the cage geometry supports the formation of normal, almost strain-free geometries in the region of the addend (Table 6). In the case of endohedral additions the considerable σ-strain of the fullerene cage would be enlarged, but if the addition is exohedral the strain of the cage would decrease. The tendency of the outside surface of fullerenes to undergo addition reactions easily is confirmed by many

Table 6. Heats of formation, spin densities, total and relative energies of the exo- and endohedral N/C_{60} complexes [196]

Compound	Position of N	Heat of formation [kcal mol^{-1}]	PM3/UHF			UB3LYB/D95*//PM3	
			Relative heat of formation[b] [kcal mol^{-1}]	Spin density	Total energy [-a.u.]	Relative energy[b] [kcal mol^{-1}]	Spin density
NC_{60} (exo)	[5,6]-aza-bridge[a] open	828.3	-83.9	6.12	2341.02492	-50.4	0.776
NC_{60} (exo)	[6,6]-aza-bridge[a] open	839.6	-72.6	5.74	2340.99040	-28.8	0.763
NC_{60} (exo)	[6,6]-aza-bridge[a] closed	844.3	-67.9	5.53	2341.02170	-48.4	0.754
$N@C_{60}$ (endo)	[5,6]-aza-bridge[a] closed	954.6	42.5	6.50	2340.84762	60.8	0.776
$N@C_{60}$ (endo)	[6,6]-aza-bridge[a] closed	944.8	32.6	6.22	2304.89144	33.3	0.760
$N@C_{60}$ (endo)	in center[a] doublet	944.5	32.4	1.75			
$N@C_{60}$ (endo)	in center[a] quartet	922.7	10.5	3.75	2340.94597	-0.9	3.753
$N@C_{60}$ (transition state)	[5,6]-center/endo bound	974.8	62.7	6.18			
$N@C_{60}$ (transition state)	[6,6]-center/endo bound	973.9	61.7	6.08			
$N@C_{60}$ (transition state)	[5,6]-endohedral–exohedral	993.7	81.5	6.70	2340.80645	86.7	1.549
$N@C_{60}$ (transition state)	[6,6]-endohedral–exohedral	985.7	73.6	6.19	2340.83958	65.9	0.761

a Geometry optimized.
b Relative to free C_{60} and N in the ground state (DH°_f).

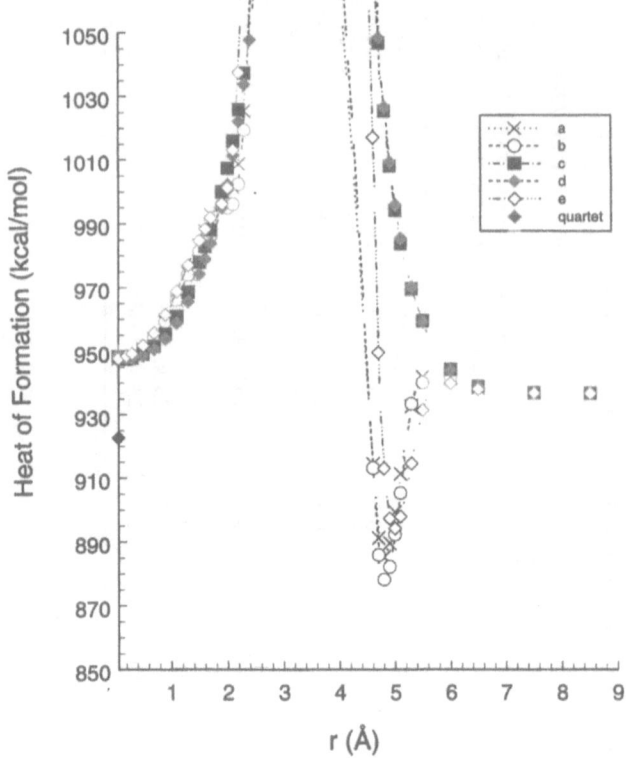

Fig. 36. Calculated PM3-UHF heats of formation for N/C_{60} complexes as a function of the distance r from N to the cage center with retention of a rigid cage geometry. Displacement of the cage-center to (*a*) a [5,6]-bond, (*b*) a [6,6]-bond, (*c*) the center of a pentagon, (*d*) the center of a hexagon and (*e*) a C-atom

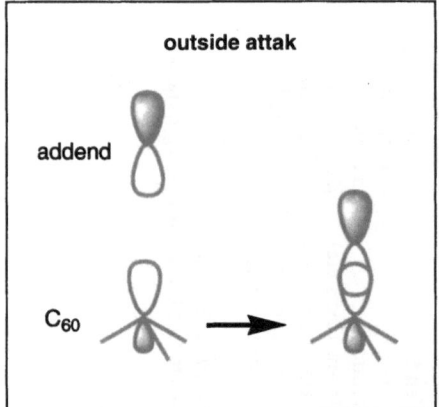
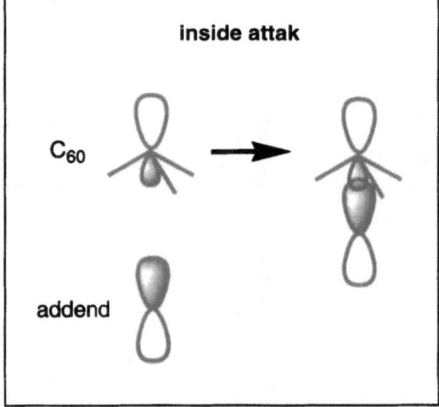

Fig. 37. Schematic representation of exohedral and endohedral addition of the p orbital of the addend to the fullerene with fixed cage symmetry

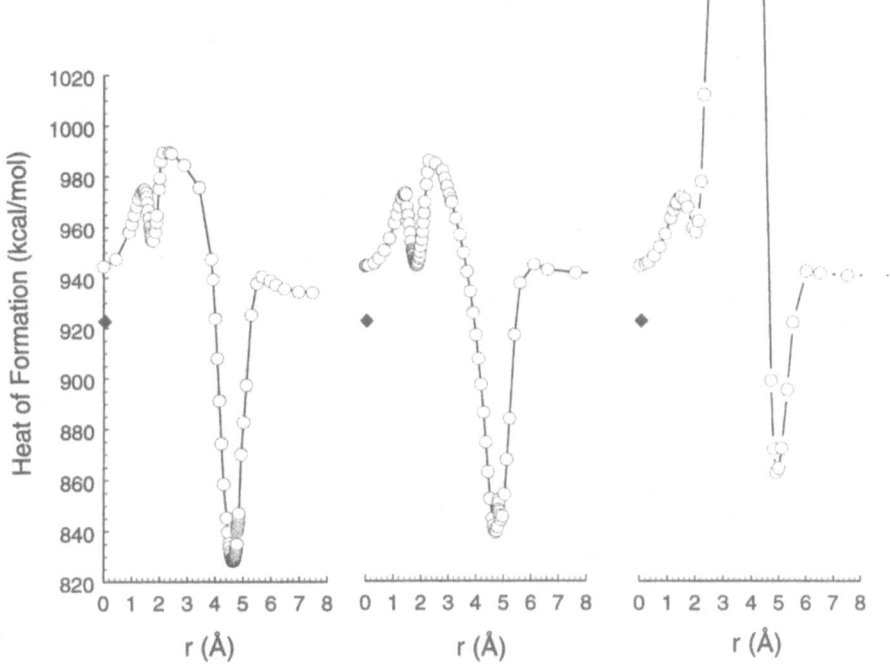

Fig. 38 a–c. Enthalpies of formation ΔH^0_f(PM3–UHF) of N/C_{60} complexes with relaxed (optimized) geometry as a function of the distance r from N to the cage center. Displacement of N from the cage center towards: **a** the center of a [5,6] bond; **b** the center of a [6,6] bond; **c** a C atom: ♦ represents the quartet minimum. The calculations were performed for N in the doublet state. Analysis of the reaction coordinates with configuration-interaction (CI) calculations showed that the ground state is very close to the excited states of the same multiplicity. Thus, points of discontinuity of the reaction coordinates in **a** and **b** are due to internal crossing between different energy surfaces. The discontinuity of the reaction coordinate in **b** next to the exohedral minima corresponds to the opening of the bridged [6,6] bond. This opening does not occur for the analogous bridged C_{60}NH compounds

experimental examples of exohedral adduct formation as demonstrated in the preceding paragraphs.

The PM3-UHF energy barriers for the penetration of nitrogen from the interior through a [5,6]- or a [6,6]-bond in the doublet state are significantly lower than the calculated most favorable value (225 kcal mol^{-1}) for the penetration of He out of He@C_{60} assuming a one- or two-window mechanism. Even the escape of N starting at the quartet state costs only little more than 60 kcal mol^{-1}. This is reasonable because, in the case of N@C_{60}, nitrogen penetrates after formation of covalently bound adducts with endohedral [5,6]- or [6,6]-aza-bridges. This bridging elongates the C-C bond, so that the penetration of the N-atom through the cluster and the following dissociation is favored. This autocatalytic breaking of the bond requires less energy than the considerable cage distortion needed for penetration by He. In the experiment He is able to leave C_{60} only after thermal treatment for hours at $600-850\,°C$ and is accompanied by an irreversible destruction of the fullerene cage. However, an N-atom escapes from $N@C_{60}$ under considerably milder conditions at $260\,°C$ [196a].

The corresponding calculations on the endohedral and exohedral complexes of C_{60} with H, F, or the methyl radical as guest predict the same behavior (Table 7) [197]. In all cases the formation of endohedral covalent bonds is energetically unfavorable due to the analogous strain arguments even if such reactive species as F-atoms are exposed to the inner surface of C_{60}.

The computational results were summarized as follows. (1) The covalent exohedral derivatives FC_{60}, HC_{60} and MeC_{60} are considerably more stable than the endohedrals $F@C_{60}$, $H@C_{60}$, and $Me@C_{60}$, regardless of the bonding situation within the cage. (2) The most favorable encapsulations of F and Me by C_{60} are slightly endothermic, whereas that of H is slightly exothermic. (3) The structures of $F@C_{60}$ and $Me@C_{60}$ with the guests placed exactly in the center of the cage are the most stable endohedral structures. In the case of $H@C_{60}$ a slightly different behavior is predicted. Hydrogen is quite free inside C_{60} and barriers between the bound and the endohedral minimum structure of $H@C_{60}$ are very low. The location of the endohedral minimum of H is consistently slightly off-center (1.1 Å from the center with the PM3-Hamiltonian). (4) In contrast to endohedral complexes of C_{60} with electropositive metals, no charge-transfer occurs between the guest and the host in the global minimum structures of $H@C_{60}$ and $Me@C_{60}$. For the global minimum structure of $F@C_{60}$ partial charge transfer from C_{60} to F is found, which is reflected by the Mulliken charge of -0.53 on F (Becke3LYP). (5) Only local minima are found for those isomers of $H@C_{60}$, $F@C_{60}$, and $Me@C_{60}$ where the guest is covalently bound to a C-atom. (6) The covalent binding of H, F, or Me with the inner surface of C_{60} leads to a very unfavorable cage distortion. Therefore, when encapsulated by C_{60}, the most favorable situa-

Table 7. PM3 dipole moments, heats of formation and bond energies of exohedral and endohedral complexes of C_{60} with H, F and the methyl radical (Me)

Compound		Dipole-moment (Debye)	Heat of Formation (kcal/mol)	Bond energy (kcal/mol)
HC_{60} (*exo*)		1.22	783.29	-67.97
$H@C_{60}$ (*endo*)	bound	0.51	851.21	-0.05
	transition state	0.16	855.95	4.69
	global minimum	0.07	848.89	-2.37
FC_{60} (*exo*)		1.23	749.20	-68.82
$F@C_{60}$ (*endo*)	bound	0.25	838.56	20.54
	transition state	1.13	848.50	30.48
	centre	0.00	827.25	9.23
CH_3C_{60} (*exo*)		1.48	778.78	-48.34
$CH_3@C_{60}$ (*endo*)	bound	0.31	885.08	57.96
	transition state	0.25	886.15	59.03
	centre	0.00	839.17	12.05

tion for such usually extremely reactive species is to stay in the middle of the cage and to avoid interactions with the inner surface as much as possible.

These investigations focused for the first time on a new aspect of topicity that takes into account the influence of the shape of a bent structure on its reactivity. The remarkable inertness of the inner surface contrasts with the pronounced reactivity of the outer convex surface of C_{60}. Almost unperturbated atomic species or reactive molecules can be studied at ambient conditions once they are encapsulated by the C_{60}. Moreover, the wavefunction of the guest atom can be influenced by a permanent distortion of the C_{60} cage. This was impressively demonstrated by exohedral derivatization of N@C_{60} leading to a lowering of the I_h symmetry which influences the ESR spectra of the paramagnetic guest [196c]. As pointed out in Sect. 3.2.2, the cage of a 1,2-monoadduct of C_{60} exhibits a very moderate droplet-like distortion. In case of symmetrical addends the symmetry is lowered to C_{2v} going from free C_{60} to a monoadduct. The nucleophilic cyclopropanation of N@C_{60} with diethyl bromomalonate (Scheme 15), afforded the endohedral monoadduct N@$C_{61}(COOEt)_2$, which was isolated from the reaction mixture together with empty $C_{61}(COOEt)_2$ by column chromatography on silica gel/toluene.

EPR investigations on N@$C_{61}(COOEt)_2$ (Fig. 39) lead to the following results. (1) The reactivity of N@C_{60} towards nucleophilic cyclopropanation is the same as that of empty C_{60}, since the mol/mol ratio of filled and empty material didn't change proving a clean conversion of N@C_{60} to N@$C_{61}(COOEt)_2$. This reflects the absence of a significant interaction of the fullerene cage with the nitrogen guest. Moreover, the complete recovery of the paramagnetic fraction as N@C_{61} $(COOEt)_2$ clearly provides further compelling evidence, that free N-atoms are located endohedrally and are shielded from the environment by the Chemical Faraday Cage C_{60}. (2) The EPR-spectrum of N@$C_{61}(COOEt)_2$ in solution (not shown) looks very similar to that of unreacted N@C_{60} but in the solid new features appear. The three main lines remain at the same position but are considerably broadened. In addition, at least four additional lines are observed. The fact that these lines are absent in the corresponding solution spectrum shows that this new feature is averaged out by the motion in solution. The hyperfine constant A and the electronic g-factor are within the experimental uncertainties

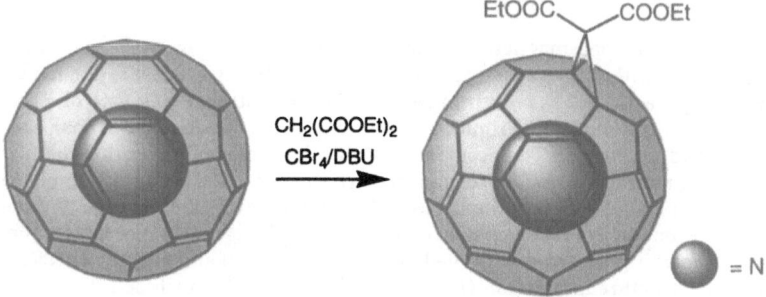

Scheme 15. Formation of the endohedral fullerene derivative N@$C_{61}(COOEt)_2$

equal to the values for unmodified $N@C_{60}$. This demonstrates that the average position of the N-atom is still on-center.

The additional lines in the solid state EPR spectrum of $N@C_{61}(COOEt)_2$ are due to a fine structure resulting from the spin-spin and spin-orbit interaction of the three unpaired electrons of the N-atom. The distortion of the cage symmetry means that the three p-orbitals of the N-atom are no longer degenerate. The expectation is corroborated by quantum mechanical calculations (PM3, UHF). Whereas in free $N@C_{60}$ all the three singly occupied p-orbitals are degenerate ($E_{px} = E_{py} = E_{pz} = -14.69108$ eV), the p_x- (-14.73942 eV), p_y- (-14.74212 eV), and p_z-orbitals (-14.74326 eV) of central N-atom in $N@C_{61}(COOEt)_2$ have slightly different energies (Fig. 40). This non-equivalency gives rise to a fine structure.

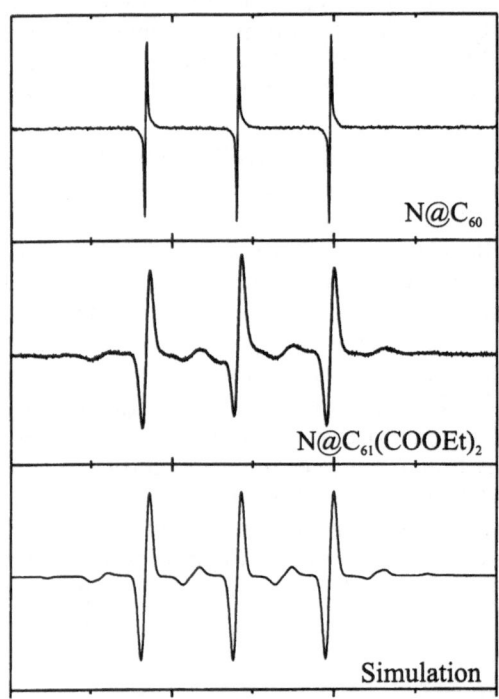

Fig. 39. EPR (electron paramagnetic resonance) spectra of: *above* $N@C_{60}$; *centre* $N@C_{61}$ $(COOEt)_2$ together with; *below* a simulation. The triplet splitting (*above*) is due to the isotropic hyperfine interaction of the electron systems with the nuclear spin $l = 1$ of ^{14}N (natural abundance 99.6%). Since the electronic spin is $S = 3/2$ (three unpaired electrons), each of the lines is three-fold degenerate. The occurrence of this degeneracy implies that the fine structure, quadrupole interaction and anisotropic hyperfine interaction are zero (complete spherical symmetry of nitrogen). In the adduct $N@C_{61}(COOEt)_2$ the icosahedral cage symmetry and therefore the degeneracy of nitrogen p orbitals is broken giving rise to new lines (*centre*). The simulation (*below*) is performed with the hyperfine interaction and g factor of $N@C_{60}$ but in addition a fine structure interaction ($D_{zz} = 2$ G and $E = 0.13$ G) is included. The effect of the deviation from spherical symmetry on the quadrupole or anisotropic hyperfine interaction is too small to be detected

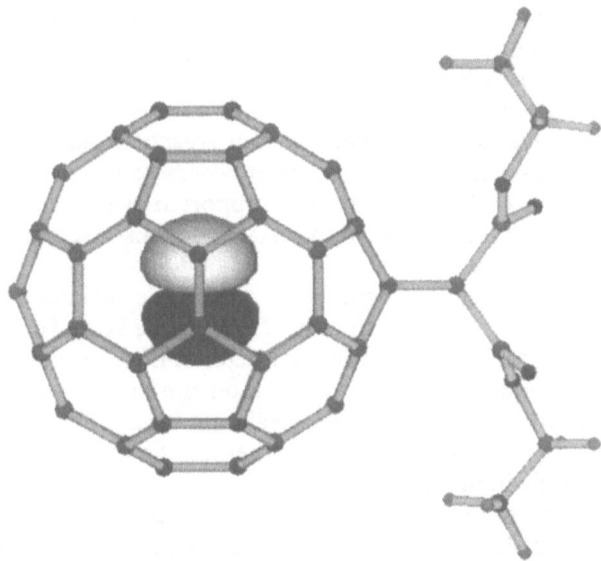

Fig. 40. Visualization of the calculated p_y orbital of nitrogen inside $C_{61}(COOEt)_2$. The x- and z-coordinates used in the text are perpendicular to the y-direction indicated there. The z-axis is in the direction from the center of $N@C_{60}$ towards the addend

Since the fine structure splitting in the ^2P excited state of nitrogen is 12,970 MHz, small admixtures of this state can already produce dramatic changes in the EPR spectra.

The simulation of the EPR-spectrum of $N@C_{61}(COOEt)_2$ (Fig. 39) taking into account a fine structure interaction is in nice accordance with the experimental spectrum. The simulation was carried out with the hyperfine interaction and g-factor of unmodified $N@C_{60}$ and the fine structure interaction $D_{zz} = 2$ G and $E = 0.13$ G (non-axial term). The shape of the extra lines definitely requires the inclusion of a non-axial term E, indicating that the molecular structure of the adduct induces some non-axiality which is not averaged out by fast axial rotation. The non-axial term was also observed at a measurement at 100 °C showing that axial motional averaging does not take place even at this temperature. These results show that like $^3He@C_{60}$ the new endohedral compound $N@C_{60}$ can be used as a probe to monitor exohedral chemical addition reactions. Due to higher sensitivity, EPR requires less material than NMR. $3He@C_{60}$ and $N@C_{60}$ are complementary probes since different interactions are measured.

5
Bent Graphitic Sheets

In the fullerenes the pyramidalization of the "sp²"-C-atoms is provided by the introduction of 12 pentagons (Eulers theorem). In a graphitic sheet which consists exclusively of a network of hexagons pyramidalization could in principle be achieved by nano-mechanically bending the sheet. Folded graphitic sheets along axes displayed by multiples of 30° were observed by scanning tunneling microscopy (STM) [198]. Similar reactivity principles that are observed for the surfaces of fullerenes are here predicted for bent graphitic sheets. In particular, bending causes the outer convex surface to become more reactive and the inner concave surface to become less reactive, for example, towards additions of radicals compared to a planar sheet. This expectation is supported by the finding of Siegel et al. [199] that complexation of transition metal fragments to corannulene is preferred at the outer convex surface.

Moreover, it is predicted that bending of a graphitic sheet also continuously changes the electronic properties, since for example the LUMO energies decrease with bending due to the increase of s character of the π-orbitals. Since carbanions prefer pyramidalized geometries and, on the other hand, carbocations prefer planar situations, a negative charge should preferably be located in the bent regions of charged graphitic sheet, whereas a positive charge would be found predominantly in the flat regions.

In conclusion the following effects associated with bending of sp^2-carbon networks (graphitic sheets) including the bent structures of buckytubes and fullerenes are predicted:

1. continuous change of the chemical reactivity upon bending – reactive convex surface and inert concave surface;
2. continuous change of the electronic properties, for example, lowering the LUMO energies upon bending;
3. continuous change of the charge distribution upon bending (shape dependence) – stabilization of negative charge in bent regions and stabilization of positive charge in flat regions.

This dependence of chemical and physical properties on the shape could be of importance for future nano-mechanics and nano-electronics.

6
Summary and Conclusion

Reactivity principles of C_{60} which were deduced in preceding reviews [3, 5] are still valid. However, the progress in the field of fullerene chemistry continuously increases our knowledge and allows us to refine the rules for the reactivity of this spherical carbon cluster. Based on the present state-of-the-art the following principles of C_{60} chemistry are summarized.

1. The spherical shape causes a pyramidalization of the C-atoms and therefore a large amount of strain energy. The addition chemistry is driven by the strain relief introduced by the formation of almost strain free sp^3-C-atoms. At a certain degree of addition this strain relief mechanism has to compete with new strain introducing processes like the increasing introduction of eclipsing interactions and the formation of planar cyclohexane substructures.
2. Due to the s character of the π-orbitals caused by the pyramidalization and the resulting repulsion of valence electron pairs (rehybridization) C_{60} is a

comparative electronegative molecule which is easily accessible for reductions and for the addition of nucleophiles.

3. In neutral C_{60} the h shell is incompletely filled resulting in a distortion corresponding to the only internal freedom that the C_{60} molecule has without breaking the I_h symmetry. This is the introduction of bond length alternation between [6,6]- and [5,6]-bonds. Since in the occupied h_u states (HOMOs) bonding interactions are predominantly located at the [6,6]-sites and the antibonding interactions (nodes) at the [5,6]-sites, a localization of the π-electrons at the [6,6]-sites accompanied by a contraction of the [6,6]-bonds causes an energy lowering. Hence, the lowest energy VB structure of neutral C_{60} contains only [6,6]-double bonds and [5,6]-single bonds. Since in the lowest lying unoccupied t_{1u} (LUMOs) and t_{1g} states the nodes are predominantly located at the [6,6]-sites and bonding interactions at the [5,6]-sites further filling of the outer shell causes bond length equalization.

4. The regiochemistry of additions to C_{60} is driven by the maintenance of the MO structure and the minimization of energetically unfavorable [5,6]-double bonds. Adducts regioselectively formed by one or several successive 1,2-additions to [6,6]-bonds have an MO structure which is closely correlated with that of free C_{60}. Their lowest energy VB structures contain only [6,6]-double bonds. 1,2-Additions occur always at [6,6]-double bonds. 1,4-Additions (introduction of one [5,6]-double bond) or 1,6-additions to position 1 and 16 (introduction of two [5,6]-double bonds) take place only with sterically demanding addends. Complexations with transition metals take place in a $^2\eta$-mode but not in $^5\eta$- or $^6\eta$-modes.

5. Due to complete orthogonality of the three LUMOs and a pronounced orthogonality of the five HOMOs the frontier orbitals of adducts with high coefficients in the *trans*-1 positions are the least accessible for attacks of nucleophiles or electrophiles, respectively (*trans*-1 effect). The most easy accessible sites in the HOMOs and LUMOs of adducts are the *e*- and *cis*-1 positions. Subsequent attacks occur preferably into *e*-positions for sterically demanding and *e*- and *cis*-1-positions for sterically less demanding addends. The regioselectivity of additions into equatorial sites increases with the number of addends already bound. The formation of thermodynamically stable hexaadducts with an octahedral addition pattern is unlikely for irreversibly binding sterically non-demanding addends (*cis*-1 additions).

6. Due to the pyramidalization of the C-atoms and the rigid cage structure of C_{60} the outer convex surface is very reactive towards addition reactions but at the same time the inner concave surface is inert (chemical Faraday cage). This allows the encapsulation, the observation, and the tuning of the wave function of extremely reactive species. This pronounced shape dependence of reactivity which was revealed for the first time with C_{60} as ambido-shaped system is predicted to be a general topicity principle.

7
References

1. Kroto HW, Heath JR, O'Brien SC, Curl RF, Smalley RE (1985) Nature 318:162
2. Krätschmer W, Lamb LD, Fostiropoulos K, Huffman DR (1990) Nature 347:354
3. Hirsch A (1994) The chemistry of the fullerenes. Thieme, Stuttgart
4. Diederich F, Isaacs L, Philip D (1994) Chem Soc Rev 243
5. Hirsch A (1995) Synthesis 895
6. Diederich F, Thilgen C (1996) Science 271:317
7. Rubin Y (1997) Chem Eur J 3:1009
8. Hebard AF, Rosseinsky MJ, Haddon RC, Murphy DW, Glarum SH, Palstra TTM, Ramirez AP, Kortan AR (1991) Nature 350:600
9. Holczer K, Klein O, Huang S-M, Kaner RB, Fu KJ, Whetten RL, Diederich F (1991) Science 252:1154
10. Hummelen JC, Prato M, Wudl F (1995) J Am Chem Soc 117:7003
11. Murphy TA, Pawlik T, Weidinger A, Höhne M, Alcala R, Spaeth JM (1996) Phys Rev Lett 77:1075
12. Mauser H, Hommes NvE, Clark T, Hirsch A, Pietzak B, Weidinger A, Dunsch L (1997) Angew Chem Int Ed Engl 36:2835
13. Averdung J, Luftmann H, Schlachter I, Mattay J (1995) Tetrahedron 51:6977
14. (a) Lamparth I, Nuber B, Schick G, Skiebe A, Grösser T, Hirsch A (1995) Angew Chem 107:2473; (b) Lamparth I, Nuber B, Schick G, Skiebe A, Grösser T, Hirsch A (1995) Angew Chem Int Ed Engl 34:2257
15. Hummelen JC, Knight B, Pavlovich J, Gonzalez R, Wudl F (1995) Science 269:1554
16. Nuber B, Hirsch A (1996) J Chem Soc Chem Commun 1421
17. Sygula A, Rabideau PW (1994) J Chem Soc Chem Commun 2271
18. Djojo F, Hirsch A (1998) Chem Eur J 2:344
19. Schmalz TG, Seitz WA, Klein DJ, Hite GE (1986) Chem Phys Lett 130:203
20. Kroto HW (1987) Nature 329:529
21. (a) Beckhaus HD, Rüchardt C, Kao M, Diederich F, Foote CS (1992) Angew Chem 104:69; (b) Beckhaus HD, Rüchardt C, Kao M, Diederich F, Foote CS (1992) Angew Chem Int Ed Engl 31:63
22. Haddon RC (1993) Science 261:1545
23. Haddon RC (1986) J Am Chem Soc 108:2837
24. Haddon RC (1987) J Am Chem Soc 109:1676
25. Haddon RC (1988) Acc Chem Res 21:243
26. Haddon RC (1990) J Am Chem Soc 112:3385
27. Haddon RC (1992) Acc Chem Res 25:127
28. Haddon RC, Brus LE, Raghavachari K (1986) Chem Phys Lett 125:459
29. Haddon RC, Brus LE, Raghavachari K (1986) Chem Phys Lett 131:165
30. Schulman M, Disch RL, Miller MA, Peck RC (1987) Chem Phys Lett 141:45
31. Lüthi HP, Almöf J (1987) Chem Phys Lett 135:357
32. Dunlap DI (1988) Int J Quantum Chem Symp 22:257
33. Scuseria GE (1991) Chem Phys Lett 176:423
34. Häser M, Almhöf J, Scuseria GE (1991) Chem Phys Lett 181:497
35. Yannoni CS, Bernier PP, Bethune DS, Meijer G, Salem JR (1991) J Am Chem Soc 113:3190
36. David WIF, Ibberson RM, Mathewman JC, Prassides K, Dennis TJS, Hare JP, Kroto HW, Taylor R, Walton DRM (1991) Nature 353:147
37. Hedberg K, Hedberg L, Bethune DS, Brown CA, Dorn HC, Johnson RD, de Vries M (1991) Science 254:410
38. Liu S, Lu YJ, Kappes MM, Ibers JA (1991) Nature 254:408
39. (a) Andreoni W, Gygi F, Parinello M (1992) Phys Rev Lett 68:823; (b) Kohanoff J, Andreoni W, Parinello M (1992) Chem Phys Lett 198:472
40. Stollhoff G (1991) Phys Rev B 44:10,998
41. Stollhoff G, Scherrer H (1995) Mater Sci Forum 191:81

42. Dresselhaus MS, Dresselhaus G, Eklund PC (1995) Science of fullerenes and carbon nanotubes. Academic Press, New York
43. Xie Q, Perez-Cordero E, Echegoyen L (1992) J Am Chem Soc 114:3978
44. Ohsawa Y, Saji T (1992) J Chem Soc Chem Commun 781
45. Fhou F, Jehoulet C, Bard AJ (1992) J Am Chem Soc 114:11,004
46. Allemand P-M, Khemani KC, Koch A, Wudl F, Holczer K, Donovan S, Grüner G, Thompson JD (1991) Science 253:301
47. Stephens PW, Cox D, Lauher JW, Mihaly L, Wiley JB, Allemand P-M, Hirsch A, Holczer K, Li Q, Thompson JD, Wudl F (1992) Nature 355:331
48. Bossard C, Rigaut S, Astruc D, Delville M-H, Felix G, Fevrier-Bouvier A, Amiall J, Flandrois S, Delhaes P (1993) J Chem Soc Chem Commun 333
49. Penicaud A, Hsu J, Reed C, Koch A, Khemani KC, Allemand P-M, Wudl F (1991) J Am Chem Soc 113:6699
50. Xie Q, Arias F, Echegoyen L (1993) J Am Chem Soc 115:9818
51. Haddon RC (1995) Nature 378:249
52. Elser V, Haddon RC (1987) Nature 325:1987
53. Bühl M (1998) Chem Eur J 4:734
54. Saunders M, Jiminez-Vazquez HA, Cross RJ, Mroczkowski S, Freedberg DI, Anet FAL (1994) Nature 367:256
55. Saunders M, Cross RJ, Jiminez-Vazquez HA, Shimshi R, Khong A (1996) Science 271:1693
56. Wudl F, Hirsch A, Khemani KC, Suzuki T, Allemand P-M, Koch A, Eckert H, Srdanov HG, Webb H (1992) In: Hammond GS, Kuck VJ (eds) Fullerenes. American Chemical Society Symposium Series 481:161
57. (a) Hirsch A, Li Q, Wudl F (1991) Angew Chem 103:1339; (b) Hirsch A, Li Q, Wudl F (1991) Angew Chem Int Ed Engl 30:1309
58. (a) Hirsch A, Soi A, Karfunkel HR (1992) Angew Chem 104:808; (b) Hirsch A, Soi A, Karfunkel HR (1992) Angew Chem Int Ed Engl 31:766
59. Hirsch A, Grösser T, Skiebe A, Soi A (1993) Chem Ber 126:1061
60. Fagan PJ, Krusic PJ, Evans DH, Lerke SA, Johnston E (1992) J Am Chem Soc 114:9697
61. Bingel C (1993) Chem Ber 126:1957
62. Kampe K-D, Egger N, Vogel M (1993) Angew Chem 105:1203
63. (a) Schick G, Kampe K-D, Hirsch A (1995) J Chem Soc, Chem Commun 2023; (b) Camps X, Hirsch A (1997) J Chem Soc, Perkin Trans 1 1595
64. Yamogo S, Yanagawa M, Nakamura E (1994) J Chem Soc Chem Commun 2093
65. Davey SN, Leigh DA, Moody AE, Tetler LW, Wade FA (1994) J Chem Soc Chem Commun. 397
66. Nagashima H, Terasaki H, Kimura E, Nakajima K, Itoh K (1994) J Org Chem 59:1246
67. Keshawarz-K M, Knight B, Srdanov G, Wudl F (1995) J Am Chem Soc 117:11,371
68. Timmermann P, Witschel LE, Diederich F, Boudon C, Gisselbrecht J-P, Gross M (1996) Helv Chim Acta 79:6
69. Osterodt J, Vögtle F (1996) Chem Commun 547
70. Mikami K, Matsumoto S, Ishida A, Takamuku S, Suenobu T, Fukuzumi S (1995) J Am Chem Soc 117:11,134
71. Murata Y, Komatsu K, Wan TSM (1996) Tetrahedron Lett 37:7061
72. Nagashima H, Terasaki H, Saito Y, Jinno K, Itoh K (1995) J Org Chem 60:4966
73. Murata Y, Shiro M, Komatsu K (1997) J Am Chem Soc 119:8117
74. Sawamura M, Iikura H, Nakamura E (1996) J Am Chem Soc 118:12,850
75. Okamura H, Murata Y, Minoda M, Komatsu K, Miyamoto T, Wan TSM (1996) J Org Chem 61:8500
76. Paulus EF, Bingel C (1995) Acta Cryst C 51:143
77. Kampe K-D, Egger N (1995) Liebigs Ann 115
78. Komatsu K, Murata Y, Takimoto N, Mori S, Sugita N, Wan TSM (1994) J Org Chem 59:6101
79. Murata Y, Motoyama K, Komatsu K, Wan TSM (1996) Tetrahedron 52:5077
80. Murata Y, Komatsu K, Wan TSM (1996) Tetrahedron Lett 37:7061

81. Komatsu K, Takimoto N, Murata Y, Wan TSM, Wong T (1996) Tetrahedron Lett 37:6153
82. Okamura H, Minoda M, Komatsu K, Miyamoto T (1997) Macromol Chem Phys 198:777
83. Wang G-W, Komatsu K, Murata Y, Shiro M (1997) Nature 387:583
84. Tanaka T, Kitagawa T, Komatsu K, Takeuchi K (1997) J Am Chem Soc 119:9313
85. Kitagawa T, Tanaka T, Takata Y, Takeuchi K (1995) J Org Chem 60:1490
86. Ganapathi PS, Friedmann SH, Kenyon GL, Rubin Y (1995) J Org Chem 60:2954
87. Kusukawa T, Ando W (1996) Angew Chem 108:1416
88. Morton JR, Preston KF, Krusic PJ, Hill SA, Wasserman E (1992) J Phys Chem 96:3576
89. Morton JR, Preston KF, Krusic PJ, Hill SA, Wasserman E (1992) J Am Chem Soc 114:5454
90. Morton JR, Preston KF, Krusic PJ, Wasserman E (1992) J Chem Soc, Perk Trans 2, 1425
91. Krusic PJ, Roe DC, Johnston E, Morton JR, Preston KF (1993) J Phys Chem 97:1736
92. Morton JR, Preston KF, Krusic PJ, Knight LB (1993) J Chem Phys Lett 204:481
93. Keizer PN, Morton JR, Preston KF (1992) J Chem Soc Chem Commun 1259
94. Johnson CS Jr (1964) J Chem Phys 41:3277
95. Krusic P, Meakin P, Jesson JP (1971) J Phys Chem 75:3438
96. Fagan PJ, Krusic PJ, Evans DH, Lerke SA, Johnston E (1992) J Am Chem Soc 114:9697
97. Krusic PJ, Wasserman E, Parkinson BA, Malone B, Holler ER, Keizer PN, Morton JR, Preston KF (1991) J Am Chem Soc 113:6274
98. Krusic PJ, Wasserman E, Keizer PN, Morton JM, Preston KF (1991) Science 254:1183
99. Fagan PJ, Krusic PJ, McEven CN, Lazar L, Holmes Parker D, Herron N, Wasserman E (1993) Science 262:404
100. Loy DA, Assink R (1992) J Am Chem Soc 114:3977
101. Skiebe A, Hirsch A, Klos H, Gotschy B (1994) Chem Phys Lett 220:138
102. Djojo F, Herzog A, Lamparth I, Hampel F, Hirsch A (1996) Chem Eur J 2:1537
103. Matsuzawa N, Dixon DA, Fukunaga T (1992) J Phys Chem 96:7594
104. Tsuda M, Ishida T, Nogami T, Kurono S, Ohashi M (1993) J Chem Soc, Chem Commun 1296
105. Schlueter JA, Seaman JM, Taha S, Cohen H, Lykke KR, Wang HH, Williams JM (1993) J Chem Soc Chem Commun 972
106. Komatsu K, Murata Y, Sugita N, Takeuchi K, Wan TSM (1993) Tetrahedron Lett 34:8473
107. (a) Lamparth I, Maichle-Mössmer C, Hirsch A (1995) Angew Chem 107:1755; (b) Lamparth I, Maichle-Mössmer C, Hirsch A (1995) Angew Chem Int Ed Engl 34:1607
108. (a) Kräutler B, Müller T, Maynollo J, Gruber K, Kratky C, Ochsenbein P, Schwarzenbach D, Bürgi H-B (1996) Angew Chem 108:1294; (b) Kräutler B, Müller T, Maynollo J, Gruber K, Kratky C, Ochsenbein P, Schwarzenbach D, Bürgi H-B (1996) Angew Chem Int Ed Engl 35:1203
109. Rotello VM, Howard JB, Yadev T, Conn MM, Viani E, Giovane LM, Lafleur AL (1993) Tetrahedron Lett 34:1561
110. Gügel A (1993) Doctoral Thesis, Max-Plank Institut, Mainz
111. Tsuda M, Ishida T, Nogami T, Kurono S, Ohashi M (1992) Chem Lett 2333
112. Hoke SH II, Molstad J, Dilattato D, Jay MJ, Carlson D, Kahr B, Cooks RG (1992) J Org Chem 57:5069
113. Prato M, Lucchini V, Maggini M, Stimpfl E, Scorrano G, Eiermann M, Suzuki T, Wudl F (1993) J Am Chem Soc 115:8479
114. Prato M, Li Q, Wudl F, Lucchini V (1993) J Am Chem Soc 115:1148
115. Banks MR, Cadogan JIG, Gosney I, Hodgson PKG, Langridge-Smith PRR, Rankin DWH (1994) J Chem Soc Chem Commun 1365
116. Banks MR, Cadogan JIG, Gosney I, Hodgson PKG, Langridge-Smith PRR, Millar JRA, Taylor AT (1994) Tetrahedron Lett 35:9067
117. Ishida T, Tanaka K, Nogami T (1994) Chem Lett 561
118. Hawker CJ, Wooley KL, Frechet JMJ (1994) J Chem Soc Chem Commun 925
119. Yan M, Cai SX, Keana JFW (1994) J Org Chem 59:5951
120. Averdung J, Luftmann H, Mattay J, Claus KU, Abraham W (1995) Tetrahedron Lett 36:2957
121. Banks MR, Cadogan JIG, Gosney I, Hodgson PKG, Langridge-Smith PRR, Millar JRA, Parkinson JAS, Rankin DWH, Taylor AT (1995) J Chem Soc Chem Commun 887

122. Banks MR, Cadogan JIG, Gosney I, Hodgson PKG, Langridge-Smith PRR, Millar JRA, Taylor AT (1995) J Chem Soc Chem Commun 88
123. Schick G, Grösser T, Hirsch A (1995) J Chem Soc Chem Commun 2289
124. Schick G, Hirsch A, Mauser H, Clark T (1996) Chem Eur J 2:935
125. (a) Grösser T, Prato M, Lucchini V, Hirsch A, Wudl F (1995) Angew Chem 107:1462; (b) Grösser T, Prato M, Lucchini V, Hirsch A, Wudl F (1995) Angew Chem Int Ed Engl 34:1343
126. Schick G, Hirsch A (1998) Tetrahedron 54:4283
127. Haldiman RF, Klärner F-G, Diederich F (1997) J Chem Soc Chem Commun 237
128. Shen CF-F, Yu H-H, Juo CG, Chien KM, Her G-R, Luh T-Y (1997) Chemistry Eur J 3:744
129. (a) Wudl F (1992) Acc Chem Res 25:157; (b) Diederich F, Isaacs L, Philp D (1994) Chem Soc Rev 243
130. Janssen RAJ, Hummelen JC, Wudl F (1995) J Am Chem Soc 117:544
131. Isaacs L, Wehrsig A, Diederich F (1993) Helv Chim Acta 76:1231
132. (a) Smith AB III, Strongin RM, Brard L, Furst GT, Romanow WJ, Owens KG, King RC (1993) J Am Chem Soc 115:5829; (b) Smith AB III, Strongin RM, Brard L, Furst GT, Romanow WJ, Owens KG, Goldschmidt RJ, King RC (1995) J Am Chem Soc 117:5492
133. Saunders M (1991) Science 253:330
134. Banks MR, Dale MJ, Gosney I, Hodgson PKG, Jennings RCK, Jones AC, Lecoultre J, Langridge-Smith PRR, Maier JP, Scrivens JH, Smith MJC, Smyth CJ, Taylor AT, Thorburn P, Webster AS (1993) J Chem Soc Chem Commun 1149
135. Rüchhardt C, Gerst M, Ebenhoch J, Beckhaus H-D, Campbell EEB, Tellgmann R, Schwarz H, Weiske T, Pitter S (1993) Angew Chem 105:609 (1993) Angew Chem Int Ed Engl 32:584
136. Vassallo AM, Wilson MA, Attalla MI (1988) Energy Fuels 2:539
137. Attalla MI, Wilson MA, Quezada RA, Vassallo AM (1989) Energy Fuels 3:59
138. Attalla MI, Vassallo AM, Tattam BN, Hanna JV (1993) J Phys Chem 97:6329
139. Henderson CC, Cahill PA (1993) Science 259:1885
140. Ballenweg S, Gleiter R, Krätschmer W (1993) Tetrahedron Lett 34:3737
141. Guarr TF, Meier MS, Vance VK, Clayton M (1993) J Am Chem Soc 115:9862
142. (a) Henderson CC, Assink RA, Cahill PA (1994) Angew Chem 106:803; (b) Henderson CC, Assink RA, Cahill PA (1994) Angew Chem Int Ed Engl 33:786
143. Meier MS, Weedon BR, Spielmann HP (1996) J Am Chem Soc 118:11,682
144. Fagan PJ, Calabrese JC, Malone B (1992) In: Hammond GS, Kuck VS (eds) Fullerenes. ACS Symosium Series 481, Am Chem Soc, Washington DC, p 177
145. (a) Balch AL, Catalano VJ, Lee JW (1991) Inorg Chem 30:3990; (b) Balch AL, Catalano VJ, Lee JW, Olmstead MM (1992) J Am Chem Soc 114:5455; (c) Balch AL, Lee JW, Noll BC, Olmstead MM (1992) J Am Chem Soc 114:10,984; (d) Balch AL, Catalano VJ, Lee JW, Olmstead MM, Parkin SR (1991) J Am Chem Soc 113:8953
146. (a) Koefod RS, Hudgens MF, Shapley JR (1991) J Am Chem Soc 113:8957; (b) Rasinkangas M, Pakkanen TT, Pakkanen TA, Ahlgren M, Rouvinen J (1993) J Am Chem Soc 115:4901
147. Yagupsky G, Brown CK, Wilkinson GJ (1970) J Chem Soc A 1392
148. (a) Diederich F, Ettl R, Rubin Y, Whetten RL, Beck R, Alvarez M, Anz S, Sehsharma D, Wudl F, Khemani KC, Koch A (1991) Science 252:548; (b) Wood JM, Kahr B, Hoke SH II, Dejarme L, Cooks RG, Ben-Amotz D (1991) J Am Chem Soc 113:5907
149. Thorp HH, Karlsbeck WA (1991) J Electroanal Chem 314:363
150. (a) Taylor R, Parsons JP, Avent AG, Rannard SP, Dennis TJ, Hare JP, Kroto HW, Walton DRM (1991) Nature 351:277; (b) Creegan KM, Robbins JL, Robbins WK, Millar JM, Sherwood RD, Tindall PJ, Cox DM, Smith AB III, McCauley JP Jr, Jones DR, Gallagher RT (1992) J Am Chem Soc 114:1103; (c) Vassalo AM, Pang LSK, Coke-Clarke PA, Wilson MA (1991) J Am Chem Soc 113:7820
151. (a) Elemes Y, Silverman SK, Sheu C, Kao M, Foote CS, Alvarez MM, Whetten RL (1992) Angew Chem 104:364; (b) Elemes Y, Silverman SK, Sheu C, Kao M, Foote CS, Alvarez MM, Whetten RL (1992) Angew Chem Int Ed Engl 31:351
152. Cataldo F, Ori O (1995) Polymer Degradation and Stability 48:291

153. (a) Hawkins JM, Lewis TA, Loren SD, Meyer A, Heath LR, Shibato Y, Saykally RJ (1990) J Org Chem 55:6250; (b) Hawkins JM, Meyer A, Lewis TA, Loren S, Hollander FJ (1991) Science 252:312; (c) Hawkins JM, Loren S, Meyer A, Nunlist R (1991) J Am Chem Soc 113:7770; (d) Hawkins JM (1992) Acc Chem Res 25:150; (e) Hawkins JM, Meyer A, Lewis TA, Bunz U, Nunlist R, Ball GE, Ebbesen TW, Tanigaki K (1992) J Am Chem Soc 114:7954

154. Holloway JH, Hope EG, Taylor R, Langley GL, Advent AG, Dennis TJ, Hare JH, Kroto HW, Walton DRM (1991) J Chem Soc Chem Commun 966

155. Selig H, Lifshitz C, Peres T, Fischer JE, McGhie AR, Romanov WJ, McCauley JP Jr, Smith AB III (1991) J Am Chem Soc 113:5475

156. Kniaz K, Fischer JE, Selig H, Vaughan GBM, Romanov WJ, Cox D, Chowdhury K, McCauley JP, Strongin RM, Smith AB III (1993) J Am Chem Soc 115:6060

157. Gakh AA, Tuinman AA, Adcock JL, Sachleben RA, Compton RN (1994) J Am Chem Soc 116:819

158. Boltalina OV, Borschevskii AY, Sidorov LN, Street JM, Taylor R (1996) Chem Commun 529

159. Boltalina OV, Markov VY, Taylor R, Waugh MP (1996) Chem Commun 2549

160. Tuinman AA, Gakh AA, Adcock JL, Compton RN (1993) J Am Chem Soc 115:5885

161. Olah GA, Bucsi I, Lambert C, Aniszfeld R, Trivedi NJ, Sensharma DK, Prakash GKS (1991) J Am Chem Soc 133:9385

162. Tebbe FN, Becker JY, Chase DB, Firment LE, Holler ER, Malone BS, Krusic PJ, Wasserman E (1991) J Am Chem Soc 113:9900

163. Birkett PR, Avent AG, Darwish AD, Kroto HW, Taylor R, Walton DRM (1993) J Chem Soc Chem Commun 1230

164. Tebbe FN, Harlow RL, Chase DB, Thorn DL, Campbell GC Jr, Calabrese JC, Herron N, Young RJ, Wasserman E (1992) Science 256:822

165. Birkett PR, Hitchcock PB, Kroto HW, Taylor R, Walton DRM (1992) Nature 357:479

166. Birkett PR, Avent AG, Darwisch AD, Kroto HW, Taylor R, Walton DRM (1993) J Chem Soc Chem Commun 1230

167. Taylor R (1995) The chemistry of fullerenes. World Scientific, Singapore

168. Miller GP (1992) Mat Res Soc Symp Proc 247:293

169. Thoman H, Bernardo M, Miller GP (1992) J Am Chem Soc 114:6593

170. Chiang LY, Swirczewski JW, Hsu CS, Chowdhury DK, Cameron S, Cressgan K (1992) J Chem Soc Chem Commun 1791

171. Chiang LY, Upasani RB, Swirczewski JW Soled S (1993) J Am Chem Soc 115:5453

172. Olah GA, Bucsi I, Lambert C, Aniszfeld R, Trivedi NJ, Sensharma DK, Prakash GK (1991) J Am Chem Soc 113:9387

173. Taylor R, Langley GL, Meidine MF, Parsons JP, Abdul-Sada AK, Dennis TJ, Hare JP, Kroto HW, Walton DRM (1992) J Chem Soc Chem Commun 667

174. Olah GA personal communication

175. Hawkins JM, Meyer A, Lewis TA, Bunz U, Nunlist R, Ball GE, Ebbesen T, Tanigaki K (1992) J Am Chem Soc 114:7954

176. Hawkins JM, Meyer A, Nambu M (1993) J Am Chem Soc 115:9844

177. (a) Hirsch A, Lamparth I, Karfunkel HR (1994) Angew Chem 106:453; (b) Hirsch A, Lamparth I, Karfunkel HR (1994) Angew Chem Int Ed Engl 33:437

178. Hirsch A, Lamparth I, Grösser T, Karfunkel HR (1994) J Am Chem Soc 116:9385

179. Avent AG, Darwish AD, Heimbach DK, Kroto HW, Meidine MF, Parsons JP, Remars C, Roers R, Ohashi O, Taylor R, Walton DRM (1994) J Chem Soc, Perkin Trans 2 15

180. (a) Henderson CC, Assink RA, Cahill PA (1994) Angew Chem 106:803; (b) Henderson CC, Assink RA, Cahill PA (1994) Angew Chem Int Ed Engl 33:786

181. Hamano T, Mashino T, Hirobe M (1995) J Chem Soc Chem Commun 1537

182. Cardullo F, Isaacs L, Diederich F, Gisselbrecht J-P, Boudon C, Gross M (1996) J. Chem Soc Chem Commun 797

183. (a) Kräutler B, Maynollo J (1995) Angew Chem 107:69; (b) Kräutler B, Maynollo J (1995) Angew Chem Int Ed Engl 34:87

184. Fagan PJ, Calabrese JC, Malone B (1991) J Am Chem Soc 113:9408

185. (a) Isaacs L, Haldimann RF, Diederich F (1994) Angew Chem 106:2435; (b) Isaacs L, Haldimann RF, Diederich F (1994) Angew Chem Int Ed Engl 33:2434
186. (a) Isaacs L, Haldimann RF, Diederich F (1995) Angew Chem 107:1636; (b) Isaacs L, Haldimann RF, Diederich F (1995) Angew Chem Int Ed Engl 34:1466
187. (a) Diederich F, Thilgen C, Herrmann A (1996) Nachr Chem Tech Lab 44:9; (b) Thilgen C, Herrmann A, Diederich F (1997) Helv Chem Acta 80:183
188. (a) Lamparth I, Herzog A, Hirsch A (1996) Tetrahedron 52:5065; (b)Camps X, Schön-berger H, Hirsch A (1997) Chem Eur J 3:561
189. Hirsch A, Lamparth I, Schick G (1996) Liebigs Ann 1725
190. (a) Anderson HL, Boudon C, Diederich F, Gisselbrecht J-P, Gross M, Seiler P (1994) Angew Chem 106:1691; (b) Anderson HL, Boudon C, Diederich F, Gisselbrecht J-P, Gross M, Seiler P (1994) Angew Chem Int Ed Engl 33:1628
191. Arce M-J, Viado AL, Khan S, Rubin Y (1996) Organometallics (in press)
192. Reuther U, Hirsch A (1998) J Chem Soc Chem Comm 1401
193. Bethune DS, Johnson DR, Salem JR, de Vries MS, Yannoni CS (1993) Nature 366:123
194. Tellgmann R, Krawez N, Lin SH, Hertel IV, Campbell EFB (1996) Nature 382:407
195. Knapp C, Dinse K-P, Pietzak B, Waiblinger M, Weidinger A (1997) Chem Phys Lett 272:433
196. (a) Mauser H, Hommes NvE, Clark T, Hirsch A, Pietzak B, Weidinger A, Dunsch L (1997) Angew Chem 109:2858; (b) Mauser H, Hommes NvE, Clark T, Hirsch A, Pietzak B, Weidinger A, Dunsch L (1997) Angew Chem Int Ed Engl 36:2835; (c) Pietzak B, Waib-linger M, Murphy TA, Weidinger A, Höhne M, Dietel E, Hirsch A (1997) Chem Phys Lett 279:259
197. Mauser H, Hirsch A, Hommes NvE, Clark T (1997) J Mol Model 3:415
198. (a) Ebbesen W (1994) In: Kuzmany H, Fink J, Mehring M, Roth S (eds.) Progress in ful-lerene research. World Scientific, Singapore, p 116; (b) Hirsch A, Mauser H, Clark T (1997) In: Kuzmany H, Fink J, Mehring M, Roth S (eds.) Molecular nanostructures. World Scientific Publishing, p 55
199. Seiders TJ, Beldridge KK, O'Connor JM, Siegel JS (1997) J Am Chem Soc 119:4781

Ring Opening Reactions of Fullerenes: Designed Approaches to Endohedral Metal Complexes

Yves Rubin

Department of Chemistry and Biochemistry, University of California, Los Angeles, CA 90095–1569 USA. *E-mail: rubin@chem.ucla.edu*

Endohedral complexes of fullerenes are some of the most attractive targets of research in fullerene chemistry and physics. An overview of the field is provided with a focus on endohedral complexes incorporating lanthanide metals, noble gases, and atomic nitrogen. Although the present approaches have been successful in providing the first members of this class of fullerenes, their preparation is still difficult and limited to a few elements. It is argued that ring opening reactions within fullerene shells have the potential to provide easy access to endohedral metallofullerenes on a large scale, including those with transition metals which are potentially the most interesting elements in regard to their materials properties. Current ring opening reactions within fullerene shells are reviewed. An outlook on ring opening reactions leading to effectively large apertures is presented.

Keywords: Fullerene, Endohedral Metallofullerene, Transition Metals, Helium, $[2+2+2]$ Ring Opening.

Topics in Current Chemistry, Vol. 199
© Springer Verlag Berlin Heidelberg 1999

1
Introduction

Fullerene chemistry is filled with exciting and formidable prospects. It is still a very young field in constant discovery and experiencing rapid developments. So far, most of the synthetic organic efforts initiated since C_{60} has become available in macroscopic quantities have been concentrated on reactivity issues combining regio- and stereochemical aspects [1,2]. However, there is still a number of ambitious goals to be reached as well as there are inspiring projects to be conceived. One of them, the stimulating prospect of introducing metal atoms or any other elements inside the cage of C_{60} and higher fullerenes [3], was initiated several years ago by carrying out the arc-evaporation of metal oxides or carbides mixed with graphite in a manner similar to the generation of the empty fullerenes. The first endohedral metallofullerenes thus obtained (endohedral ≡ inside the cage) brought immediate excitement to the scientific community, as evidenced by the number of experimental and theoretical papers that have dealt with them [4].

The goal of inserting metals within fullerene cages is partly inspired by the phenomenal diversity of desirable physical properties that C_{60} alone has demonstrated [5]. By placing a transition metal, distinguished by its diversity of redox states, inside a fullerene, one can justifiably argue that the physical properties displayed by C_{60} can be significantly altered and, even perhaps, result in higher temperature superconductors, redox active systems, ferromagnetic compounds, and many other interesting materials. Moreover, having either high spin or radioactive metals inside these structures could constitute a convenient way to "transport" these properties into biological systems without leaching out metals that are often highly cytotoxic [e.g. Gd(III)]. This would allow convenient magnetic resonance imaging (MRI) or other diagnostics to be carried out for medical purposes [6]. The utilization of fullerenes, endohedral metallofullerenes, or their functionalized derivatives in medicine will of course depend on their biological and pharmacokinetic properties [7].

Reaching the goal of synthesizing endohedral fullerene complexes at will and on a significant scale is undoubtedly bound to have a tremendous impact in the field and in science and technology in general. However, this relies critically on the achievement of an exquisite control of the regioselectivity of additions to C_{60} or other fullerenes [2], and on the understanding and design of subsequent ring-opening reactions within the shell of these molecules.

2
Endohedral Fullerene Complexes

2.1
Endohedral Metallofullerenes

Almost immediately after the discovery of C_{60} in the gas phase [8], it was suggested that atoms could be encapsulated into the hollow structure of this fullerene [9]. This concept was in fact used to probe whether the structure of the fullerenes are spherical as originally proposed [8], when no other methods other than mass spectrometry were available. In "shrink-wrap" experiments, C_2-units were "evaporated" one-by-one by laser-induced fragmentation of potassium, cesium, or lanthanum–C_{60} ions [10]. Past a critical size (C_{44} for potassium, C_{48}–C_{50} for cesium, and C_{42}–C_{44} for lanthanum), complete disappearance of the metal–C_{2n} ions was observed. This was taken as proof of the sudden disintegration of the carbon shells arising from the high strain generated by the shrinking fullerene cages surrounding these endohedrally trapped metal atoms.

The significant cavity size present in C_{60} (~ 3.5 Å diameter) and other fullerenes allows practically any element of the periodic table to be incorporated inside them (Fig. 1). Endohedral complexes of fullerenes are represented as $M_m@C_{2n}$, where the symbol @ denotes that the metal M is encapsulated within the carbon cage, in contrast to being attached outside (exohedral) or incorporated within the shell (heterohedral, see pp. 93 this volume.

Endohedral metallofullerenes were first prepared in macroscopic yields by laser or arc vaporization of graphite impregnated with rare-earth metal oxides [11]. More generally, they are prepared by evaporation of hollow carbon rods filled with metal oxides or carbides. As early as 1993, pristine samples of metallofullerenes became available [12,13]. The isolation and purification of endohedral complexes is usually carried out by high performance liquid chromatography (HPLC) through a tedious and repetitive process often necessitating strictly airless conditions. Characteristic isolated yields for pure endohedral metallofullerenes can be up to 1% of the produced soot, amounting to microgram up to milligram quantities. Curiously, the metals incorporated by these methods are essentially limited to rare-earth elements (Sc, Y, La, Ce, Pr, Nd, Sm, Eu, Gd, Tb, Dy, Ho, Er, Tm, Yb, Lu) [12–14], although a few examples in the alkaline-earth metals have become available recently ($M@C_{72}$, $M@C_{74}$, $M@C_{80}$, $M@C_{82}$, $M@C_{84}$; M = Ca, Sr, Ba) [15]. The reasons for why only certain metal atoms can be

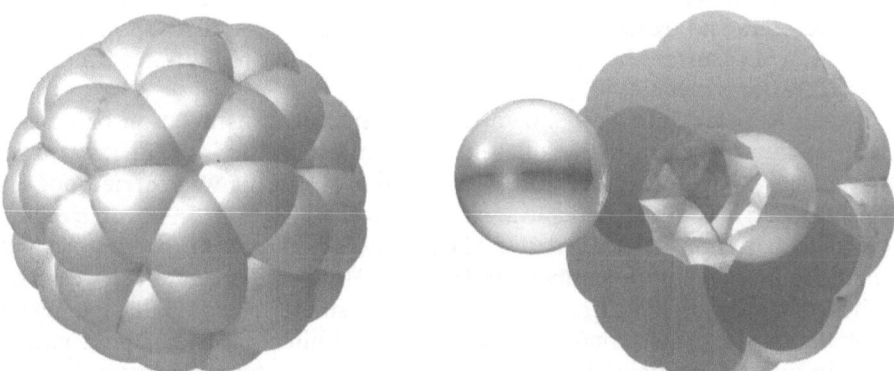

Fig. 1. Space filling representation of C_{60} and the hollow cavity within it, shown next to a transition metal atom (radius = 2.0 Å) for size comparison. The structures were ray-traced with POV-Ray Tracer 3 downloadable at: http://www.povray.org. Note that the atoms are not hard as perhaps suggested here by the metallic surfaces; metal atoms can "bury" themselves significantly within the π-cloud of the carbon shell through metal-to-carbon bonding

incorporated within a fullerene by the arc-evaporation method are not well understood.

Several pure endohedral metallofullerenes have been characterized, including Ln@C_{82} (Ln=La, Y, Ce, Nd, Gd, Pr) [12], as well as some incorporating two or even three metal atoms such as La$_2$@C_{72}, La$_2$@C_{80}, Sc$_2$@C_{74}, Ce$_2$@C_{80}, Sc$_2$@C_{82}, Sc$_2$@C_{84}, and Sc$_3$@C_{82} [13]. The main obstacles to their study are their low production yields and, for some of them, their high air sensitivity. There are two important questions regarding these species: (1) Are the metal atoms within the cages, and (2) what are the charge states of the metals?

X-ray synchrotron powder diffraction of solid Y@C_{82} confirmed for the first time that the metal is indeed within this structure, as is the case for Sc$_2$@C_{84}, and the other lanthanides by inference [13c,16]. Most interestingly, the metal in Y@C_{82} is bound in an off-center position inside the cage. This creates a large, permanent dipole, which can be observed for example in oriented head-to-tail clusters of Y@C_{82} by scanning tunneling microscopy (STM) [17]. In fact, calculations had predicted this behavior [18]. In addition, both theory and experiment confirm the existence of an interesting dynamic circular motion of the metal atoms inside some of these compounds. In the case of La$_2$@C_{80}, ^{13}C and ^{139}La NMR investigations indicate that the lanthanum atoms are bound loosely to the cage (5 kcal/mol activation barrier) and circulate inside the cage at a fast rate [19].

The charge state of the metals inside the fullerenes has been studied by electron spin resonance (ESR). The lanthanum atom inside La@C_{82} was first determined to be in the +3 state, with scandium and yttrium inside Sc@C_{82} and Y@C_{82} proving to be in the same state [11b,c,20]. However, from more recent ESR experiments, there is a standing question on the amount of charge transfer in these compounds, i.e. whether these structures can be best described as either Ln^{3+}C$_{82}^{3-}$ or Ln^{2+}C$_{82}^{2-}$ [4b]. Most interesting is the fact that these structures can be formally compared to "superatoms" in that the positive metal acts like a nucleus and the carbon shell as the "electron shell" of an atom [4b]. In this regard, dimers and trimers of Y@C_{82} have been observed to form spontaneously on a Cu(111) surface by STM in a way perhaps similar to lithium forming Li$_2$ and Li$_3$ cluster molecules [17].

Elegant experiments have shown that endohedral transmutation of an element can take place within the fullerene cage while it is practically unaffected by high-energy neutron capture and β-emission events. This was carried out on non-radioactive Gd@C_{82}, which transmutes into Tb@C_{82} under neutron irradiation with emission of a β-particle [21]. Monoisotopic ^{165}Ho in macroscopic quantities (mg) of pure samples of Ho@C_{82}, Ho$_2$@C_{82}, or Ho$_3$@C_{82} was also activated with a high neutron flux to produce stable ^{166}Er [22]. These initial results strongly suggest that endohedral metallofullerenes may find a useful place in nuclear medicine in a way similar to boron neutron capture therapy (BNCT) [23].

Several reports on the chemistry of endohedral lanthanide fullerene complexes indicate that they are quite similar in reactivity to the empty fullerenes. However, their redox chemistry is very rich since they can also get easily oxidized in contrast to C_{60} and C_{70} [13b,24]. Reactions with a disilirane [*cyclo-*

($Mes_2Si-CH_2-SiMes_2$), Mes = mesityl] or germanium analogs [25], and with diphenyldiazomethane [26], give well-defined addition products in a manner similar to the empty fullerenes. This is very encouraging in that the rich chemistry developed for empty fullerenes should be applicable to these novel species [2].

2.2
Endohedral Metallofullerenes Based on C_{60}

In analogy to the spectacular physical properties found for C_{60} and its deriva-
tives, endohedral metallofullerenes have great potential in the materials and biological sciences. The most interesting endohedral metallofullerenes are like-
ly to be those based on icosahedral C_{60} because this high symmetry confers its special properties to the empty fullerene [5], whereas higher fullerenes like C_{70}, C_{76}, and C_{84} have not yet demonstrated the same trends. It is therefore imperative that these structures are isolated.

In this regard, it appears that the endohedral lanthanide–C_{60} complexes are formed abundantly in the resistive heating method, but they cannot be isolated because they are either too reactive or insoluble. Reports on the extraction of endohedral complexes of C_{60} in aniline or pyridine solutions have recently appeared [27]. The extracts were characterized by LD-TOF mass spectroscopy, indicating the presence of substantial amounts of the $M@C_{60}$ species among higher metallofullerenes. Unfortunately, well-characterized endohedral complexes of C_{60} have not yet been isolated [15b], and it is not certain whether the species observed by mass spectroscopy are the intact fullerenes or derivatives of them. It is possible that aniline or hydroxylated adducts of $M@C_{60}$ are formed, which would not be necessarily distinguishable from the parent $M@C_{60}$ by mass spectrometry if the addends fragment easily. This is the case for many C_{60} adducts where the C_{60} ions make by far the most intense peak in FAB and especially LD-TOF mass spectra. Nevertheless, the above results are important in that they indicate that the C_{60} complexes are viable entities, which was not at all cer-
tain for a long time. It is useful to point out that these compounds suffer the same preparative limitations as those mentioned above.

Reports on the formation of $Li@C_{60}$ or $Li@C_{70}$ by ion beam implantation of the metals into thin films of the fullerenes have appeared [28]. The ~400-nm thick films were characterized by LD-TOF, IR and Raman spectroscopy. However, until recently, there was not much support for the *molecular* integrity of these materials. The very recent CS_2-extraction and HPLC isolation of $Li@C_{60}$-
and $Li@C_{70}$-containing fractions constitutes an important development [28b]. Intriguingly, two fractions of $Li@C_{60}$-containing material are isolated, sug-
gesting that there are different chemical species in solution, perhaps generated by post-incorporation exohedral functionalization with water or oxygen. This observation is supported by the fact that empty C_{60} is more easily extracted from the thin films by CS_2 than $Li@C_{60}$-containing material. The latter is probably polymeric to a large degree in the solid state and only gets extracted upon reaction with air, water, or the solvent CS_2.

In an innovative experiment, high-energy ^7Be atoms, generated in the nuclear recoil following neutron capture of lithium-7 [$^7Li(p,n)^7Be$], were incorporated

into C_{60} molecules in their trajectory [29]. The incorporation was monitored by γ-ray measurement during elution of the soluble extracts with HPLC showing high radioactivity within the eluting C_{60} fraction. However, the amount of ^7Be incorporation inside C_{60} is presumably low.

A number of achievements have been made in endohedral metallofullerene research and the outlook on their physical and chemical properties is very exciting. However, it is imperative to find better approaches to their preparation. A synthetic-organic approach to these attractive compounds should bring a practical answer to the limitations of the graphite evaporation method. Subsequent sections in this chapter present some of the first results towards this goal.

2.3
Noble Gases Trapped Inside Fullerenes

Reports of noble gas incorporation inside C_{60} and C_{70} appeared as early as 1991 from the Schwarz group [30]. These first experiments were carried out by gas phase collisions of helium with high-energy C_{60} and C_{70} radical cations in a mass spectrometer. Consequently, the endohedral complexes could not be isolated in macroscopic amounts in these experiments. In an important development, Saunders and Cross et al. reported that heating crystalline fullerenes at 650°C under 3000 atmospheres of the noble gases helium, neon, argon, krypton, or xenon results in fractional incorporation of these atoms inside the fullerene structures [31]. This remarkable transformation occurs with relatively low efficiency, however, to give materials with a composition of endohedral complex up to ~0.1% of the total fullerene mass. In the process of incorporation, a significant portion of C_{60} is lost through polymerization. These compounds have found an interesting application in the ^3He NMR spectroscopy of fullerene derivatives [32]. Organic derivatives of the $^3He@C_{60}$, $^3He@C_{70}$, and $^3He@C_{76}$ complexes have characteristic ^3He chemical shifts due to differential distortions and loss of ring currents of the fullerene framework by the added fragment, strongly influencing the shielding of ^3He inside those compounds [32]. This property has the potential of becoming a facile method for the characterization of regioisomers, for example, of bisadducts [32d].

Recently, remarkable enrichment of the endohedral component up to 400 times has been realized by HPLC of the $C_{60}/Ar@C_{60}$ mixture, giving µg quantities of $Ar@C_{60}$ samples with greater than 40% purity [33]. The current degree of incorporation achieved during synthesis (0.1%) has been calculated to be far from equilibrium [34]. Ab initio MP2 calculations with an extended basis set give a $2\,kcal \cdot mol^{-1}$ thermodynamic advantage to helium incorporation inside C_{60}. The computed equilibrium (RHF/EXT or BLYP/EXT) at 650°C and 3000 atmospheres exceeds 10%. It is clear that suitable catalysts should considerably increase the experimental incorporation yields.

An important question relevant to this chapter lies in the mechanism of this incorporation reaction, which can also be realized in a preparative way by beam implantation [35]. The prevalent theory for the solid-state incorporation of noble gases inside fullerenes, which is so far unique to these unreactive atoms, has been that one or two bonds are broken temporarily to allow the noble gas

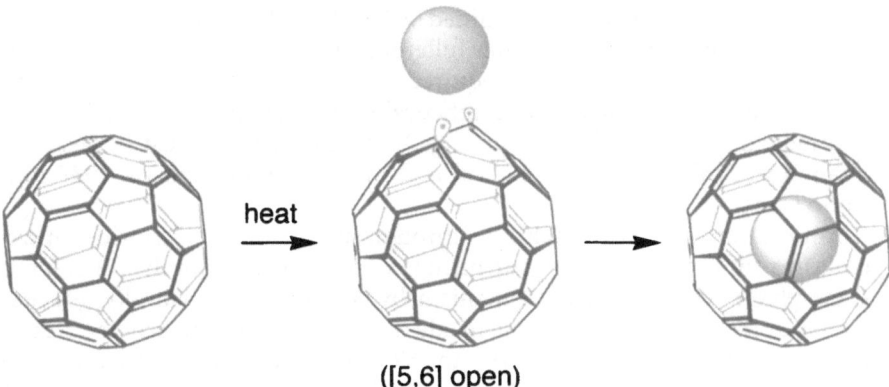

([5,6] open)

Scheme 1. Schematic representation of one of the possible mechanisms involved in the incorporation of noble gases inside C_{60}

atom to slip in (Scheme 1). Under the high-temperature and high-pressure conditions of this reaction, vibrational distortions of the cage must favor a temporary lengthening of the narrow distance separating two non-bonded carbon atoms. This process has been calculated at good levels of theory and was given a lower estimate of ~200 kcal mol^{-1} for the activation barrier [36]. However, experimental evidence based on noble gas release under heating suggests that this barrier may be much lower at 90 kcal mol^{-1} [37]. This number, supported to some extent by BLYP//MNDO calculations, was taken as an indication that a catalyst must be involved [38]. Reversible radical additions loosen the C–C bonds adjacent to the addition site or temporary dimerization of the fullerene increases the orifice size. In any case, conclusive proof of this mechanism has to be brought forward. Its possible application in the preparation of endohedral fullerene complexes with other elements could be crucial.

As noted above, endohedral noble gas–fullerene compounds have found a special niche as NMR probes of magnetic ring currents within fullerene cages and, as a result, provide a diagnostic tool for chemical addition reactions. If the incorporation of xenon can be increased to the same or higher levels as the other noble gases, ^{129}Xe NMR could provide dramatically larger chemical shift sensitivities adding to the utility of these compounds. On the other hand, the fact that the noble gases do not interact significantly with the inside π-orbital lobes of the cage implies that the chemical and physical properties of these compounds will not differ much from those of empty fullerenes. This is one of the reasons why the formation of endohedral metal–C_{60} complexes remains an important quest.

2.4
Atomic Nitrogen Trapped Inside Fullerenes

A recent surprising development is the partial incorporation of the first highly reactive element, atomic nitrogen, inside C_{60} [39]. This new species, N@C_{60}, has astounding stability considering that the nitrogen atom inside it has the atomic quartet ground state ($^4S_{3/2}$) as determined by ESR [39,40]. It implies that the interior of C_{60}, and presumably of higher fullerenes, is exceptionally inert. This can be understood by the high energy required to hybridize a carbon atom inwards in order to form a bond with the endohedral nitrogen atom, resulting in a very unfavorable distortion of the carbon cage [41]. The energetic cost of this bond formation was calculated to be 62 kcal mol^{-1} by the semi-empirical UHF-PM3 method [41a].

The formation of N@C_{60} is achieved through sublimation of C_{60} inside a nitrogen plasma tube, whereby energetic nitrogen cations react with the fullerene in the gas phase. The probable mechanism by which the exohedrally attached nitrogen atoms are able to penetrate inside the cages is shown in Scheme 2. First, a nitrogen atom reacts with C_{60} to give the 5,6-open structure 1 or the analogous 6,6-open structure. Under the plasma conditions, these atoms have enough energy to pass through the barrier of insertion to form the covalently bound endohedral compound 2, which as pointed out above has very unusual hybridization and must therefore undergo immediate homolytic scission of both bonds to give the endohedral product 3 (N@C_{60}). This mechanism is supported by the fact that the ESR signal at 260 °C for endohedral nitrogen decreases with time, suggesting that it escapes from the cage unless other quenching pathways are operating. The barrier of this endohedral-to-exohedral expulsion was estimated to be substantially lower (40 kcal · mol^{-1} at 260 °C) than that of the ^{3}He

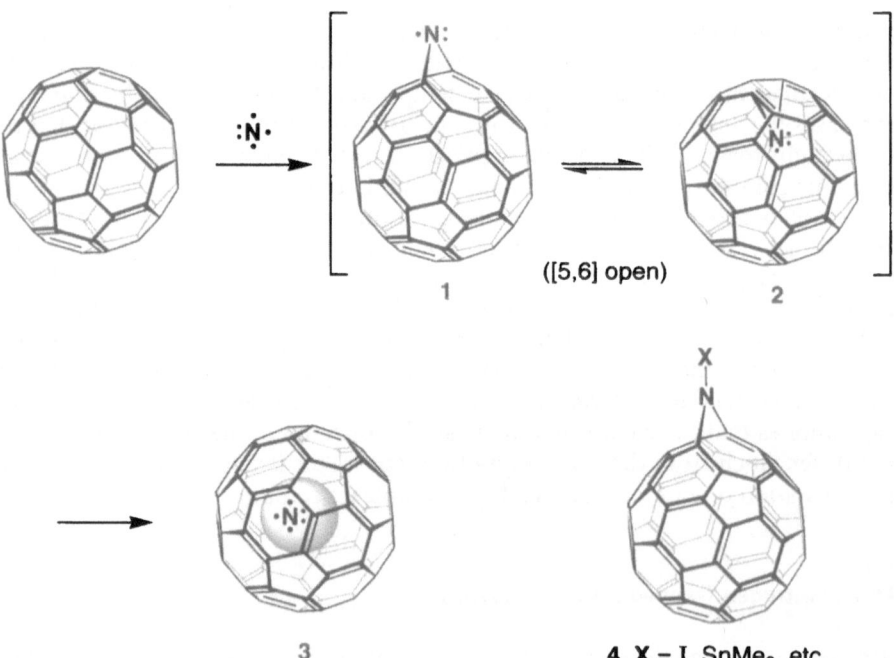

Scheme 2. Schematic representation of the probable mechanism for the incorporation of a nitrogen atom inside C_{60} in plasma experiments

atom release (90 kcal·mol^{-1} at 630 °C) [37, 41a]. There is indeed a large difference between this "covalently" driven penetration compared to the high-energy process required for the incorporation of the noble gases in which covalent bonds cannot be formed temporarily. Correspondingly, larger amounts of N@C$_{60}$ may become accessible through appropriate generation of 1 from defined chemical precursors 4.

As inferred from calculations, most light elements should behave similarly to atomic nitrogen inside C$_{60}$ and form unusual endohedral compounds [41b]. Their formation is one more hot topic of pursuit in current fullerene research.

3
Ring-Opening Reactions of Fullerene Shells

The formation of a large orifice within a fullerene framework could constitute a powerful method of introducing practically any atom inside these hollow structures. However, the task of opening an effective aperture on the surface of fullerenes has proven very challenging. Should this be achieved, reforming the original fullerene shell through ring-closing reactions will also need to be addressed, ideally by exploiting the reversibility of the original ring-opening reaction(s). This whole game can be compared – in a humorous manner – to the "Pacman" game illustrated in Fig. 2.

Fig. 2. The "Pacman" analogy to metal insertion inside fullerenes

3.1
First Observation of an Opening Reaction

The [5,6]-fulleroids, and corresponding [5,6]-*aza*fulleroids such as 5 (Scheme 3), are the first compounds to embody an opening within their shell in the form of a one-bond scission (see pp. 93 this volume). However, the cavity is completely constrained by the proximity of the attached bridge and bridgehead carbons. A breakthrough in this search was achieved when Wudl's group observed that azafulleroid 5 reacts in a regioselective [2+2]-cycloaddition reaction with singlet oxygen ($^1O_2$) [42]. This induces the preferential formation of a dioxetane intermediate 6 spontaneously rearranging to the ketolactam 7. The 11-membered ring opening in 7 – although real as shown by the space-filling model in Scheme 3 – is rather small and sterically hindered. In an attempt to estimate the effective size of the cavity, expulsion experiments – the reverse of insertion – were made with endohedrally ^{3}He-labeled 7 [43]. Heating of labeled 7 in solution up to 200 °C did not show any escape of ^{3}He (van der Waals radius = 1.2 Å) as revealed by ^{3}He NMR.

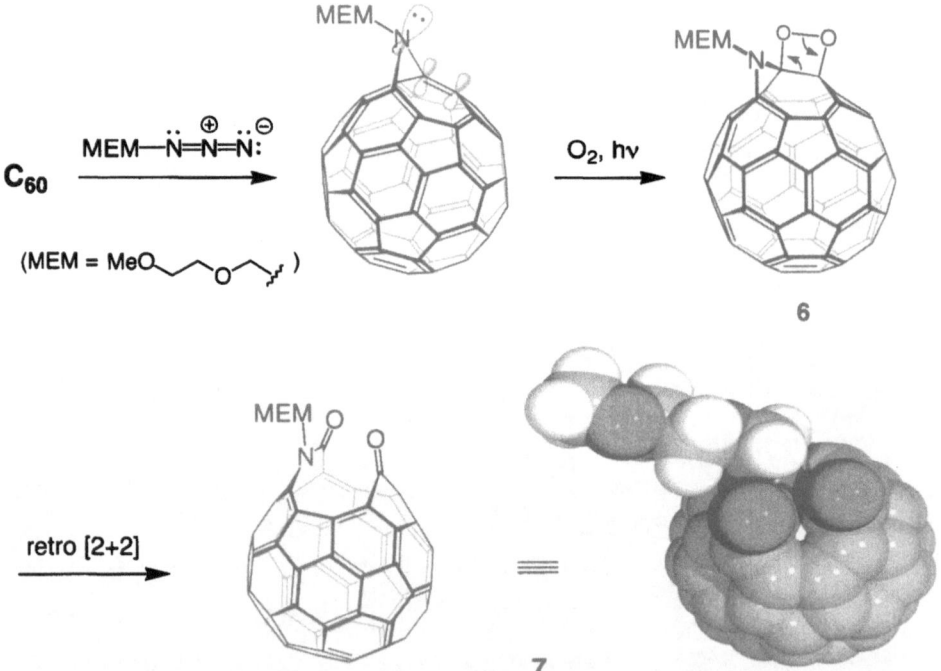

Scheme 3. Formation and space-filling representation of the first ring-opened fullerene 7. The van der Waals radius of all carbon atoms was reduced to 88% to accentuate the size of the opening; the inside of the cavity is illuminated for easier visualization

3.2
Other Restricted Openings

Several reports on the formation of openings within fullerene shells have appeared since Wudl's first communication [44]. Unfortunately, none of them – except for 14 – have matched the effective opening size of ketolactam 7 because they are all bridged by one or two atoms.

In a recent example, Hirsch et al. found that the regioselectivity of two azide additions to C_{60} can be controlled to give the opened bisiminofullerene 9b (Scheme 4) [44a]. After the initial monoadduct 8 formation (6,6-closed), the second addition of azide occurs primarily at the *cis*-1 position, which is unusual in respect to the inaccessibility of this site in other cycloadditions. The first aziridine nitrogen is small enough to allow addition of the incoming nitrene or azide at the *cis*-1 site, helped in part by favorable HOMO coefficients at this double bond. Remarkably, the two [6,6] bonds in the primary bisiminofullerene 9a open up to afford the first example of a system (9b) having [6,6] ring junctions broken. The corresponding 14-atom opening in 9b is quite large (not count-

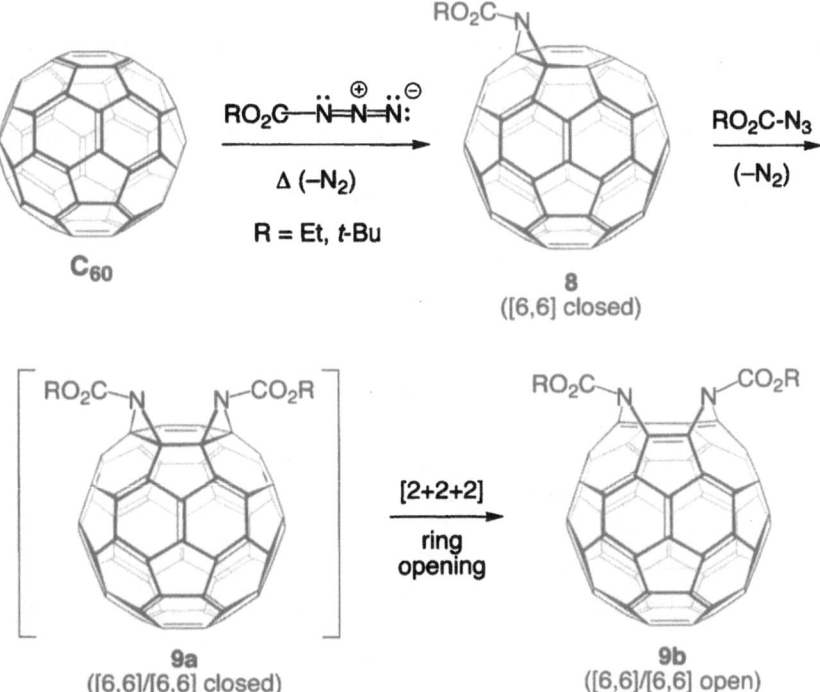

Scheme 4. Addition of two azides to C_{60} *via* the intermediate [6,6]-closed aziridine 8 to give the first [6,6]/[6,6]-open structure 9b

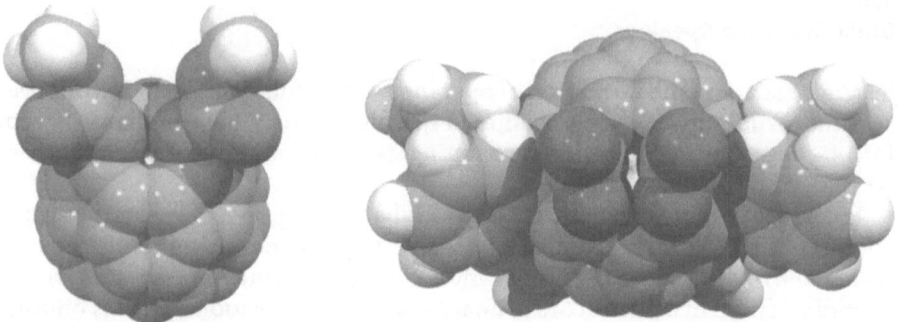

Fig. 3. Space-filling representations of the ring-opened fullerenes **9b** and **14**. Rendering options are the same as in Scheme 3

ing the bridging N-CO$_2$R moieties), but unfortunately it is obstructed by the two imino bridges (Fig. 3).

Luh et al. [44b, c], as well as Chan et al. [44d], have carried out similar bisadditions involving two azide groups that are tied together with a short alkyl chain as in compound **10** (Scheme 5). In this case, another type of bisadduct is formed (**12**) with a 13-atom orifice (not including the bridging -NCH$_2$C(Bn)$_2$CH$_2$N-moiety) next to the [6,6]/[6,6]-closed diaziridine **11**. Direct conversion of **11** to the [5,6]/[5,6]-open system **12** by heating was not observed. Consequently, the [5,6]/[5,6] ring opened bis-azafulleroid **12** must be formed directly from the intermediate bis-triazoline adduct in a mechanism similar to the one related by Wudl in chap 5.

Taylor et al. found that spontaneous oxidation of Ph$_8$C$_{70}$ (**13**) in air leads to Ph$_8$C$_{70}$O$_4$ (**14**), a luminescent compound which was assigned the bis-lactone structure shown in Scheme 6 based on NMR and IR data (13-atom orifice not

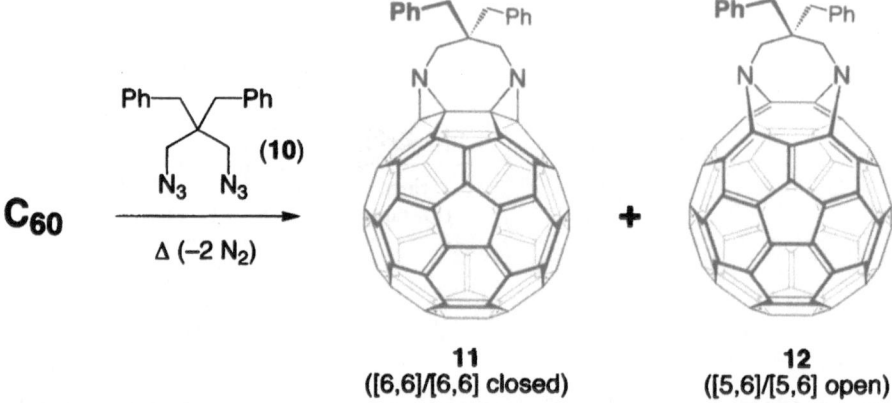

Scheme 5. Addition of a bisazide to C$_{60}$ to give the [5,6]/[5,6] open bisazafulleroid **12**

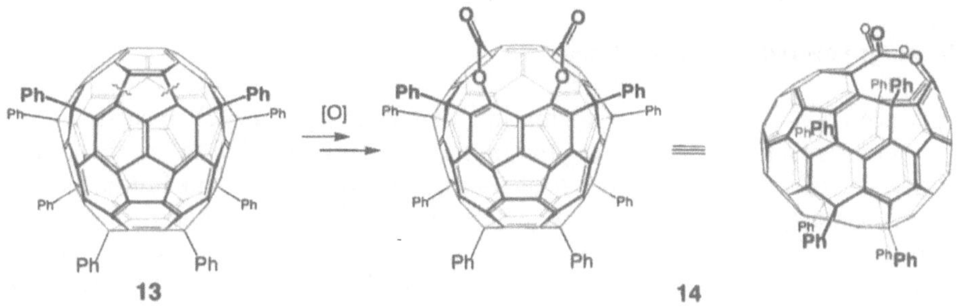

Scheme 6. Oxidation of Ph_8C_{70} (**13**) to $Ph_8C_{70}O_4$ (**14**)

counting the bridging -COO- moieties) [45]. Its precursor **13** has an energetically unfavorable [5,6]-double bond, which would favor oxygen addition at this location [45b]. Ring opening of a putative dioxetane intermediate and Bayer–Williger-type oxidation of the resulting diketone presumably affords the bis-lactone regioisomer **14** having a significant cavity (Fig 3). However, unequivocal proof of this structure awaits a single crystal X-ray structure determination since there is presently no certainty for the exact location of the two lactone bridges.

Both types of restricted openings in compounds **9b** and **14** (Fig. 3) provide illustrative examples of the current difficulties encountered in opening reactions. In principle, an additional imino bridge can be placed at the third double bond of bisiminofullerene **9a** to give the larger opening of structure 15 (Fig. 4). However, here too, the nitrogen bridges constrain the cavity to a very large extent. Similarly, in the bis-lactone **14**, possible elaboration of the structure by ester cleavage and decarboxylation/oxidation reactions may provide access to more widely opened fullerene shells. These and many other possibilities offer an exciting playground for imaginative chemical explorations.

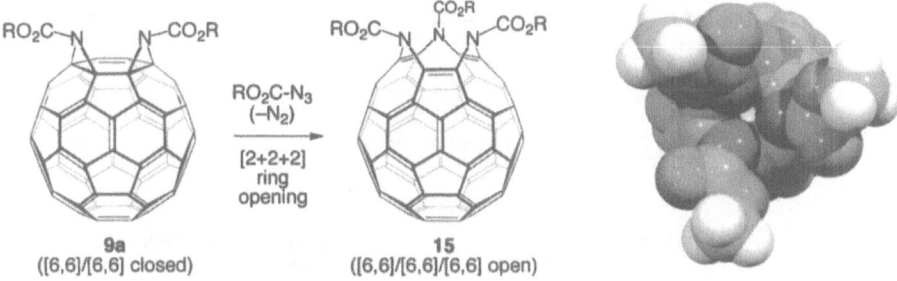

Fig. 4. Structure and space-filling representation of the hypothetical ring-opened fullerene 15. Rendering options are the same as in Scheme 3

3.3
Openings Covered by a Metal Atom

In a scenario similar to the formation of N@C_{60} (Scheme 2), insertion into fullerenes is facilitated by having the metal covalently bonded to the rim of the cavity. In addition, the formation of the metal–carbon covalent bonds can be advantageously effected by oxidative insertion of the metal. This principle was demonstrated recently by our group with the discovery of a very facile sequence of bond scissions involving pericyclic and metal insertion reactions [46]. The so-far largest formal ring opening – 15 atoms without counting the C_4H_4 and Co bridges – was generated as the cobalt(III) complex 19a having the metal attached on top of the opening (Scheme 7). This complex is the final product of an overall sequential triple scission of a 6-membered ring on C_{60} affording the 15-membered ring opening, whereby the conversion of diene 16 to the bisfulleroid 18 involves a retro [2 + 2 + 2] cyclization step (Scheme 8). Two of the three 5-membered rings surrounding the original 6-membered ring open up during the conversion of intermediate 17 to the bisfulleroid 18, while the last bond scission occurs during the oxidative insertion step with cobalt(I) to afford 19a.

A general difficulty in fullerene chemistry resides in the structural characterization of new compounds. Contrary to most aliphatic or aromatic compounds where connectivities can be deduced readily from H–H coupling data, carbon–carbon connectivities in fullerene frameworks cannot be directly ascertained on a routine basis from ^{13}C NMR experiments unless recourse is made to

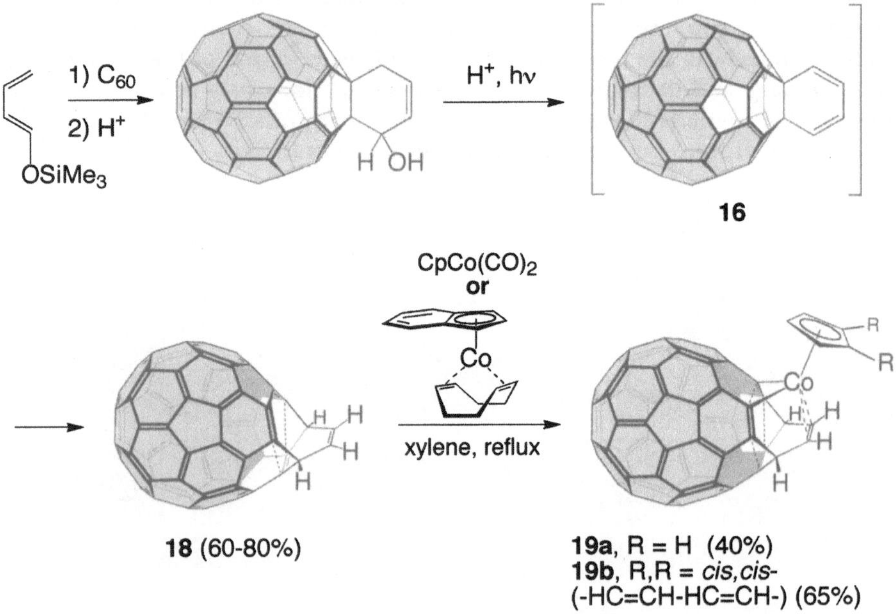

Scheme 7. Synthesis of the cobalt(III) complexes 19a and 19b

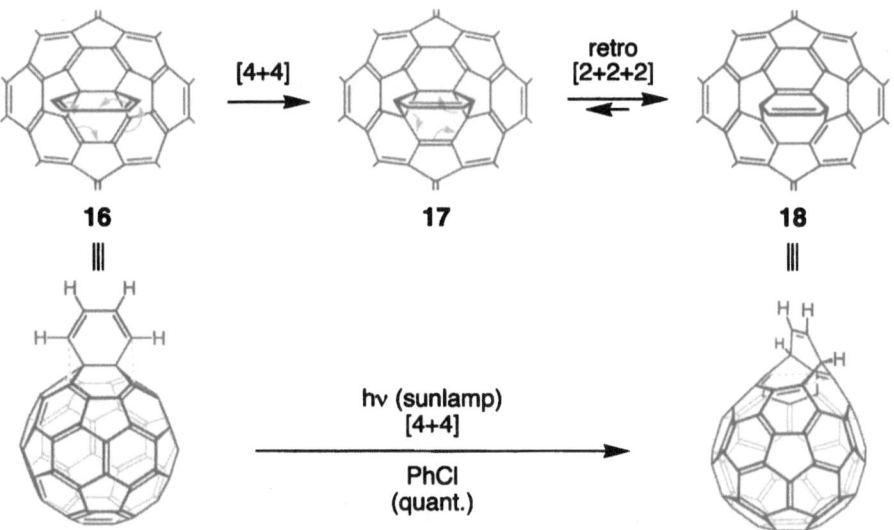

Scheme 8. Mechanism of the conversion of compound 16 to the bisfulleroid 18

expensive INADEQUATE or HOHAHA experiments using ^{13}C-labeled material [47]. As a consequence, symmetry and reactivity are commonly used as circumstantial evidence of structural integrity. The structure of complex **19a** could be unequivocally confirmed by single crystal X-ray crystallography (Fig. 5). Even though the C_4H_4 bridge (C1–C4) constrains its cavity, this structure has its cobalt atom ideally seated on top of the orifice of the cage, waiting for insertion into the fullerene pending proper activation. Several attempts to effect cobalt insertion have been performed, including high-temperature heating of the crystals up to 400°C [3] or under high pressure [48]. From the head-to-tail arrangement of molecules in the crystal (Fig. 5, bottom), it was hypothesized that crystal vibrations, in addition to intramolecular vibrations, could promote the metal insertion process at higher temperatures and pressures. However, this was not found to be effective under the applied conditions [3,48].

Alternatively, we have explored oxidation, reduction, or photochemical reactions to induce the cyclopentadienyl ligand (Cp) to detach itself from the metal. It appears – not unexpectedly – that the cyclopentadienyl ligand is a critical parameter to play with because it is strongly bound to the cobalt atom. More recent work in our laboratory has allowed the preparation of compound **19b** (Scheme 7) having the more labile indenyl ligand [3b,49]. The latter underwent facile loss of indene upon heating in the crystalline state as found by DSC and TGA. Unfortunately, loss of indene was also accompanied by rapid solid state polymerization affording intractable material. Examination of the thermal chemistry of **19b** by laser desorption mass spectroscopy did not indicate the generation of $Co@C_{64}H_4$ or $Co@C_{60}$.

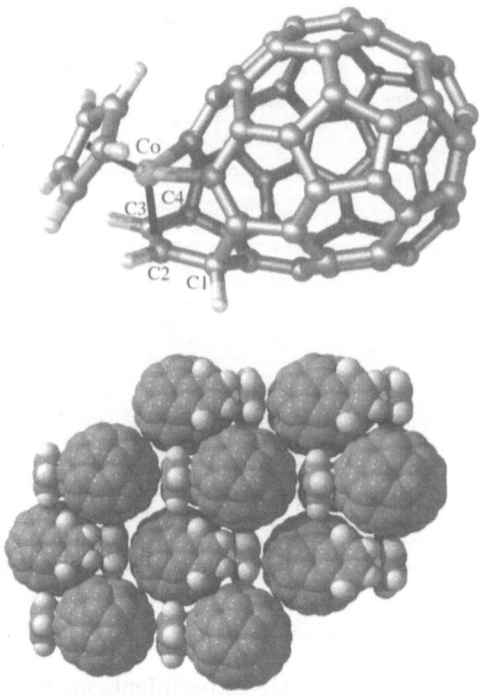

Fig. 5. *Top* X-ray structure of cobalt complex **19a** in a shaded ball-and-stick representation. Note that the carbon atoms C1–C4 were part of the original cyclohexadiene ring in compound **16**. The isolated 5-membered ring is the cobalt-complexed cyclopentadienyl ligand. *Bottom* Packing structure of **19a** in the crystal showing the close proximity between the Cp ligands and the fullerene $C_{64}H_4$ moieties

4
Potential Mechanisms for the Formation of Effectively Large Openings

One of the principal reasons for the difficulty of inserting cobalt into the fullerene cage of complexes **19a** and **19b** has to do with the small effective size of the orifice as can be seen readily from the space-filling representation of Fig. 6. Because the C_4H_4 bridge in complex **19b** hinders full expansion of the opening in the same way as in compounds **12**, **14**, and **15**, we have started to investigate more effective mechanisms for ring-opening reactions.

An intriguing mechanism was recently proposed by Shijun Zheng in our group, whereby the ligand is used as a template to direct radical or electrophilic additions of two groups at the fullerene sp^2-hybridized carbons directly attached to cobalt (Figs 6 and 7) [49]. In concurrence with this hypothesis, the HOMO coefficients at these carbons are significant as found by PM3 calculations. Furthermore, linking the ligand to the fullerene core effectively locks the metal into place, having no other way to go than inside the fullerene if heat and/or pressure are applied. Approaches to this type of complexes are currently being pursued in our group.

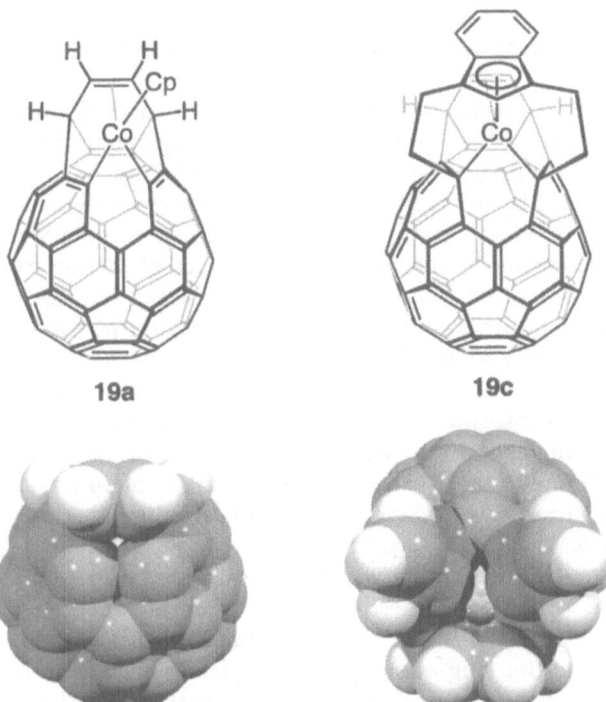

Fig. 6. Structures and space-filling representations of the cobalt complex **19a** from its X-ray structure coordinates, and the PM3 calculated cobalt complex **19c**, for which the Cp or indenyl ligands and cobalt atoms have been removed to show the openings. The van der Waals radius of all carbon atoms was reduced to 85 % to accentuate the size of the openings; the inside of the cavity is illuminated for easier visualization

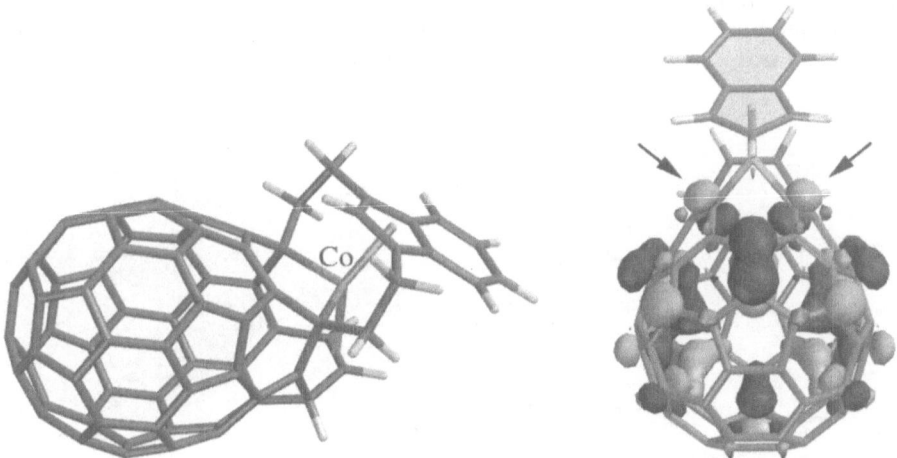

Fig. 7. PM3 calculated structure of cobalt complex **19 c** and HOMO orbitals of **19 b** (PM3)

4.1
Openings Resulting from Unhindered Retro [2 + 2 + 2] Reactions

The mechanism operating in practically all known cases of opening reactions consists of a concerted [2 + 2 + 2] six-membered ring opening illustrated by the transformations leading to compounds **5**, **9b**, **12**, and **18**. Accordingly, this mechanism appears to be the most powerful tool in opening rings within fullerene shells, and its exploitation in forming *unhindered* openings is indispensable. This mechanism operates at various degrees of carbon oxidation states (sp^2 or sp^3) provided that highly bent alkyne moieties are not formed in the bond rearrangement process. If one examines the latter possibility on C_{60} itself (Scheme 9), it is readily deduced that this opening reaction is largely unfavorable as calculated by PM3. This mechanism was not examined in the papers by Thiel et al. in regard to the insertion mechanism of noble gases inside fullerenes [36,38]. However, it is unlikely to operate in view of its high energetic cost, unless presaturation of one or more carbons occurs at the 6-membered ring via reversible additions of radicals or oxygen as hypothesized for the one- or two-bond mechanisms [36,38].

In contrast to the above process, partially saturated 6-membered rings as in compounds **9a** or **17** (Schemes 4 and 8) undergo energetically favorable [2 + 2 + 2] ring openings. If one can go as far as saturating all carbons of a 6-membered ring, the ring-opening reaction leads to an unhindered orifice, provided that there are no bridging groups to constrain it (Scheme 10).

This situation was examined theoretically with the $C_{60}H_6$ model compound **21 a** and its methylated analog **22 a** (Table 1). As can be readily seen from the computed differences in heats of formation, opening of the 1,2,3,4,5,6-hexahydro[60]fullerene **21 a** in two different ways shows that it should be an endothermic process. The two possible modes of opening correspond to bond-breaking events at either [6,6] or [5,6] ring junctions, termed 6- or 5-open for the purpose of this chapter. The latter is not unexpectedly more favorable because it involves breaking σ-bonds at fused 5-membered rings which are more strained than 6-membered rings. Accordingly, opening energies for the 5-open structure **21 b** range from 9.0 to 29.2 kcal·mol^{-1} at AM1, PM3, and ab initio RHF/3–21G

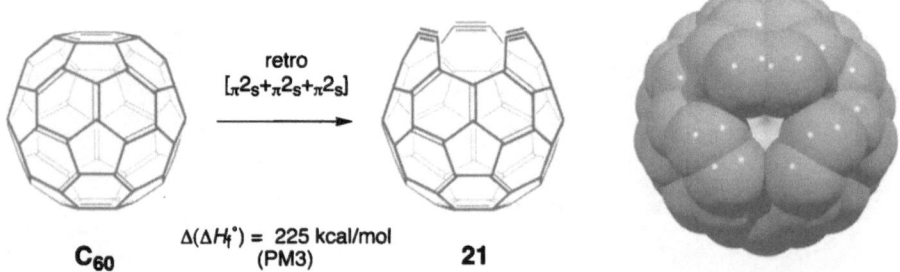

$$\Delta(\Delta H_f^\circ) = 225 \text{ kcal/mol (PM3)}$$

C$_{60}$ **21**

Scheme 9. [2 + 2 + 2] ring opening of C_{60} leading to triyne **21** and space-filling representation of the PM3-calculated structure of triyne **21**. Rendering options are the same as in Scheme 3

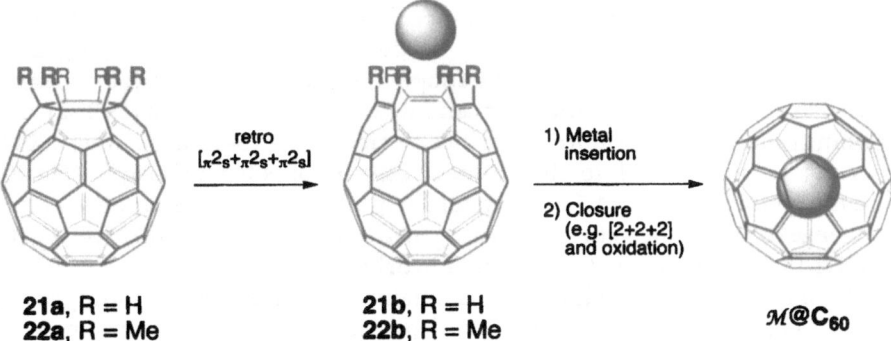

21a, R = H
22a, R = Me

21b, R = H
22b, R = Me

$\mathcal{M}@C_{60}$

Scheme 10. [2 + 2 + 2] ring opening of a 1,2,3,4,5,6-hexahydro-derivative of C_{60} as a potential entry into endohedral metallofullerenes

Table 1. Relative semi-empirical (AM1, PM3) and ab initio (RHF 3-21G) heats of formation for 1,2,3,4,5,6-hexahydro derivatives **21a–c**, 1,2,3,4,5,6-hexamethyl derivatives **22a–c**, and structures **16–18** involved in the [4 + 4] cyclization of Scheme 8 [52]. All energies are given in kcal mol^{-1}. The ab initio calculated geometries for **21a–c** are shown below

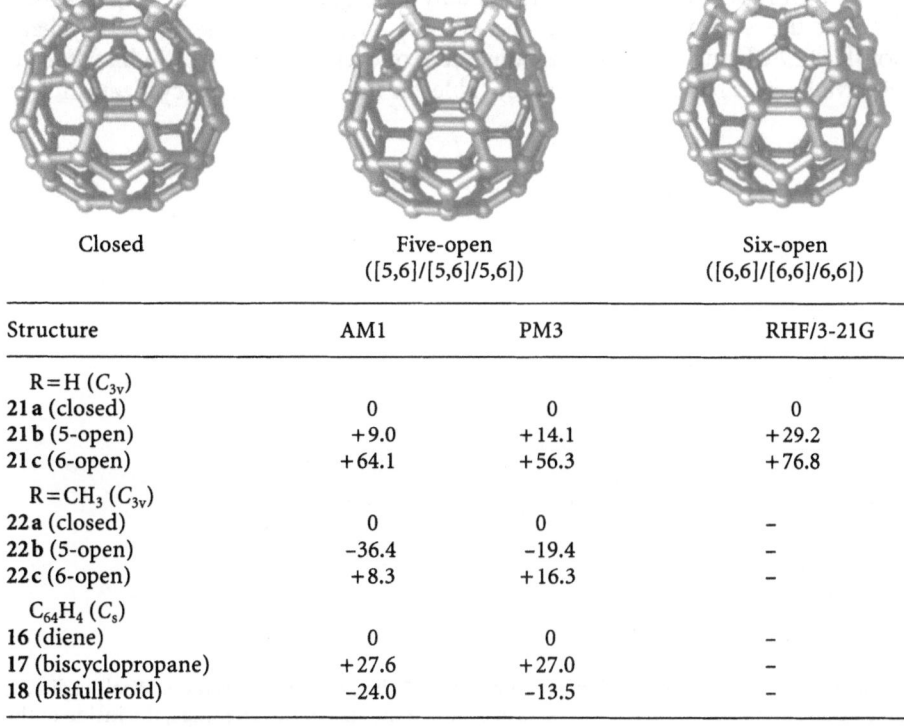

	Closed	Five-open ([5,6]/[5,6]/5,6])	Six-open ([6,6]/[6,6]/6,6])
Structure	AM1	PM3	RHF/3-21G
R=H (C_{3v})			
21a (closed)	0	0	0
21b (5-open)	+9.0	+14.1	+29.2
21c (6-open)	+64.1	+56.3	+76.8
R=CH$_3$ (C_{3v})			
22a (closed)	0	0	–
22b (5-open)	–36.4	–19.4	–
22c (6-open)	+8.3	+16.3	–
C$_{64}$H$_4$ (C_s)			
16 (diene)	0	0	–
17 (biscyclopropane)	+27.6	+27.0	–
18 (bisfulleroid)	–24.0	–13.5	–

levels, while opening energies for the 6-open structure **21c** are 30–55 kcal · mol^{-1} higher. Although these energies have to be treated with caution [50], their magnitude already indicates that the simple 1,2,3,4,5,6-hexahydro[60]fullerene **21a** will presumably not be the right candidate for the unhindered [2+2+2] ring opening process. Furthermore, the orifice created by the 15-atom opening in **21b** may not be wide enough to allow a metal atom to pass through easily (Fig. 8), and the synthesis of 1,2,3,4,5,6-hexahydro[60]fullerene **21a** poses a considerable challenge owing to the general dehydrogenative tendency of its 1,2,33,41,42,50- and 1,2,18,22,23,36-isomers and similar hydrides [51].

Considerably larger cavities can be obtained by replacing the hydrogens of **21b** with more sterically demanding addends, for example, methyl groups as in **22b** (Scheme 10, Fig. 8). While ab initio methods correctly predict small diffe- rences in energies between the isomers of $C_{60}H_4$ or $C_{60}(CH_2)_2$ [51,53], they are computationally demanding and have not yet been carried out on the methyl- ated series **22a–c**. In any case, the PM3 semi-empirical method appears to give reliable results within ± 2 kcal mol$^-$ [51,53] and can be taken as an indication of general trends. As expected, the AM1 and PM3 energies of the open structures **22b** and **22c** are considerably lower (~40–55 kcal · mol^{-1}) than their counter- parts **21b** and **21c** and, here again, the 5-open structure **22b** is much more favor- ed than the 6-open (**22c**) by at least 36 kcal · mol^{-1}. As a consequence, the 5-open structure **22b** lies 36.4 and 19.4 kcal · mol^{-1} lower, respectively, than the closed structure **22a**.

For calibration with an experimentally confirmed [2 + 2 + 2] ring opening process, the structures of the compounds involved in the [4 + 4] cycliza- tion/[2 + 2 + 2] ring opening rearrangement of cyclohexadiene **16** to bisfulleroid **18** have been calculated (Scheme 8, Table 1). The [4 + 4] photocyclization step

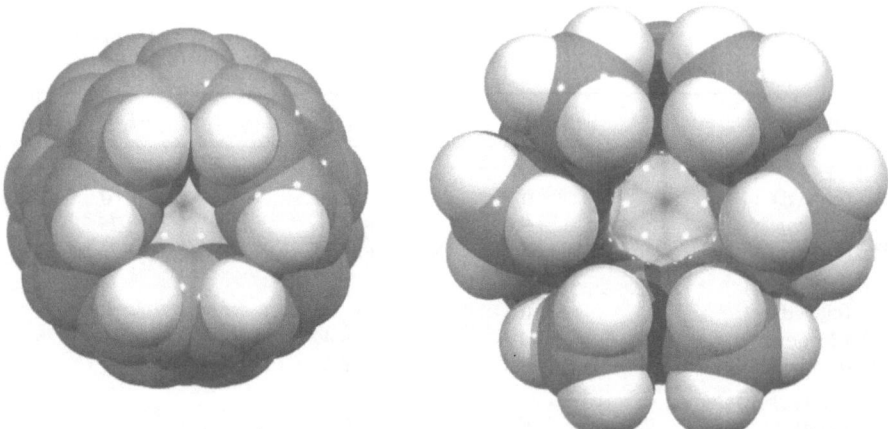

Fig. 8. Space-filling representations of compounds **21b** and **22b**. The van der Waals radius of all carbon atoms was reduced to 88% to accentuate the size of the openings; the inside of the cavities are illuminated for easier visualization

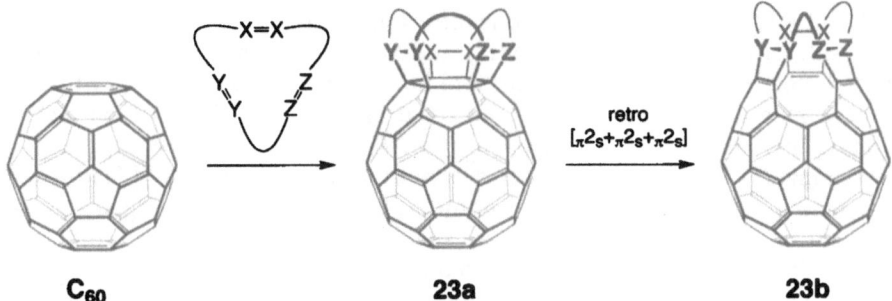

Scheme 11. [2 + 2 + 2] ring opening reaction leading to unhindered orifices

(**16** to **17**) goes uphill at both semi-empirical levels, and largely downhill to the [2 + 2 + 2] ring opened product **18**, in good agreement with the experimental observation. Since intermediate **17** has never been observed experimentally, it is clear that it undergoes [2 + 2 + 2] ring opening with a large thermodynamic advantage, corroborating the calculated values.

It can be inferred from these calculations that 5-open structures with substituents larger than hydrogen will be formed from the corresponding 1,2,3, 4,5,6-hexahydro derivatives with a solid thermodynamic advantage. However, there is a considerable challenge in synthesizing such strained systems. The six substituents of the closed structures **21a** and **22a** are all eclipsed, amounting to substantial steric clashes. If six groups are to be placed on a single 6-membered ring in one or through multistep reactions by addition to the three double bonds, strong "incentives" for these groups not to react with adjacent double bonds (e.g. *cis*-2, *cis*-3) have to be provided. An example following this precept is shown in Scheme 11. Here, all three adding groups (diene, 1,3-dipole, [2 + 2] partner, etc. [2]) are tied together in a macrocyclic structure, resulting in good preorganization in all of the three transition states leading to the addition reactions. We are currently examining several systems utilizing this principle.

5
Conclusion

The aspiration of preparing the first macroscopic amounts of endohedral metallofullerenes has been partly fulfilled by the application of the arc-evaporation method or by insertion of noble gases or atomic nitrogen in fullerenes under high-energy conditions. In spite of these important results, the study of their chemical and physical properties is just beginning to emerge owing to the scarce availability of these compounds. Most significantly, endohedral metallofullerenes based on C_{60} are not yet characterized. As a consequence, finding more practical methods to obtain these compounds in high incorporation or preparative yields is of utmost importance.

[2 + 2 + 2] Ring-opening reactions offer some of the potentially most important entries into preparatively useful and versatile ways to form endohedral metallofullerenes. Several problems still have to be addressed in order to form widely open orifices, but they do not seem insurmountable. Practical realizations of the retro [2 + 2 + 2] ring opening on the strained bis-cyclopropane **17** or bisziridine **9a** indicate that proper design of adding groups could lead to unhindered openings. Even smaller unhindered orifices can be enlarged through thermal activation, in all likeliness assisted by intermediate η^2,η^2,η^2-complexation of the three orifice C=C bonds to the metal. Such intermediate complexation could expand the orifice to a point that allows the metal to slip inside the fullerene under proper thermal activation and/or pressure. Furthermore, most metals should bind more or less strongly to the inside fullerene cavity [16–19], providing thermodynamic incentive for metal insertion.

6
References

1. (a) Krätschmer W, Lamb LD, Fostiropoulos K, Huffman DR (1990) Nature 347:354; (b) Ajie H, Alvarez MM, Anz SJ, Beck RD, Diederich F, Fostiropoulos K, Huffman DR, Krätschmer W, Rubin Y, Schriver KE, Sensharma D, Whetten RL (1990) J Phys Chem 94:8630
2. For reviews, see: (a) Taylor R, Walton DRM (1993) Nature 363:685; (b) Hirsch A (1994) The chemistry of the fullerenes. Thieme, Stuttgart; c) Diederich F, Isaacs L, Philp D (1994) Chem Soc Rev 23:243; (d) Hirsch A (1995) Synthesis 895; (e) Taylor R (1995) The chemistry of fullerenes. World Scientific, River Edge, NJ; (f) Diederich F, Thilgen C (1996) Science 271:317; (g) Diederich F (1997) Pure Appl Chem 69:395; (h) Thilgen C, Herrmann A, Diederich F (1997) Angew Chem Int Ed Engl 36:2269; (i) Prato M, Maggini M (1998) Acc Chem Res, ASAP Article: Web release date, June 3
3. (a) Rubin Y (1997) Chem Eur J 3:1009; (b) Rubin Y (1998) Chimia 52:118
4. For reviews, see: (a) Edelmann FT (1995) Angew Chem Int Ed Engl 34:981; (b) Nagase S, Kobayashi K, Akasaka T (1996) Bull Chem Soc Jpn 69:2131; (c) Nagase S, Kobayashi K, Akasaka T (1997) Theochem J Mol Struct 398:221; (d) Nagase S, Kobayashi K, Akasaka T (1998) J Comput Chem 19:232
5. (a) Dresselhaus MS, Dresselhaus G, Eklund PC (1995) Science of fullerenes and carbon nanotubes. Academic Press, San Diego; (b) Holczer K, Klein O, Huang SM, Kaner RB, Fu KJ, Whetten RL, Diederich F (1991) Science 252:1154; (c) Holczer K, Whetten RL (1992) Carbon 30:1261; (d) Allemand PM, Khemani KC, Koch A, Wudl F, Holczer K, Donovan S, Grüner G, Thompson JD (1991) Science 253:301; (e) Nalwa HS (1993) Adv Mater 5:341; (f) Sun YP, Riggs JE, Liu B (1997) Chem Mater 9:1268
6. For reviews on the applications of lanthanide(III) ions in MRI, see: (a) Aime S, Botta M, Fasano M, Terreno E (1998) Chem Soc Rev 27:19; (b) Lauffer RB (1987) Chem Rev 87:901
7. Jensen AW, Wilson SR, Schuster DI (1996) Bioorg Med Chem 4:767
8. Kroto HW, Heath JR, O'Brien SC, Curl RF, Smalley RE (1985) Nature 318:162
9. Heath JR, O'Brien SC, Zhang Q, Liu Y, Curl RF, Kroto HW, Tittel FK, Smalley RE (1985) J Am Chem Soc 107:7779
10. Weiss FD, Elkind JL, O'Brien SC, Curl RF, Smalley RE (1988) J Am Chem Soc 110:4464
11. (a) Chai Y, Guo T, Jin C, Haufler RE, Chibante LPF, Fure J, Wang L, Alford JM, Smalley RE (1991) J Phys Chem 95:7564; (b) Johnson RD, de Vries MS, Salem J, Bethune DS, Yannoni CS (1992) Nature 355:239; (c) Shinohara H, Sato H, Saito Y, Ohkohchi M, Ando Y (1992) J Phys Chem 96:3571

12. (a) Kikuchi K, Suzuki S, Nakao Y, Nakahara N, Wakabayashi T, Shiromaru H, Saito K, Ikemoto I, Achiba Y (1993) Chem Phys Lett 216:67; (b) Kikuchi K, Nakao Y, Suzuki S, Achiba Y, Suzuki T, Maruyama Y (1994) J Am Chem Soc 116:9367; (c) Ding JQ, Weng LT, Yang SH (1996) J Phys Chem 100:11120; (d) Ding JQ, Lin N, Weng LT, Cue N, Yang SH (1996) Chem Phys Lett 261:92; (e) Funasaka H, Sakurai K, Oda Y, Yamamoto K, Takahashi T (1995) Chem Phys Lett 232:273; (f) Ding JQ, Yang SH (1996) J Am Chem Soc 118:11254

13. (a) Stevenson S, Burbank P, Harich K, Sun Z, Dorn HC, van Loosdrecht PHM, de Vries MS, Salem JR, Kiang CH, Johnson RD, Bethune DS (1998) J Phys Chem A 102:2833; (b) Suzuki T, Maruyama Y, Kato T, Kikuchi K, Nakao Y, Achiba Y, Kobayashi K, Nagase S (1995) Angew Chem Int Ed Engl 34:1094; (c) Shinohara H, Yamaguchi H, Hayashi N, Sato H, Ohkohchi M, Ando Y, Saito Y (1993) J Phys Chem 97:4259; (d) Takata M, Nishibori E, Umeda B, Sakata M, Yamamoto E, Shinohara H (1997) Phys Rev Lett 78:3330; (e) Ding J, Yang S (1996) Angew Chem Int Ed Engl 35:2234; (f) Shinohara H, Inakuma M, Hayashi N, Sato H, Saito Y, Kato T, Bandow S (1994) J Phys Chem 98:8597; (g) van Loosdrecht PHM, Johnson RD, de Vries MS, Kiang CH, Bethune DS, Dorn HC, Burbank P, Stevenson S (1994) Phys Rev Lett 73:3415

14. (a) Weaver JH, Chai Y, Kroll GH, Jin C, Ohno TR, Haufler RE, Guo T, Alford JM, Conceicao J, Chibante LPF, Jain A, G. P, Smalley RE (1992) Chem Phys Lett 190:460; (b) Gillan EG, Yeretzian C, Min KS, Alvarez MM, Whetten RL, Kaner RB (1992) J Phys Chem 96:6869; (c) Kirbach W, Dunsch L (1996) Angew Chem Int Ed Engl 35:2380

15. (a) Wan TSM, Zhang H-W, Nakane T, Xu Z, Inakuma M, Shinohara H, Kobayashi K, Nagase S (1998) J Am Chem Soc 120:6806; (b) Dennis TJS, Shinohara H (1998) J Chem Soc Chem Commun 883; (c) Dennis TJS, Shinohara H (1997) Chem Phys Lett 278:107

16. Takata M, Umeda B, Nishibori E, Sakata M, Saito Y, Ohno M, Shinohara H (1995) Nature 377:46

17. Shinohara H, Inakuma M, Kishida M, Yamazaki S, Hashizume T, Sakurai T (1995) J Phys Chem 99:13769

18. (a) Andreoni W, Curioni A (1996) Phys Rev Lett 77:834; (b) Laasonen K, Andreoni W, Parrinello M (1992) Science 258:1916

19. Akasaka T, Nagase S, Kobayashi K, Wälchli M, Yamamoto K, Funasaka H, Kako M, Hoshino T, Erata T (1997) Angew Chem Int Ed Engl 36:1643

20. (a) Suzuki S, Kawata S, Shiromaru H, Yamauchi K, Kikuchi K, Kato T, Achiba Y (1992) J Phys Chem 96:7159; (b) Hoinkis M, Yannoni CS, Bethune DS, Salem JR, Johnson RD, Crowder MS, de Vries MS (1992) Chem Phys Lett 198:461

21. Kikuchi K, Kobayashi K, Sueki K, Suzuki S, Nakahara H, Achiba Y, Tomura K, Katada M (1994) J Am Chem Soc 116:9775

22. Cagle DW, Thrash TP, Alford M, Chibante LPF, Ehrhardt GJ, Wilson LJ (1996) J Am Chem Soc 118:8043

23. (a) Soloway AH, Tjarks W, Barnum BA, Rong F-G, Barth RF, Codogni IM, Wilson JG (1998) Chem Rev 98:1515; (b) Hawthorne MF (1993) Angew Chem Int Ed Engl 32:950

24. (a) Suzuki T, Maruyama Y, Kato T, Kikuchi K, Achiba Y (1993) J Am Chem Soc 115:11006; (b) Suzuki T, Kikuchi K, Oguri F, Nakao Y, Suzuki S, Achiba Y, Yamamoto K, Funasaka H, Takahashi T (1996) Tetrahedron 52:4973

25. (a) Akasaka T, Nagase S, Kobayashi K, Suzuki T, Kato T, Kikuchi K, Achiba Y, Yamamoto K, Funasaka H, Takahashi T (1995) Angew Chem Int Ed Engl 34:2139; (b) Akasaka T, Kato T, Nagase S, Kobayashi K, Yamamoto K, Funasaka H, Takahashi T (1996) Tetrahedron 52:5015; (c) Akasaka T, Nagase S, Kobayashi K, Suzuki T, Kato T, Yamamoto K, Funasaka H, Takahashi T (1995) J Chem Soc Chem Commun 1343

26. Suzuki T, Maruyama Y, Kato T, Akasaka T, Kobayashi K, Nagase S, Yamamoto K, Funasaka H, Takahashi T (1995) J Am Chem Soc 117:9606

27. (a) Kubozono Y, Maeda H, Takabayashi Y, Hiraoka K, Nakai T, Kashino S, Emura S, Ukita S, Sogabe T (1996) J Am Chem Soc 118:6998; (b) Kubozono Y, Noto T, Ohta T, Maeda H, Kashino S, Emura S, Ukita S, Sogabe T (1996) Chem Lett 453; (c) Kubozono Y, Hiraoka K, Takabayashi Y, Nakai T, Ohta T, Maeda H, Ishida H, Kashino S, Emura S, Ukita S, Sogabe T

(1996) Chem Lett 1061; (d) Kubozono Y, Ohta T, Hayashibara T, Maeda H, Ishida H, Kashino S, Oshima K, Yamazaki H, Ukita S, Sogabe T (1995) Chem Lett 457

28. (a) Campbell EEB, Tellgmann R, Krawez N, Hertel IV (1997) J Phys Chem Solids 58:1763; (b) Gromov A, Krätschmer W, Krawez N, Tellgmann R, Campbell EEB (1997) J Chem Soc Chem Commun 2003

29. Ohtsuki T, Masumoto K, Ohno K, Maruyma Y, Kawazoe Y, Sueki K, Kikuchi K (1996) Phys Rev Lett 77:3522

30. Weiske T, Bohme DK, Hrusak J, Krätschmer W, Schwarz H (1991) Angew Chem Int Ed Engl 30:884

31. (a) Saunders M, Jiménez-Vázquez HA, Cross RJ, Poreda RJ (1993) Science 259:1428; (b) Saunders M, Jiménez-Vázquez HA, Cross RJ, Mroczkowski S, Gross ML, Giblin DE, Poreda RJ (1994) J Am Chem Soc 116:2193; (c) Saunders M, Cross RJ, Jiménez-Vázquez HA, Shimshi R, Khong A (1996) Science 271:1693

32. (a) Saunders M, Jiménez-Vázquez HA, Bangerter BW, Cross RJ, Mroczkowski S, Freedberg DI, Anet FAL (1994) J Am Chem Soc 116:3621; (b) Saunders M, Jiménez-Vázquez HA, Cross RJ, Billups WE, Gesenberg C, Gonzalez A, Luo W, Haddon RC, Diederich F, Herrmann A (1995) J Am Chem Soc 117:9305; (c) Smith AB, Strongin RM, Brard L, Furst GT, Atkins JH, Romanow WJ, Saunders M, Jiménez-Vázquez HA, Owens KG, Goldschmidt RJ (1996) J Org Chem 61:1904; (d) Cross RJ, Jiménez-Vázquez HA, Lu Q, Saunders M, Schuster DI, Wilson SR, Zhao H (1996) J Am Chem Soc 118:11454; (e) Ruttimann M, Haldimann RF, Isaacs L, Diederich F, Khong A, Jiménez-Vázquez H, Cross RJ, Saunders M (1997) Chem Eur J 3:1071; (f) Jensen AW, Khong A, Saunders M, Wilson SR, Schuster DI (1997) J Am Chem Soc 119:7303

33. DiCamillo BA, Hettich RL, Guiochon G, Compton RN, Saunders M, Jiménez-Vázquez HA, Khong A, Cross RJ (1996) J Phys Chem 100:9197

34. Patchkovskii S, Thiel W (1997) J Chem Phys 106:1796

35. Shimshi R, Cross RJ, Saunders M (1997) J Am Chem Soc 119:1163

36. Patchkovskii S, Thiel W (1996) J Am Chem Soc 118:7164

37. Shimshi R, Khong A, Jiménez-Vázquez HA, Cross RJ, Saunders M (1996) Tetrahedron 52:5143

38. (a) Patchkovskii S, Thiel W (1997) Helv Chim Acta 80:495; (b) Patchkovskii S, Thiel W (1998) J Am Chem Soc 120:556

39. (a) Murphy TA, Pawlik T, Weidinger A, Höhne M, Alcala R, Spaeth JM (1996) Phys Rev Lett 77:1075; (b) Pietzak B, Waiblinger M, Murphy TA, Weidinger A, Hohne M, Dietel E, Hirsch A (1997) Chem Phys Lett 279:259

40. Knapp C, Dinse KP, Pietzak B, Waiblinger M, Weidinger A (1997) Chem Phys Lett 272:433

41. (a) Mauser H, van Eikema Hommes NJR, Clark T, Hirsch A, Pietzak B, Weidinger A, Dunsch L (1998) Angew Chem Int Ed Engl 36:2835; (b) Mauser H, Hirsch A, Hommes N, Clark T (1997) Journal of Molecular Modeling 3:415

42. Hummelen JC, Prato M, Wudl F (1995) J Am Chem Soc 117:7003

43. Wudl F, Saunders, M, personal communication; see also pp. 93 this volume

44. (a) Schick G, Hirsch A, Mauser H, Clark T (1996) Chem Eur J 2:935; (b) Shen CKF, Yu H, Juo C-G, Chien K-M, Her G-R, Luh T-Y (1997) Chem Eur J 3:744; (c) Shiu LL, Chien KM, Liu TY, Lin TI, Her GR, Luh TY (1995) J Chem Soc Chem Commun 1159; (d) Dong GX, Li JS, Chan TH (1995) J Chem Soc Chem Commun 1725

45. (a) Birkett PR, Avent AG, Darwish AD, Kroto HW, Taylor R, Walton DRM (1995) J Chem Soc Chem Commun 1869; (b) Avent AG, Benito AM, Birkett PR, Darwish AD, Hitchcock PB, Kroto HW, Locke IW, Meidine MF, Odonovan BF, Prassides K, Taylor R, Walton DRM, van Wijnkoop M (1997) J Mol Struct 437:1

46. Arce MJ, Viado AL, An YZ, Khan SI, Rubin Y (1996) J Am Chem Soc 118:3775

47. (a) Hawkins JM, Loren S, Meyer A, Nunlist R (1991) J Am Chem Soc 113:7770; (b) Hawkins JM, Meyer A, Lewis TA, Bunz U, Nunlist R, Ball GE, Ebbesen TW, Tanigaki K (1992) J Am Chem Soc 114:7954

48. Edwards CM, Butler IS, Qian WY, Rubin Y (1998) J Mol Struct 442:169

49. Zheng S, Rubin Y, unpublished results

50. (a) Cahill PA (1996) Chem Phys Lett 254:257; (b) Henderson CC, Rohlfing CM, Assink RA, Cahill PA (1994) Angew Chem Int Ed Engl 33:786
51. (a) Bergosh RG, Meier MS, Cooke JAL, Spielmann HP, Weedon BR (1997) J Org Chem 62:7667; (b) Meier MS, Weedon BR, Spielmann HP (1996) J Am Chem Soc 118:11682
52. AM1/PM3 semi-empirical and RHF/3-21G ab initio calculations were performed with Spartan 4.0 for Silicon Graphics computers.
53. Rohlfing CM, Cahill PA (1997) Molec. Phys. 91:561

Heterofullerenes

Jan C. Hummelen[1] · Cheryl Bellavia-Lund[2] · Fred Wudl[3]

[1] Stratingh Institute & Materials Science Center, University of Groningen, Nijenborgh 4, NL-9747 AG Groningen, The Netherlands. *E-mail: j.c.hummelen@chem.rug.nl*
[2] SYMYX Technologies, 3100 Central Expressway, Santa Clara, CA 95051, USA. *E-mail: clund@symyx.com*
[3] Department of Chemistry and Biochemistry, 4505A Molecular Science Building, University of California, 405 Hilgard Ave, Los Angeles, CA 90095-1569, USA. *E-mail: wudl@chem.ucla.edu*

The state of the art in heterofullerene chemistry and physics is reviewed with emphasis on azafullerenes. The macroscopic synthetic methods that have been developed for aza[60] fullerene compounds since 1995 have led to a whole new and rich area in the science of fullerenes: cage modification chemistry. The synthetic routes towards aza[60]fullerene and its derivatives are reviewed in Sect. 2. The synthetic routes for aza[70]fullerene and its derivatives are summarized in Sect. 3. Section 4 comprises the theoretical and experimental work on the physicochemical properties of azafullerene compounds. Finally, in Sect. 5, the literature regarding heterofullerenes other than monoazafullerenes is reviewed.

Keywords: Heterofullerene, Azafullerene, Borafullerene, Fullerene Dimer, Azirenofullerene, Azafulleroid, Fullerene Salt, Boron, Holeyball, Dopeyball, Dopyball.

Abbreviations

MEM   2-methoxyethoxymethyl
SEM   2-(trimethylsilyl)ethoxymethyl
ODCB   1,2-dichlorobenzene
pTsOH *para*-toluenesulfonic acid

1
Introduction

Heterofullerenes ("dopeyballs", "heterohedral" fullerenes) are fullerenes in which one or more carbon atoms that form the cage structure is replaced by a non-carbon atom, i.e. a heteroatom. Fullerenes are defined by IUPAC as poly-hedral closed cages made up entirely of n three-coordinate carbon atoms and having 12 pentagonal and $(n/2-10)$ hexagonal faces, where n is equal to or greater than 20. Carbon cages that do not fulfill the requirement are referred to as *quasi*-fullerenes; the hetero-analogues of such structures being hetero-*quasi*-fullerenes and they will not be reviewed in this chapter. Since most inorganic (non-carbon) cage and cluster structures do not fulfill the requirement, these compounds are also not considered here. Thus, this review concerns mainly hetero-analogues of [60]fullerene (Buckminsterfullerene C_{60}) and [70]fullerene.

In 1991, 6 years after the experimental discovery of the fullerenes [1] and one year after the discovery of a method for bulk preparation of fullerenes [2], the first spectroscopic observation of gas-phase formation of heterofullerene ions was reported by the Smalley group [3]. Very soon thereafter, other groups found spectroscopic evidence for gas-phase heterofullerene ion formation and some theoretical studies concerning the electronic structure and thermodynamic stability of heterofullerenes emerged (see Sect. 5). From the beginning, hetero-fullerenes have been considered as interesting molecules and materials, mainly because it was envisioned that through modification of the cage structure, sig-nificant modification of geometry, chemical functionality and electronic char-acter of the fullerenes would occur. Consequently, it was anticipated that these heterofullerenes should exhibit a variety of properties, e.g. superconductivity (leading to superconducting materials with improved thermodynamic stability) or altered electronegativity (making them interesting candidates for photo-diode-based devices).

In 1995, three chemistry research groups found independently that, during mass spectroscopic analysis, certain exohedrally nitrogen-containing [60]fuller-ene *derivatives* showed efficient gas-phase rearrangements to aza[60]fullerene ions [4–6]. Based on the gas-phase observation and after rationalization of the events that take place during the rearrangement process, the first and highly efficient synthetic method for the bulk preparation of aza[60]fullerene was developed in Santa Barbara (USA) in 1995 [7]. Soon thereafter, the Hirsch group succeeded in converting their gas-phase process into a preparative method [8]. These developments have enabled scientists to synthesize and study a variety of aza[60]- and aza[70]fullerene compounds. Since there are no such synthetic methods for other fullerenes to date, azafullerenes are by far the most studied heterofullerenes until now. Hence, this review concerns mainly the preparation of aza[60]fullerene and its derivatives (Sect. 2), aza[70]fullerene and derivatives (Sect. 3), and the properties of aza[60]- and aza[70]fullerene compounds (Sect. 4). In Sect. 5, we will summarize the work concerning presently observed or considered (mainly [60]-) heterofullerenes other than aza[60]- and aza[70]fullerene.

Although the field is very young, a number of reviews concerning heterofullerenes and heteroatom-containing versions of carbon nanotubes have appeared recently. In some of these, with varying success, efforts have been made to cover the whole area [9–12], in others only part of the field is concerned [13–15], while in some articles the authors summarize their own work on heterofullerenes almost exclusively [16–21].

The nomenclature that has been used for fullerenes, fullerene derivatives, and heterofullerenes, has been until now rather haphazard. In this review, we will generally try to comply with the latest IUPAC nomenclature, as described in the article by Godly and Taylor [22].

2
Synthesis of Aza[60]fullerene

2.1
Introduction

The potential influence of substitution of a carbon atom by a nitrogen atom in the C_{60} cage structure on its electronic and optical properties has attracted the theoretical interest of a considerable number of scientific groups since 1992. Neutral aza[60]fullerene (CAS: $2H$-1-aza[5,6]fulleren-C_{60}-I_h-yl) is an open shell molecule due to the trivalency of nitrogen leaving a "dangling bond" on an adjacent carbon atom in the cage (structure 1 in Fig. 1). Aza[60]fullerene in the first oxidized state is sometimes called aza[60]fulleronium (CAS: azonia[5,6]fullerene-C_{60}-I_h), for which one resonance structure is shown (2, see Fig. 1).

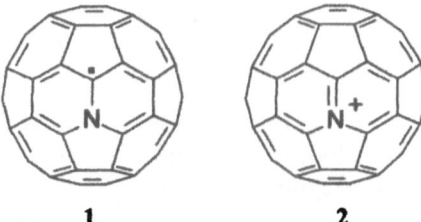

1   **2**

Fig. 1. Important resonance structures of aza[60]fullerene radical (1) and aza[60]fulleronium (2)

The first preparative efforts, using the arc-discharge method in an atmosphere of helium and nitrogen, ammonia, or methylamine, followed by MS analysis of the toluene extraction of the resultant soot, pointed towards the formation of several nitrogen-containing heterofullerenes, but not $C_{59}N$ [23,24]. Gas-phase formation of $C_{59}N$ (as $C_{59}N^+$) was first reported by Christian et al. in 1992 [25]. An experiment in which a N^+ ion beam was collided with an evaporated mixture of C_{60} and C_{70} ($\sim 85:15$) and the collision products were collected by the ion guide and mass analyzed by a double-focusing mass spectrometer and counted by an on-axis Daly detector, yielded evidence for the formation of $C_{59}N^+$. It was reported that the $C_{59}N^+$ thus formed decayed primarily by CN elimination (analogous to the "shrink-wrap" mechanism found for fullerenes) to C_{58}^+. In hindsight, this proves that their $C_{59}N^+$ was almost certainly aza[60]fulleronium, since this shrink-wrap pattern was also observed later in mass spectra of pure samples of $(C_{59}N)_2$ [7]. Arc vaporization of graphite in the presence of pyrrole vapor was reported to yield "N-doped" fullerenes. The products were analyzed by a series of spectroscopic methods, yielding little structural evidence for $C_{59}N$ [26, 27]. Negative mode mass spectroscopic analysis of products from arc vaporization of graphite in an atmosphere of pure nitrogen showed a relatively increased intensity of the peak at m/e 722, which was interpreted as proof for the presence of $C_{59}N^-$ [28]. Products from a similar experiment were also characterized by XPS [29]. Other investigators analyzed similarly obtained products by MS, UV-Vis and IR spectroscopy [30]. Mass spectroscopy, ESR, Raman scattering, and IR analysis of Soxhlet extracts of soot from arc vaporization of graphite rods of which one was filled with boron nitrite, were interpreted as if $C_{59}N$ would exist as an open shell monomer at ambient conditions [31], which is not the case. Laser ablation of graphite in a heated nitrogen gas atmosphere revealed ions with masses in accordance with $C_{59}N^+$ and $C_{69}N^+$ in FT mass spectroscopy. The presence of nitrogen in the product was confirmed by isotope labeling experiments with $^{15}N_2$. The nitrogen 1s level was found to be at 400.7 eV using XPS [32,33]. A hot plasma method for the preparation of all kinds of heterofullerenes including azafullerenes has been patented [34].

This summarizes all the theoretical and gas-phase preparation work that has been done up to now with respect to aza[60]fullerene, except for the developments that led to the realization of actual bulk preparation of [60]- and aza[70]fullerene dimers and derivatives and the investigations of these products. This work will be described in detail in the remainder of this section.

2.2
Synthesis of Aza[60]fullerene Dimer $(C_{59}N)_2$

In 1995, three different but related ways were found that first led to *efficient* gas-phase preparation of aza[60]fulleronium 2. These findings subsequently gave way to two closely related synthetic methods for bulk preparation of $C_{59}N$ in its dimeric form [4–8]. Since all of these methods start with the chemical modification of C_{60} or C_{70} using the addition of azides, we will review the work related to this reaction first in Sect. 2.2.1. A second step in the synthesis is described in Sect. 2.2.2 and the two approaches for the final synthetic step are described in Sects. 2.2.3 and 2.2.4.

2.2.1
The Addition of Azides and Nitrenes to C60

In 1993, it was found that alkyl azides react with C_{60} in a 1,3-dipolar cycloaddition manner [35]. The course of the reaction is closely analogous to the 1,3-dipolar addition of alkyl diazo compounds. Preliminary results on the reaction of C_{60} with organic azides were complicated. Treatment of C_{60} with SEM azide in refluxing chlorobenzene afforded two major products **3** and **4** (Fig. 2). While ^{15}N-^{13}C spin-spin coupling experiments supported the structure for **4**, the ^{13}C NMR spectrum of **3** contained a resonance at 160 ppm coupled to ^{15}N, which was not compatible with the expected triazoline.

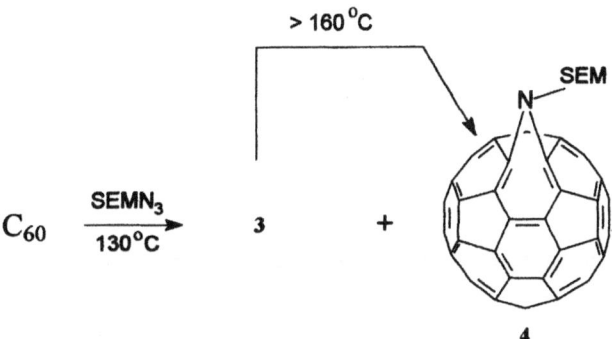

Fig. 2. The addition of $SEMN_3$ to C_{60}

The initial monoaddition product from the reaction with MEM azide across a [6,6] bond of C_{60}, the triazoline **5**, is stable enough to allow isolation at room temperature and confirmation of the structure through X-ray analysis of single crystals [36]. Thermolysis of the monoadduct, triazoline **5**, followed by concomitant loss of nitrogen, gives rise to azafulleroid (azahomofullerene; IUPAC: 1a-aza-1(6)a-homo[60]fullerene; CAS: 2a-aza-1,2(2a)-homo[5,6]fullerene-C_{60}-I_h) **6** and azirenofullerene ("fulleroaziridine", 1,2-(dihydro-1H-aziridino)-[60]fullerene; aziridinofullerene; IUPAC: 1,2-epimino[60]fullerene or 1'H-azireno[2',3':1,2][60]fullerene; CAS: 1'H-[5,6]fullereno-C_{60}-I_h-[1,9-b]azirene) **7** in 15 and 5% yield, respectively (Fig. 3) [35]. Azirenofullerenes and azafulleroids can also be obtained directly from C_{60} at higher reaction temperatures (160–180°C).

Interestingly, when MEM azide addition reactions are carried out at temperatures ranging from 60–140°C, in addition to the formation of the azafulleroid and azirenofullerene, a bisazafulleroid (diazabishomofullerene) **8** is formed and is the major product [37]. For reasons which are not well understood, the double bonds adjacent to the nitrogen bridge are more reactive towards further cyclo-

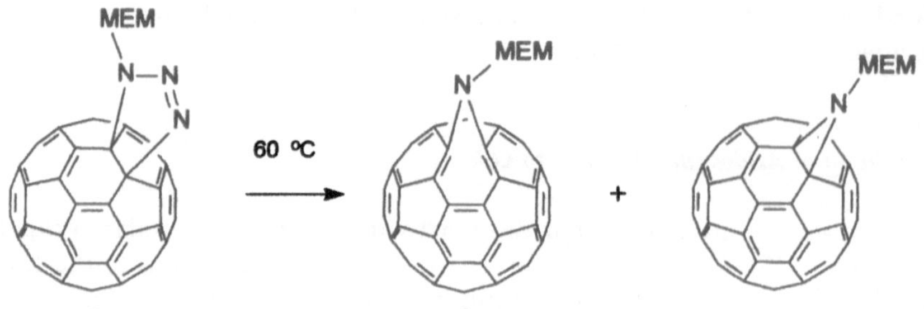

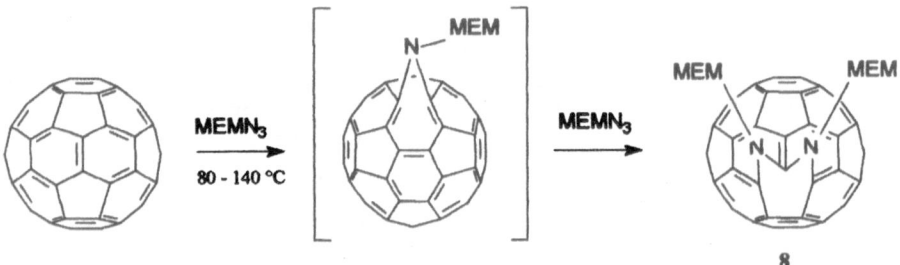

Fig. 3. Thermolysis products from MEM-substituted triazoline **5**

Fig. 4. Chemo- and regioselective formation of bisazafulleroid **8** from C_{60}

addition of alkyl azides than the other double bonds in the fullerene framework. Upon heating above 140 °C, **8** (or a MEM analogue of **3**) reverts back to the mono adduct **6**. The appearance of a carbon resonance at 160 ppm for **8** (**3**) is consistent with a carbon atom adjoined to two nitrogen atoms. (Fig. 4)

Investigation of this phenomenon led to the finding that not only is the azide addition chemoselective, it is also regioselective, as indicated by the isolation of the bisadduct **8**, and not **9** or **10** (Fig. 5). The preparation of bisazafulleroid compounds has been patented [38].

Several groups have noted similar findings for azide addition, in which the structure of the bisadduct depends on the nature of the starting alkyl azide or bisazide [39–41]. Even chiral bisazafulleroids have been prepared [42,43]. As an interesting example, it was found that subsequent addition of azido ester to N-ethoxycarbonyl-azirenofullerene resulted in the formation of bisadduct **12** and not **11** [40]. Structure **12** constitutes the first fullerene derivative having [6, 6] open bridges (Fig. 6). Before that, all [5,6] adducts were found to adopt an open bridged structure and all [6,6] adducts to adopt closed bridged structures.

In the case of **12**, electronic factors exceed the normal characteristic of fullerene derivatization; that is the tendency to avoid endocyclic double bonds in five-membered rings. Prior to this work, Banks claimed the existence of a [5,6]

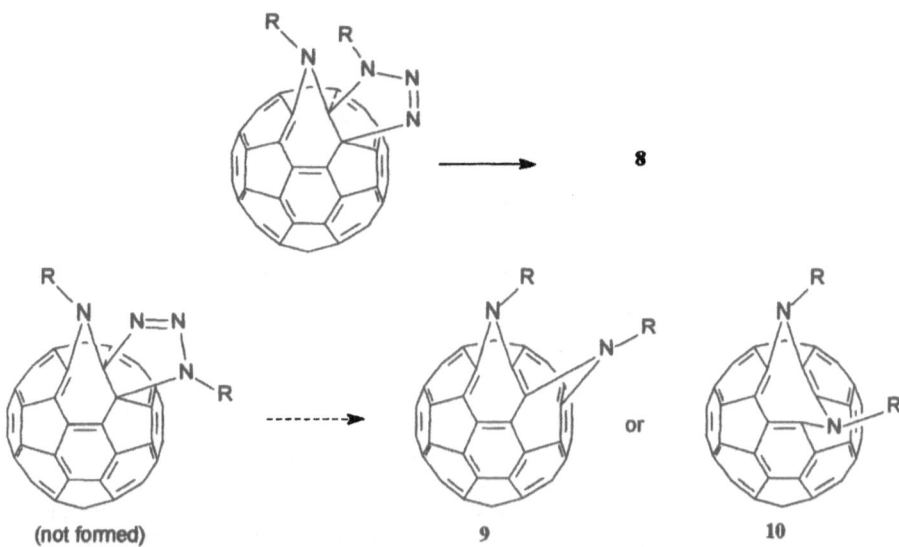

Fig. 5. Expected additional bisazafulleroid isomers **9** and/or **10** in the case of non-regioselective addition of MEMN$_3$ to MEM-azafulleroid **6**

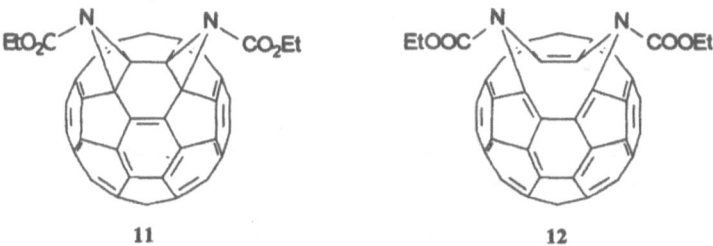

Fig. 6. Addition of azidoester to N-ethoxycarbonyl-azireno[60]fullerene results in formation of bisadduct **12** and not **11**

closed structure, but this was corrected by the Hirsch group [44, 45]. The existence of the [6, 6] open structure **12** is well established; ^{15}N labeling experiments were used to determine the bonding of these unique bisadducts [40]. Azirenofullerenes are formed in good yield by the reaction of in situ formed nitrenes (either photochemically or thermally from azides) with C$_{60}$. Averdung and Mattay and co-workers investigated the addition to C$_{60}$ of nitrenes derived from aroylazides and tert-butylazidoformate and found that the parent azirenofullerene **14** can be obtained upon chromatography of the N-t-butoxycarbonyl compound **13** over a column of neutral alumina (Fig. 7) [46–48]. Banks et al. found a similar route to **14** independently [49]. The parent azirenofullerene **14** can be derivatized on the nitrogen atom by addition of electrophiles to yield a urethane, an amide, or a sulfonamide [50].

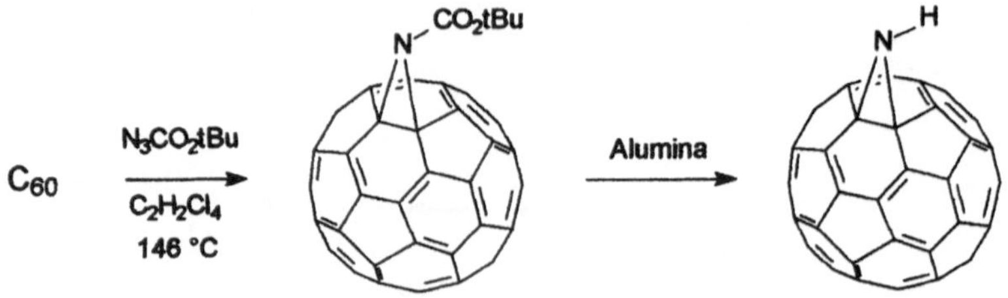

Fig. 7. Two-step synthesis of azireno[60]fullerene **14** from C_{60}

In contrast to the N-MEM (N-alkyl) analogue (see Sect. 2.2.2), N-aryl-aza-fulleroids, obtained from the thermal reaction of aryl azides with C_{60}, can be photochemically isomerized to azirenofullerenes [51]. This closely resembles the difference in the photochemical behavior of C-aryl- and C-alkyl fulleroids obtained from the reaction of diazo compounds with C_{60} (e.g. [5, 6] $C_{61}H_2$ is photochemically stable). After some initial studies (see for example [52–55]), the addition of azides and nitrenes to C_{60} has been investigated and used for the preparation of a series of functional fullerene derivatives by a number of other investigators, but their work is not relevant in relation to the preparation of aza[60]fullerene. The addition of azides to fullerenes has been briefly reviewed previously [56–58].

2.2.2
Gas-Phase Observation of Aza-dihydro[60]fullerene and Azafulleronium Ions from Aazirenofullerenes and Bisazafulleroids

Averdung et al. found in 1995 that when azirenofullerene **14** was subjected to desorptive chemical ionization (DCI) mass spectroscopy, using ammonia as reagent gas and a rhenium wire as heater, peaks at m/z 723 and 724 were found, next to $[M+H]^+$, C_{60}^+, and $C_{60}H^+$ [6]. These peaks were rationalized as dihydro-aza[60]fullerene ions $C_{59}NH^{+\cdot}$ and $C_{59}NH_2^+$. A normal exact mass determination was not possible, but the calculated centroid masses of the peaks for $C_{59}NH^+$ and $C_{59}NH_2^+$ were reported to fit the experimentally found values. Based on quantum-chemical calculations, the authors proposed a 1,2-closed structure as the most stable for both the neutral $C_{59}NH$ and the charged species $C_{59}NH_2^+$ (structures **15** and **16** in Fig. 8).

The exact mechanism of this formation of heterofullerene ions remains unclear, although a possible mechanism involving NH_3 addition to azafulleroid (presumably formed by isomerization of the azirenofullerene) was later proposed [18]. The authors reported on several occasions on this gas-phase transformation [17, 59–61]. The proposed structure for $C_{59}NH$ was later confirmed by the Wudl group upon the bulk synthesis of this compound (see Sect. 2.3).

Also in 1995, Lamparth et al. found that when bisadducts **12** and **17** (Fig. 9) were submitted to FAB-MS, aza[60]fulleronium $C_{59}N^+$ was formed in high yield,

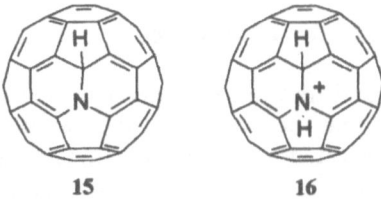

Fig. 8. Structures for neutral $C_{59}NH$ (**15**) and $C_{59}NH_2^+$ (**16**), proposed by Averdung et al.

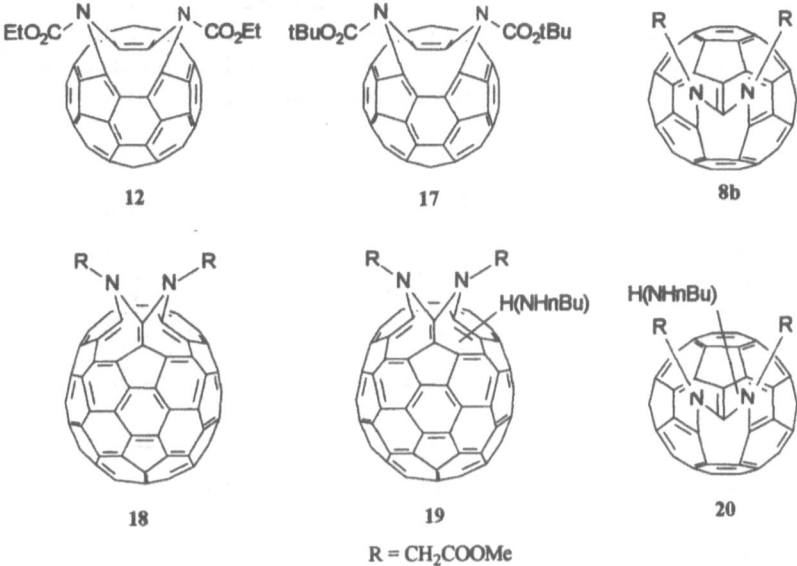

R = CH$_2$COOMe

Fig. 9. Precursors for the gas phase formation of $C_{59}N^+$ (**8b, 12, 17, 20**) and $C_{69}N^+$ (**18, 19**) in FAB-MS. Compounds **19** and **20** are the products of the reaction of butylamine/DBU with bisazafulleroids **18** and **8b**, respectively

whereas the mass spectra of the bisazafulleroids show only the typical characteristics of fullerene adducts (a relatively small M$^+$ peak and, at 720, C_{60}^+, the base peak) [5].

In the case of **12** and **17** the peak at m/z 722 ($C_{59}N^+$) was the base peak, and the 720 peak showed a relative intensity of 80%. The well-known shrink-wrap degradation was observed, supporting the fullerene structure of this material. Furthermore, it was found that treatment of bisazafulleroids **8b** and **18** with one equivalent of butylamine in the presence of DBU yielded (presumably regioselective monoaddition) products that fragmented in FAB-MS with efficient formation of aza[60]fulleronium and aza[70]fulleronium ions, respectively ($C_{60}^+/C_{59}N^+ \sim 1:1$ and $C_{70}^+/C_{69}N^+ \sim 1:2$). Exact mass determination confirmed the proposed formulae. The synthetic method for the preparation of azafullerenes, based on these findings, is described in Sect. 2.2.5.

2.2.3
Controlled Chemical Opening of the Fullerene Cage

From the fact that azide addition to fullerenes yields specific bisadducts with
high selectivity, it is clear that the two identical "enamine" C=C bonds in azafulleroids are relatively activated in certain reactions. In an attempt to expand
the scope of the chemoselectivity of the azafulleroids, other cycloadditions were
performed. Taking advantage of the fullerenes as effective triplet sensitizers,
Hummelen et al. showed that self-sensitized photo-oxygenation of N-MEM
aza[60]fulleroid 6 afforded the cage-opened [60]N-MEM ketolactam 22 in 65%
yield [62]. This was the first step towards the controlled opening of the fullerene cage. By causing oxidative rupture of a double bond adjacent to the 5,6 bridged nitrogen atom, an 11-membered ring was created. It is assumed that the
reaction proceeds through a 1,2-dioxetane intermediate 21 (Fig. 10).

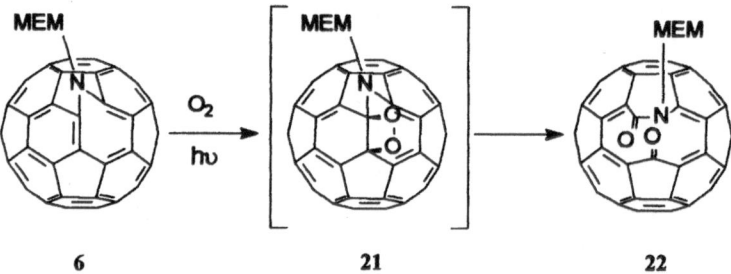

Fig. 10. Controlled opening of the fullerene cage by self-sensitized photo-oxygenation of 6

   Product 22 was fully characterized by ^{13}C and ^1H NMR, FTIR, UV-Vis and
FAB-MS spectroscopy and elemental analysis. In the course of the characterization, FAB-MS revealed a base peak at m/z=722 (and no peak at all at 720!) and
smaller peaks at m/z=780 (95%) and 856 (M+H$^+$). The exact mass of the
m/z 722 ion, 721.9991, was consistent with the molecular formula $C_{59}N^+$. Further
confirmation was found in the shrink-wrap fragmentation pattern [63]: loss of
CN, m/z = 696, followed by the successive loss of C_2, m/z 672 and 648. Examination of the FAB mass fragmentation pattern of N-MEM ketolactam 22 to azafulleronium $C_{59}N^+$ led to the proposed mechanism of formation as depicted in
Fig. 11.
   The relatively intense peak at m/z = 780 is due to loss of 2-methoxyethanol
from the protonated M to yield the N-methyl carbonium ion 23 which is, in turn,
transformed to the four-membered 1,3-oxazetidinium ring in 24. The latter
eliminates formaldehyde to yield intermediate 25, followed by the loss of carbon monoxide to yield the azafulleronium ion.
   When bisazafulleroid 8 was submitted to self-sensitized photooxygenation,
the expected symmetrical oxidation product was not observed, but instead the
asymmetric mono-oxidized product 26 was isolated in low yield (Fig. 12). The

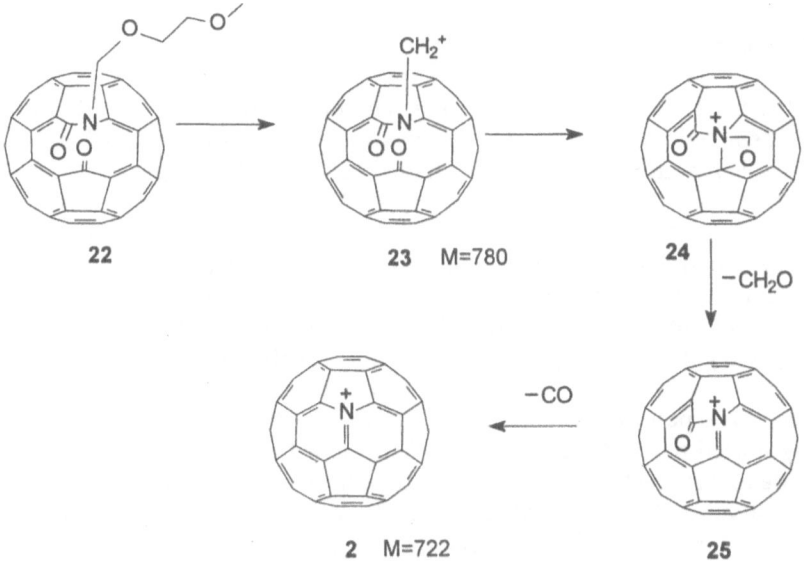

Fig. 11. Possible mechanism of gas-phase formation of $C_{59}N^+$ from ketolactam **22** in FAB-MS

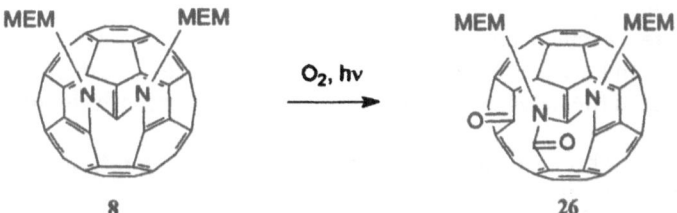

Fig. 12. Regioselective self-sensitized photo-oxygenation of bisazafulleroid **8**

FAB-MS degradation pattern of **26** gave further support for the proposed mechanism of azafulleronium formation, because, next to the M^+ peak at $m/z = 958$, and the base peak at 722, peaks at $m/z = 883$ (17%) and 853 (35%) were observed, corresponding to subsequent loss of 2-methoxyethanol and formaldehyde from **26** [7].

These gas-phase observations paved the way for the first synthetic route towards azafullerenes, as described in the next section. The cage-opened structures **22** and **26** are asymmetric. Ketolactam **22** was later resolved on a chiral HPLC column, and the circular dichroism of the enantiomers was determined [64].

2.2.4
(C₅₉N)₂ from a Cage-Opened Fullerene

Soon thereafter, during the search for a synthetic method that would mimic the events depicted above to obtain macroscopic quantities of "$C_{59}N$", a fast and remarkable reaction was observed when N-MEM ketolactam was treated with a large excess (15–20 equiv.) of pTsOH-H_2O in ODCB at reflux temperature under an atmosphere of nitrogen [7]. The color of the reaction mixture changed from red-brown to green-brown upon the formation of a non-polar major product in high yield. After purification by HPLC, the latter (green in solution) was shown by cyclic voltammetry, ^{13}C NMR, FTIR, UV-Vis, elemental analysis and mass spectrometry (electrospray) to be the azafullerene dimer $(C_{59}N)_2$. A strong argument for the dimer was obtained from the cyclic voltammogram (CV), which showed three pairs of reversible one-electron reductions. The appearance of closely spaced pairs of waves in the CV suggested that this system consisted of two (identical) weakly interacting electrophores. The ^{13}C NMR spectrum of $(C_{59}N)_2$ showed 29 lines in the region between 156 and 124 ppm, which is consistent with a highly symmetrical dimer structure. Other resonances (i.e. sp^3-hybridized carbon atoms) were not observed. The absence of a resonance in the 70–110 ppm region in the ^{13}C NMR spectrum forced the investigators to consider the [6,6]-open structure **27** as a possible structure for the azafullerene, as the only alternative to the [6,6]-closed structure **28** (symmetry arguments ruled out all other structures; Fig. 13). Electrospray mass spectroscopy was the only MS technique that allowed the detection of the dimer M⁺ peak (at m/z = 1445). In FAB-MS, MALDI, EI and CI MS, only the monomer $C_{59}N^+$ or $C_{59}N^-$ (depending on the detection mode) ions were observed, with shrink-wrap peaks in FAB-MS.

After establishing the dimeric structure, the following rationale for the molecule's formation process was proposed (Fig. 14): the azafulleronium ion is

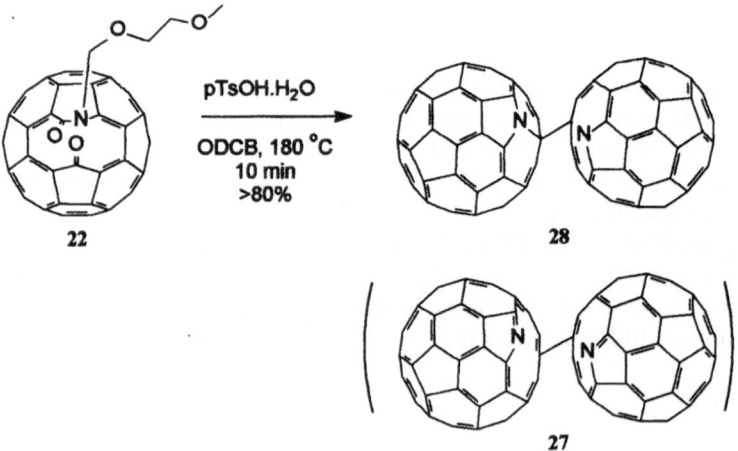

Fig. 13. Synthesis of $(C_{59}N)_2$ from MEM-ketolactam **22**

formed first in a manner that mimics the gas-phase formation of azafulleronium, i. e. the acid protonates the MEM moiety, inducing the loss of 2-methoxyethanol, followed by rearrangement and loss of formaldehyde and carbon monoxide to yield the azafulleronium ion. In the presence of the 2-methoxyethanol (or water) the azafulleronium ion can be reduced to the azafullerenyl radical. Finally, this radical dimerizes to yield $(C_{59}N)_2$.

Repeating the synthetic sequence $C_{60} \rightarrow$ **28** starting with ^{15}N-labeled MEM azide afforded ^{15}N-labeled **28**. Using various NMR techniques, the [6,6]-closed structure of $(C_{59}N)_2$ (i. e. structure **28**) was be proven unambiguously [65].

The Wudl group later optimized the three-step synthetic route from C_{60} to $(C_{59}N)_2$ to a reproducible overall yield of 15% on larger scale to yield gram quantities of the dimer. The synthetic method is potentially applicable to all fullerenes (see Sect. 3). When oxygen was not rigorously excluded during the reaction **22** → **28**, some oxidized dimeric material was formed. The structure of this material was proposed to be the N-oxide **29** (Fig. 15) [66]. Further support for

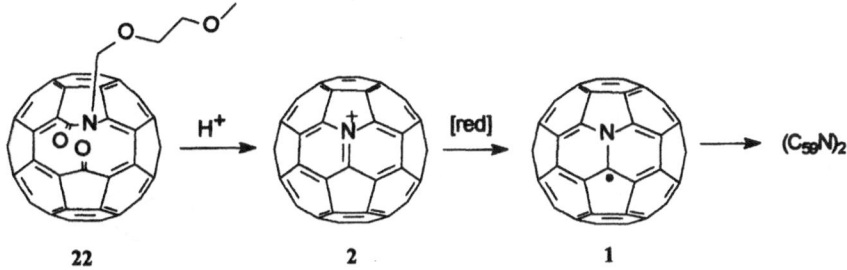

Fig. 14. Mechanism of formation of $(C_{59}N)_2$ from MEM-ketolactam **22**

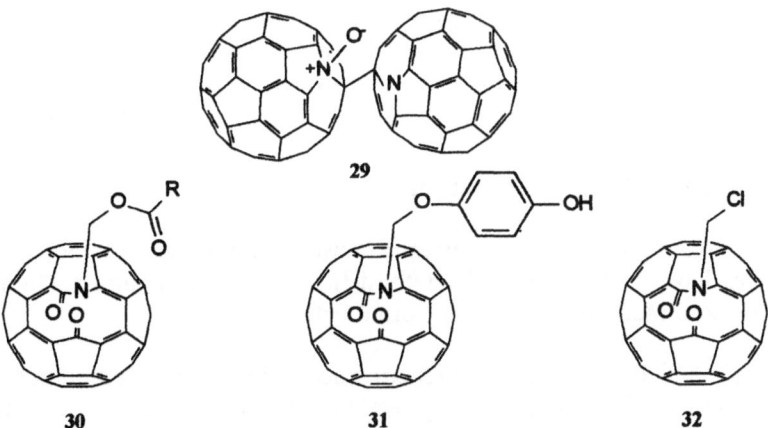

Fig. 15. N-oxide **29** is formed during the reaction **22** → **28** in the presence of traces of oxygen. Holeyballs **30**, **31**, and **32** are carbonium ion intermediate **23** trapping products from the reaction **22** → **28** in the presence of nucleophiles

the proposed mechanism for the formation of $(C_{59}N)_2$ was obtained when the reaction was carried out in the presence of nucleophiles to trap the proposed carbonium ion intermediate 23 [67]. This led to the isolation of new holeyballs 30, 31, and 32, fullerenes with a ring larger than a hexagon (Fig. 15).

2.2.5
$(C_{59}N)_2$ from a Bisazafulleroid

Based on their gas-phase observation of $C_{59}N^+$, a second route to $(C_{59}N)_2$ was developed by Hirsch and co-workers [7, 8]. When bisazafulleroid 8 was treated with butylamine and DBU in toluene at room temperature, "aminofullerides" were formed, which yielded a mixture of dimer 28 and aza[60]fullerene derivative 33 in 12–15% yield after heating with 20 equivalents of pTsOH in ODCB at reflux temperature over 8 min (Fig. 16). The reaction did not occur when 8b, the bis-NCH_2CO_2Me analogue of 8, was used as starting material. The dimer 28 obtained by this method was identical to the material obtained earlier by the Wudl group.

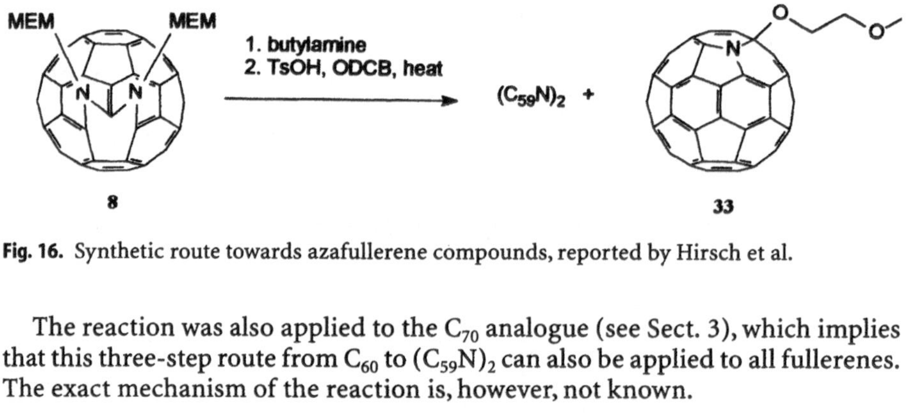

Fig. 16. Synthetic route towards azafullerene compounds, reported by Hirsch et al.

The reaction was also applied to the C_{70} analogue (see Sect. 3), which implies that this three-step route from C_{60} to $(C_{59}N)_2$ can also be applied to all fullerenes. The exact mechanism of the reaction is, however, not known.

2.3
Synthesis of Aza[60]fullerene Derivatives

Up to now, two types of synthetic methods towards monomeric aza[60]fullerene derivatives have been developed: Method (a) starting from non-azafullerene precursors and Method (b) starting from the dimer $(C_{59}N)_2$.

Method (a). The first example of a reaction according to this method has already been mentioned in the preceding section. Another example is the reaction depicted in Fig. 17. Hence, when a 15-fold excess (relative to 8) of hydroquinone was added to the reaction mixture that otherwise yielded the azafullerene dimer, the intermediate azafulleronium ion 2 was reduced to azafullerenyl radical 1, which was subsequently trapped by the hydroquinone to give the parent mono-

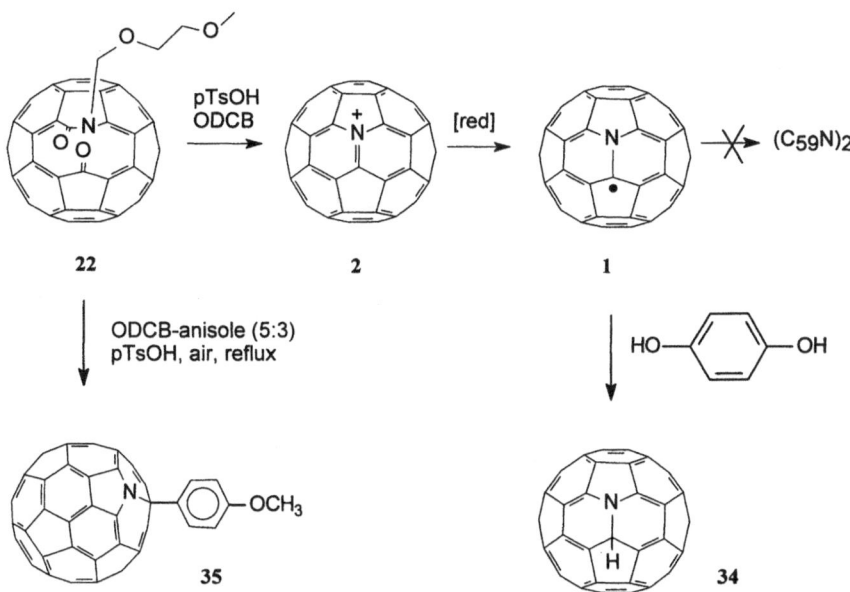

Fig. 17. More trapping products of the reaction **22→28.** Trapping of azafullerenyl radical intermediate **1** by hydroquinone leads to the formation of hydroaza[60]fullerene **34.** Trapping of intermediate azafulleronium **2** by anisole yields arylated azafullerene **35**

meric azafullerene, hydroazafullerene **34,** in a high yield [67]. As was mentioned in Sect. 2.2.4, a larger (75-fold) excess of hydroquinone resulted in the trapping of an earlier intermediate in this reaction, i.e. carbonium ion **23.**

Negative ion ES-MS of **34** showed a base peak at m/z = 723. DCI mass spectroscopy shows shrink-wrap peaks at m/z = 696, 672, 648, and 624. Compound **34,** like dimer **28** and derivative **33,** is green in common organic solvents. A clear resonance signal at 72.1 ppm in the ^{13}C NMR allowed for unambiguous assignment of the [6,6]-closed structure to hydroazafullerene **34.**

Recently, it was found that when the reaction is carried out in the presence of a large excess of anisole and air, arylated azafullerene **35** can be obtained in 38% yield [68] (Fig. 18). Compound **35** was also obtained, but in a much lower yield, directly from bisazafulleroid **8,** when the reaction shown in Fig. 16 was carried out in the presence of anisole and air. When the same reaction was carried out in pure 1-chloronaphthalene at 220°C, a mixture of arylated products **36** was obtained in 46% isolated yield [68].

Method (b). Several examples of the preparation of monomeric aza[60]fullerene derivatives from the dimer **28** have been reported. First, it was found that treatment of the dimer with excess diphenylmethane in refluxing ODCB for 48 h afforded, after chromatographic purification, 2-diphenylmethyl-azafullerene **37** in 42% yield, as depicted in Fig. 19 [69].

A short free-radical chain mechanism has been proposed to explain the result. Hence, thermal homolysis of the dimer generates $C_{59}N^{\cdot}$, which either

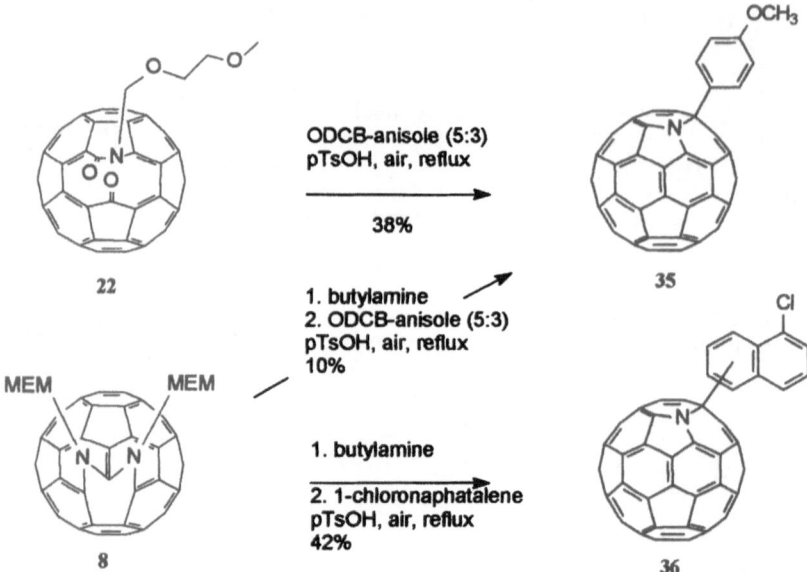

Fig. 18. Arylation products from the common azafulleronium intermediate **1**, formed in some reactions of **22** and **8**

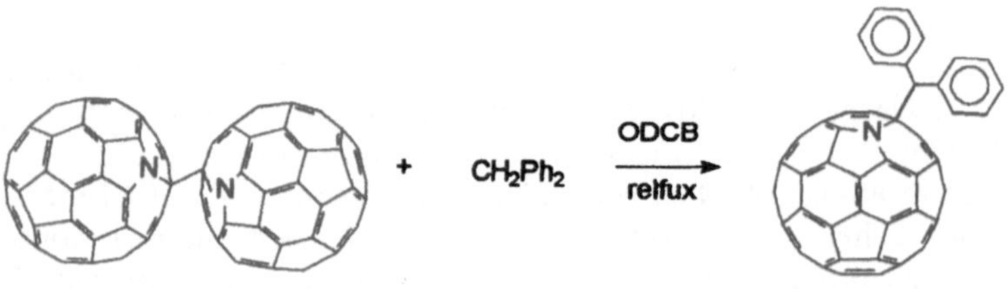

Fig. 19. Thermal homolysis of $(C_{59}N)_2$ in the presence of diphenylmethane yields azafullerenyl radical trapping product **37**

abstracts a hydrogen atom from the hydrogen donor diphenylmethane or re-combines with the thus formed diphenylmethyl radical. The hydroazafullerene is not stable under these reaction conditions and generates more dimer:

$$(C_{59}N)_2 \rightarrow 2\ C_{59}N^{\cdot}$$

$$C_{59}N^{\cdot} + Ph_2CH_2 \rightarrow C_{59}NH + Ph_2CH^{\cdot}$$

$$Ph_2CH^{\cdot} + C_{59}N^{\cdot} \rightarrow C_{59}(CHPh_2)N$$

$$2\ C_{59}NH\ (\text{with trace of } O_2) \rightarrow (C_{59}N)_2 + H_2O_2$$

This mechanism is supported by two control experiments: (a) a fresh sample of hydroazafullerene **34** was found to dimerize on standing at room temperature, even in the solid state; (b) treatment of **34** with diphenylmethane in refluxing ODCB for 24 h afforded 73 % of **37**, 7 % of dimer **28**, and 5 % of recovered starting material.

Treatment of $(C_{59}N)_2$ with diphenylmethane under photolytic conditions at room temperature was expected to produce **37**. However, after five hours of irradiation, HPLC analysis of an aliquot of the reaction mixture showed no formation of **37**, but instead revealed the formation of material with proposed N-oxide structure **38** (25 %), next to recovered $(C_{59}N)_2$ (75 %). It was assumed that trace amounts of $^1O_2$ initiate this conversion (Fig. 20). Because fullerenes are excellent sensitizers for the conversion of triplet to singlet oxygen, any oxygen in the reaction flask is converted to $^1O_2$ under these reaction circumstances [66].

Thermal treatment of the dimer with anisole, toluene, and 1-chloronaphthalene in the presence of a large excess of pTsOH and air leads to the formation of mono-arylated azafullerenes in 78–90 % isolated yields [68]. The reaction with anisole and toluene yield *para*-substitution products **35** and **39**, while 1-chloronaphthalene is substituted at various positions (Fig. 21). The reaction does not take place in the absence of air or pTsOH. The reaction is presumed to proceed through electrophilic aromatic substitution by $C_{59}N^+$, which was proposed as being formed via thermal homolysis of the dimer, followed by oxidation with O_2.

Very recently, Reuther and Hirsch have found that **35**, **39**, and yet another arylated azafullerene **40**, react with iodine monochloride in a reaction closely analogous to the one found by Taylor and co-workers, yielding tetra-chlorinated compounds **41**, **42**, and **43**, respectively, in 50–60 % yields [70,71]. An interesting point is that the five-membered ring containing the nitrogen atom adopts the

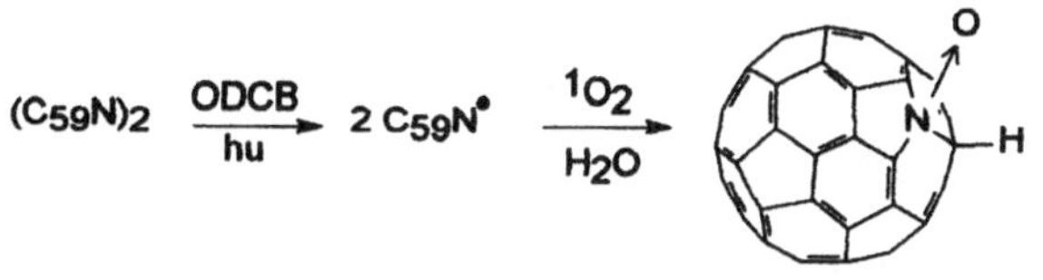

Fig. 20. Formation of N-oxide **38** by photolysis of $(C_{59}N)_2$ in the presence of traces of oxygen

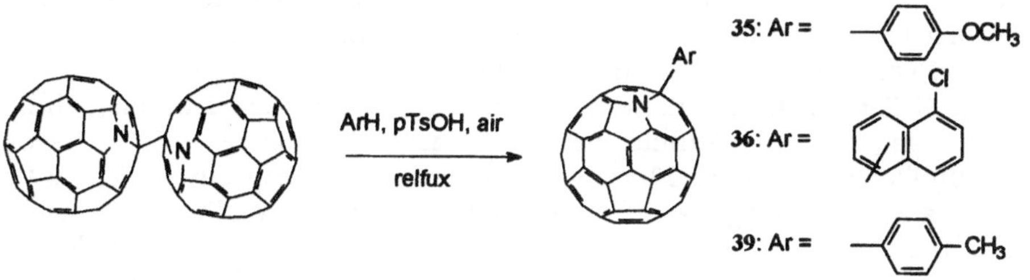

Fig. 21. Azafulleronium arylation products from oxidative thermolysis of $(C_{59}N)_2$

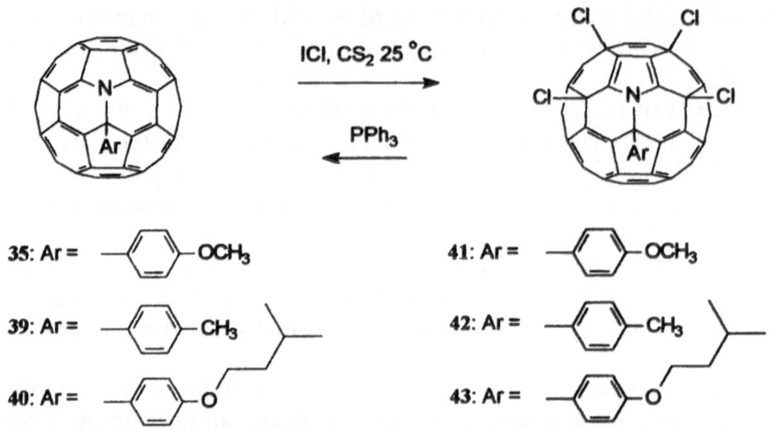

Fig. 22. Formation of tetra-chlorinated compounds 41–43 upon the regioselective reaction of 2-Aryl-azafullerenes with ICl. The reverse reaction takes place with excess PPh$_3$ in ODCB at 25 °C

pyrrole configuration (Fig. 22). As a result, the UV-Vis spectrum of these compounds is very similar to that of C$_{60}$Cl$_6$. AM1 calculations support the pyrrole configuration. The reaction can be reversed upon treatment of the tetrachloro adducts with excess triphenylphosphine in ODCB at room temperature.

The authors envision that compounds like 41, 42, and 43 could have potential as precursors for diaza[60]fullerene C$_{58}$N$_2$ derivatives by way of an azide addition/dechlorination sequence.

This summarizes all presently known chemistry towards aza[60]fullerene compounds, and with that all preparative chemistry related to hetero[60]fullerenes in general.

3
Synthesis of Aza[70]fullerene

The preparation of pure aza[70]fullerene dimers and derivatives is an even more challenging endeavor than the aza[60]fullerene case, because, in principle, there is more than one possible isomer of C$_{69}$N. This complication was recognized from the beginning [8] and has been worked out in detail by Bellavia-Lund and Wudl [20, 72]. C$_{70}$ (D$_{5h}$) has five different types of carbon atoms (a – e in Fig. 23), therefore there are five isomers of C$_{69}$N, of which the oxidized forms 44–48 are shown in Fig. 23. Simultaneous formation of all five corresponding radicals could lead to 15 distinct C$_{69}$N dimers (if bound by specific[6,6]-α-carbon atoms!).

The IUPAC names for these fulleronium compounds are 2-azonia[70]fullerene (44), 1-azonia[70]fullerene (45), 5-azonia[70]fullerene (46), 7-azonia[70]fullerene (47), and 20-azonia[70]fullerene (48), respectively (The CAS names and numbering are different again!). The corresponding most stable radicals are expected to be 2-aza[70]fulleren-1-yl (from 44), 1-aza[70]fulleren-2-yl (from 45), 5-aza[70]fulleren-6-yl (from 46), 7-aza[70]fulleren-21-yl (from 47), and 20-aza[70]fulleren-19(or 21)-yl (from 48).

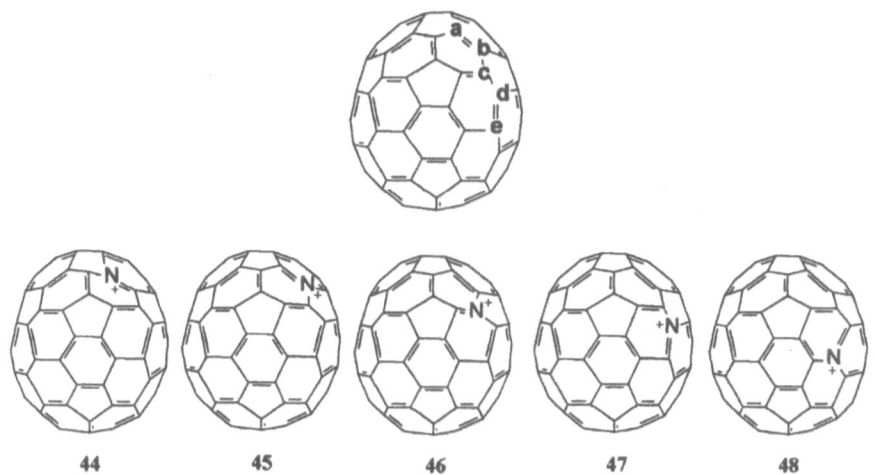

Fig. 23. C_{70} has five different types of carbon atoms a-e (*above*). Hence, there are five possible isomers of aza[70]fulleronium $C_{69}N^+$: **44–48**

Experiments leading to gas-phase observations of $C_{69}N^+$ have already been mentioned in Sects. 2.1 and 2.2.2, and will not be reviewed again here [5, 25, 32, 33].

3.1
Synthesis of Aza[70]fullerene Dimers ($C_{69}N$)$_2$ and Aza[70]fullerene Derivatives

The groups of Hirsch and Wudl both applied their own synthetic methods for the preparation of aza[60]fullerene dimer to C_{70}. Since both methods start with the preparation of aza[70]fulleroids and bisaza[70]fulleroids as precursors, we will summarize these results first.

The 1,3-dipolar addition of azide could, in theory, give rise to a maximum of six triazoline isomers (Fig. 24), where addition across a [6,6] bond between two different types of carbon atoms in general gives rise to a pair of regio-isomers [73, 74]. C_{60} gains its reactivity as a result of inherent strain in the molecule. In the case of C_{70}, having lower symmetry, the strain is not uniform over the entire molecule. Hawkin's studies on the osmylation of C_{70} revealed that most of the strain is localized at the poles rendering the [a–b] bond most reactive [75].

Treatment of C_{70} with an excess of MEMN$_3$ in ODCB at room temperature for 48 h afforded, upon chromatography, one band which contained a mixture of triazoline isomers **49, 50,** and **51** (32% yield, 60% based on consumed C_{70}) [72]. When the mixture of the three isomers was heated to 55 °C overnight, only one azirenofullerene (**55**) and one azafulleroid (**56**) were found, next to **49, 50,** and a little C_{70} (Fig. 25). Hence, **51** was converted selectively, which allowed the separation and complete characterization of both **49** and **50** as well as for separately heating of these isomers to yield the *same* azirenofullerene **57** and azafulleroids **58** and **56**, respectively.

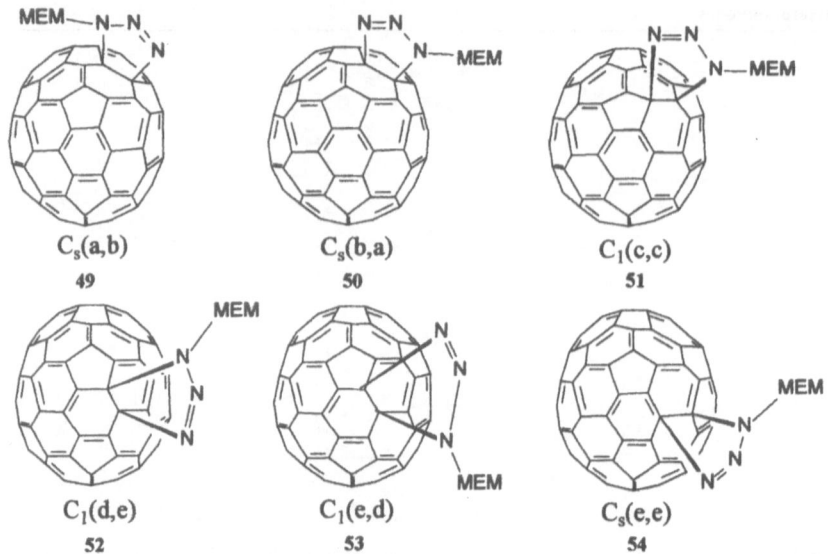

Fig. 24. The six possible triazoline isomers from the reaction of MEM-azide with C$_{70}$

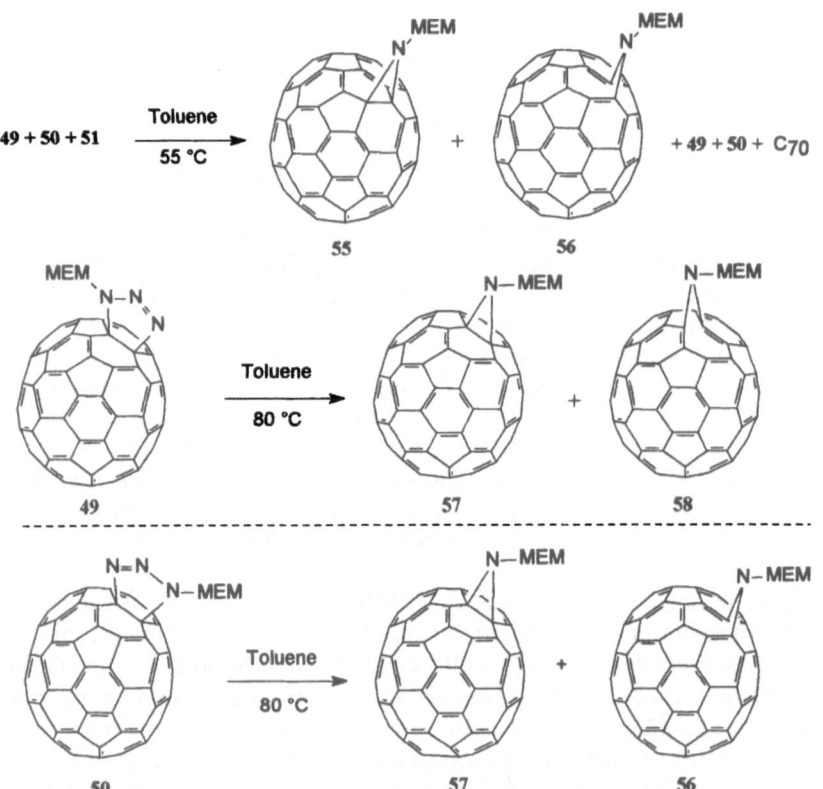

Fig. 25. Selective thermal decomposition of triazolino[70]fullerene **51**, present in a mixture of three isomers **49–51** (*above*). Separate thermal decomposition of triazolino[70]fullerene isomers **49** and **50**, leading to a common product, azireno[70]fullerene **57**, and isomeric aza[70]fulleroids **58** and **56**, respectively (*below*)

Differentiation between 49 and 50 was accomplished by thermolysis of a solution of 50 and 51, affording the azirenofullerenes 57 and 59 [supposedly the (c,c)-isomer] as well as a common azafulleroid 56 [76]. In conclusion, perseverance afforded two well-characterized aza[70]fulleroids 56 and 58, that were used for further transformation to aza[70]fullerenes (see below).

Nuber and Hirsch, in the meantime, prepared bisaza[70]fulleroid 60 in a one-step synthesis from C_{70} with MEM azide in ODCB at 120 °C. They observed that while 60 was formed preferentially, another bisadduct isomer [presumably the (1,6);(1,9)-isomer, according to *modern* IUPAC numbering] was formed as a byproduct. When this material was subjected to the reaction conditions that afforded aza[60]fullerenes from 8 (see Sect. 2.2.5), an analogous aza[70]fullerene dimer, most likely with the structure 61, and the methoxyethoxy derivative 62 could be isolated from the mixture in 10 and 15% yield, respectively (Fig. 26) [8]. The detection of a resonance at 96.4 ppm in the ^{13}C NMR spectrum of 62 is indicative for the presence of an sp^3-hybridized carbon atom next to nitrogen. Hence, the aza[70]fullerene has a [6,6]-closed structure as shown in Fig. 26.

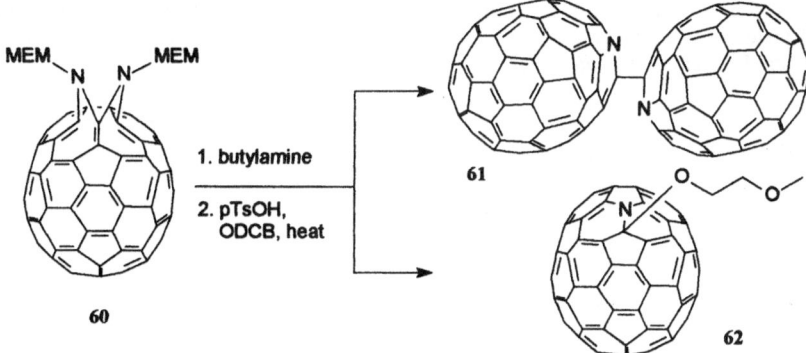

Fig. 26. Aza[70]fullerene compounds obtained from bisazafulleroid 60 by the method of Hirsch et al.

Compound 62 (as its aza[60]fullerene analogue 33) is unstable and decomposes slowly into insoluble products, which show carbonyl vibrations in IR. A rigorous proof of the structure of 61, using ^{13}C NMR, was impossible due to the low solubility of the material.

Photooxygenation of C_s symmetric aza[70]fulleroid 56 yielded a single (chiral) oxidized product, ketolactam 63, in good yield [20]. The structure of 63 was fully characterized by spectroscopic methods (Fig. 27).

The analogous photooxygenation of C_1 symmetric 58 yielded two ketolactam products 64 and 65 (4:1). Interestingly, and unlike the UV-Vis spectra of the isomeric aza[70]fulleroids, the UV-Vis spectra of the ketolactam isomers are not superimposible. Isomer 63 is brick red, 64 is orange red, and 65 is brownish green in solution. Similar to the [60] analog, the FAB-MS of all three ketolactams revealed a base peak at m/z = 842, consistent with $C_{69}N^+$.

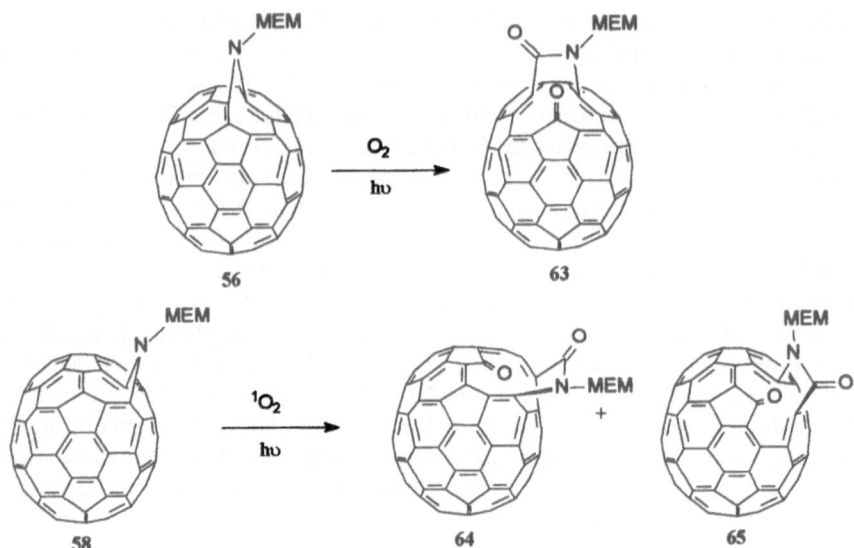

Fig. 27. Self-sensitized photo-oxygenation products from aza[70]fulleroids: symmetric fulleroid **56** yields one ketolactam **63**, while asymmetric azafulleroid **58** yields the two possible ketolactam isomers **64** and **65** (4:1)

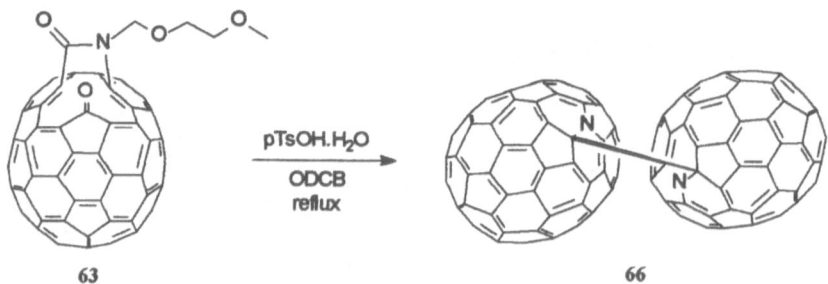

Fig. 28. Formation of the "a-a" $(C_{69}N)_2$ isomer **66** from ketolactam **63**

Treatment of ketolactam **63** with an excess of pTsOH in refluxing ODCB produced a single non-polar product, aza[70]fullerene dimer **66** (Fig. 28). Electrospray mass analysis confirmed the elemental composition of **66**, showing a molecular ion at m/z = 1685 and a strong base peak at m/z = 842. The ^{13}C NMR spectrum revealed 29 peaks in the sp^2 region, illustrating the high symmetry of the isomer. Because of the low solubility of the dimer and the tendency for sp^3 carbons adjacent to nitrogen to have long T_1 times, no resonances in the sp^3 region were detected. Dimer **66** is most likely the same dimer as **61**, described by Nuber and Hirsch, but this cannot be fully confirmed by comparison of the data presently reported for the latter [8, 76, 77].

Analysis of the decomposition of the ketolactam isomers **64** and **65** according to the proposed mechanism suggests that each ketolactam isomer should give a different single $(C_{69}N)_2$ isomer. Decomposition of **64** leads to the exchange of carbon "b" for nitrogen (Fig. 23) and decomposition of **65** leads to exchange of carbon "c". Hence, **64** should initially yield azafulleronium **45**, while **65** is expected to give azafulleronium **46** before reduction to its corresponding radical. Interestingly, the symmetries of **45** and **46** are different, allowing for the distinction between the two dimers, **67** and **68**. Ketolactams **64** and **65** were indeed reported to yield the expected aza[70]fullerene dimers, as shown in Fig. 29, but rigorous proof of the structure of **68** is still awaited from ^{13}C NMR analysis [76].

The three isomeric dimers show different retention times on a Cosmosil Buckyprep HPLC column. In the reaction **65** → **68**, a byproduct was formed, which was assigned to hydroaza[70]fullerene **69**.

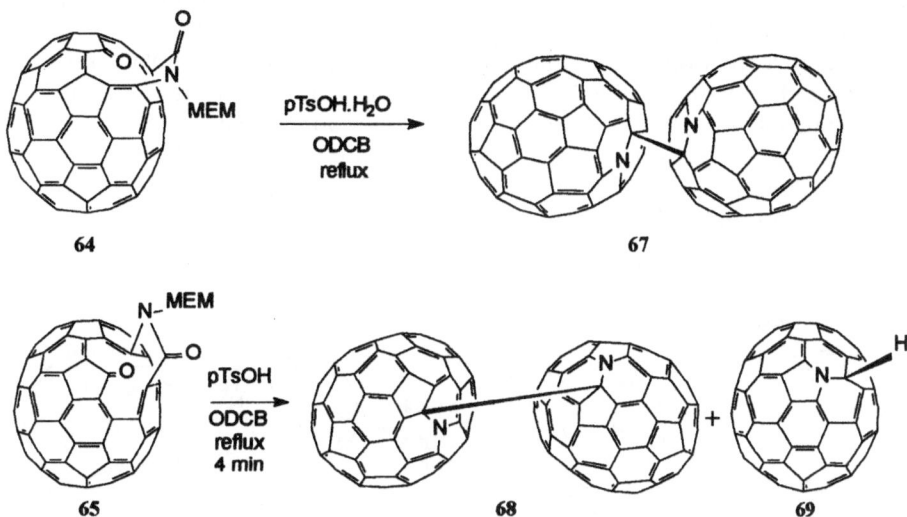

Fig. 29. Formation of the "b-b" $(C_{69}N)_2$ isomer **67** from ketolactam **64** (*above*). Formation of the "c-c" $(C_{69}N)_2$ isomer **68** and hydroaza[70]fullerene **69** from ketolactam **65** (*below*)

The $C_{69}N$ dimer, in which carbon atom "e" is exchanged with a nitrogen atom, cannot be made following the above procedure, since the nitrogen atom in position "e" is not part of a five-membered ring. As a consequence, the corresponding azafulleroid cannot exist. Formation of an isomer in which carbon "d" is replaced by nitrogen is theoretically possible, using the above mentioned sequence of steps, but not probable. Its formation would require several extra steps, i.e. blocking the more reactive sites, prior to the addition of azide.

In conclusion, three of the five possible homodimers of aza[70]fullerene and two monomeric aza[70]fullerene derivatives have been synthesized up to now.

3.2
The Mixed Dimer $C_{59}N-C_{69}N$

The inter-dimer bond in aza[60]fullerene and aza[70]fullerene dimers is by far the weakest bond in these structures, with an estimated bond energy of ~ 18 kcal mol^{-1}. This has been used in the syntheses of monomeric azafullerene derivatives as described in Sect. 2.3: heating the dimer to 180°C in ODCB in the presence of a radical scavenger yields monomeric products after thermal rupture of the dimer bond. It was found that at 80 °C no net reaction had occurred. In order to gain a better perspective on the nature of the $C_{59}N^{\cdot}$ radical, the equilibrium with a radical source which has a similar binding energy, i.e. $C_{69}N-C_{69}N$ was studied. The investigation of the formation of the heterodimer, $C_{59}N-C_{69}N$, was not only to provide information about the binding energy of the $C_{59}N$ dimer, but also to shed some light on the reactivity of the aza[60]fullerene radical towards its higher analog aza[70]fullerene radical. In addition, this study furnished the first fullerene with commingled properties from both heterofullerenes [78]. Bellavia-Lund and Wudl took two approaches to this study [76].

First, the formation of the dimers from their corresponding ketolactams was investigated. Treatment of a 1:1 molar solution of [60]ketolactam 8 and [70] ketolactam 64 with pTsOH in refluxing ODCB for 5 min resulted in the formation of $(C_{59}N)_2$, $C_{69}HN$, $(C_{69}N)_2$, and the heterodimer $C_{59}N-C_{69}N$ 70, as determined by HPLC and UV-Vis analysis (Fig. 30). (The aza[70]fullerene moiety is of the type present in dimer 67).

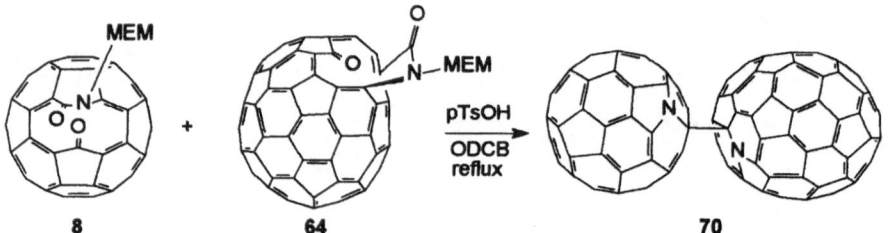

Fig. 30. Method 1 for the preparation of mixed azafullerene dimers: through generation of a mixture of azafullerene radicals from the corresponding ketolactams. The mixed dimer obtained from a mixture of 8 and 64 is proposed to have the structure 70, the $C_{59}N$-"**b**"-$C_{69}N$ isomer with the dimer bond position on the a-"**a**" carbon atom of the $C_{69}N$ moiety

Second, an equimolar solution of $(C_{59}N)_2$ (28) and aza[70]fullerene dimer 66 in ODCB was degassed and refluxed overnight. HPLC analysis of the resulting mixture revealed the presence of $(C_{59}N)_2$, $(C_{69}N)_2$, and the heterodimer $C_{59}N-C_{69}N$ 71 (Fig. 31).

Interestingly, the three dimers were formed in essentially a 1:1:2 ratio of 28, 66, and 71, respectively, suggesting that at the elevated temperature they are in equilibrium with each other, as determined by statistical mixing. The overlapping UV-Vis spectra of $(C_{59}N)_2$, $C_{59}N-C_{69}N$, and $(C_{69}N)_2$, in which the heterodimer spectrum contains equal features of both homo-dimers, are shown in Fig. 32a.

$(C_{59}N)_2$ +

ODCB
reflux

66   **71**

Fig. 31. Method 2 for the preparation of mixed azafullerene dimers: through generation of a mixture of azafullerene radicals from the two corresponding homodimers. The mixed dimer obtained from a mixture of $(C_{59}N)_2$ **28** and "a-a" $(C_{69}N)_2$ dimer **66** is proposed to have the structure **71**, the $C_{59}N$-"a"-$C_{69}N$ isomer with the dimer bond position on the **a**-"**b**" carbon atom of the $C_{69}N$ moiety

The fact that heterodimer **71** contains a different $C_{69}N$ moiety than **70** was clear from the difference in the HPLC retention times and the the UV-Vis spectra of these compounds. The UV-Vis spectra of **70** and **71** are shown in Fig. 32b.

Heterodimers **70** and **71** are the first examples of what is theoretically an enormous family of possible (hetero)fullerene heterodimers. It is expected that fullerene heterodimers will exhibit interesting electronic (optical, electron accepting) properties. Nuber and Hirsch recognized the fact that they had obtained a mixture of three aza[70]fullerene dimers upon treatment of a mixture of two isomeric bisaza[70]fulleroids. One of their products is most likely a mixed dimer [8]. This is a "hetero"-dimer in a sense that differs from **70** and **71**;

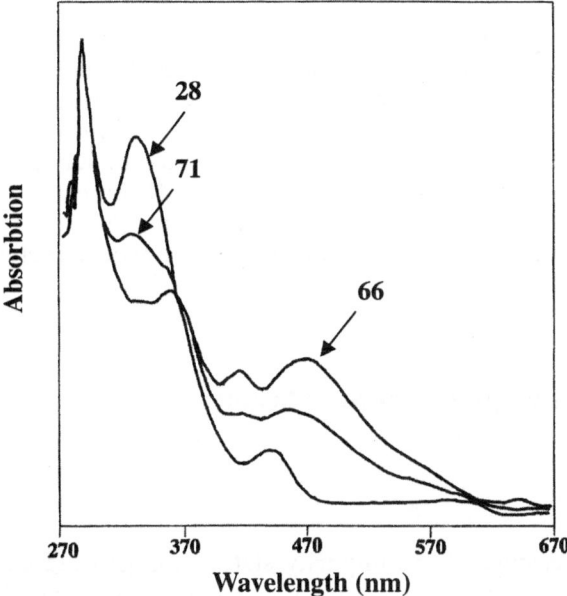

Fig. 32 a. UV-Vis spectrum of mixed $C_{59}N$-$C_{69}N$ azafullerene dimer **71** and of both corresponding homodimers $(C_{59}N)_2$ (**28**) and $(C_{69}N)_2$ isomer **66**

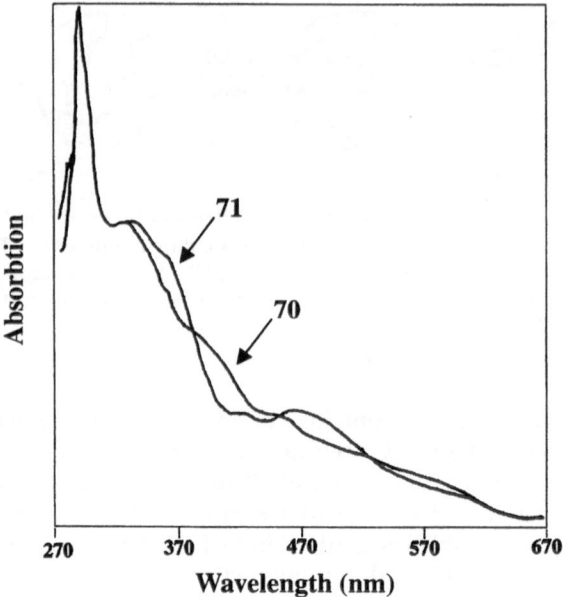

Fig. 32 b. UV-Vis spectra of $C_{59}N$-$C_{69}N$ azafullerene dimers 70 and 71

however, it could be called a "heteroisomeric" dimer, since the two parts are isomers of each other.

The theoretical possibility of a "donor-acceptor" dimer, i.e. a Lewis acid/base complex $C_{59}N - C_{59}B$, has been mentioned by Wang et al., who calculated on their hypothetical heterodimeric structures $C_{60} - C_{59}N$, $C_{60} - C_{59}B$, and $C_{59}B - C_{59}N$ [79]. These workers, however, only considered structures in which dimer bonds are formed by the heteroatoms, a highly unlikely situation in our opinion. Andreoni et al. also envisaged the possibility of a $C_{59}B - C_{59}N$ dimer, bound in part by charge exchange between donor and acceptor states [80]. We regard a C–C'-bonded heterodimer $C_{59}B - C_{59}N$, with marginal "donor-acceptor" properties, perhaps more feasible.

4
Physicochemical Properties of Azafullerenes

4.1
Theory

Andreoni et al. calculated the impurity states in $C_{59}N$ (and $C_{59}B$) using the Car–Parrinello method and found that the deformation of the fullerene cage is limited to the environment of the "impurity" [80,81]. From the calculated cohesive energy, it was concluded that $C_{59}N$ could exist. Kurita et al. calculated the molecular structure, binding energies and electronic properties of $C_{59}N$ (also of $C_{59}B$ and $C_{59}S$) using the Harris approximation [82]. Liu et al. used a tight-binding approximation method for calculating the electronic structure [83]. Rosen and Oestling did molecular cluster calculations on $C_{59}N$ within the local density approximation [84]. Chen et al. considered the possible electronic effects of metal complexation of $C_{59}N$ (and $C_{59}B$) [85]. In 1994, the geometrical structure and relative stability of $C_{59}N$ were also calculated using LCAO MO and molecular mechanics methods [86]. Jiang and Xing and co-workers calculated the structural and electronic properties [87], the third-order nonlinear polarizability [88, 89], the linear optical absorption [90], and the second-order nonlinear susceptibility [91] of $C_{59}N$, all using the Su–Schrieffer–Heeger model and the sum-over-states method. Some nonlinear optical properties had been calculated previously by Rustagi et al. using a tight-binding method [92]. It has even been speculated that $C_{59}N$ (and C_{60}) could be formed via a modified isoprene mechanism as an abiotic terpenoid [93]. Piechota and Byszewski reported on the calculation of the electronic structure of $C_{59}N^+$ [94]. These authors also claimed to have prepared samples containing $C_{59}N$, $C_{59}B$, and $C_{58}BN$, which would be at least quite remarkable, if not just plain unlikely (see Sect. 5). The reported ESR spectrum of their material does not fit the data obtained from pure azafullerene (see next section). Andreoni is unique among the theoreticians by anticipating the synthetic preparation of the *real* heterofullerenes $(C_{59}N)_2$ and $C_{59}NH$. While most of the work of her group has been published together with that of the experimentalists (see below), part of it has appeared independently in the literature [95–97].

4.2
The Dimer Bond in Azafullerene Dimers; Generation and Study of Aza[60]fulleren-2-yl and Aza[70]fullerenyl Radicals

Soon after the first preparation of $(C_{59}N)_2$, when the "6,6-closed" structure of the azafullerene moieties was not yet fully established, Andreoni et al. calculated the optimized structure and the energetically favored conformation of the dimer using Car-Parrinello molecular dynamics [98]. It was found that the two balls are linked in a *trans*-configuration by one bond and that the molecule has C_{2h} symmetry. The link is made by the carbon atoms 2 and 2' (making the hexagon-hexagon fusions together with the nitrogen atoms), with the C^2–N bond length being 1.520 Å (i.e. the "6,6-closed" configuration). The *trans*-conformation minimizes the repulsion of the nitrogen electron clouds. The C–C' dimer bond was further calculated to be 1.609 Å, i.e. 0.05 Å longer than that between two average sp^3 carbons. The binding energy was calculated to be ~ 18 kcal mol^{-1}, a value in the range of the formation enthalpies of dimers of C_{60} with monoalkyl radical adducts.

Synthetically, the weakness of the dimer bond was used to prepare azafullerene monomer derivatives from the dimer and mixed dimers from two different dimers, as was discussed in Sects. 2.3 and 3.2, respectively. The dimer can be cleaved both thermally and photochemically, yielding two $C_{59}N^•$ radicals 1 [69]. $C_{59}N^•$ and $C_{69}N^•$ radicals are isoelectronic with C_{60}^- and C_{70}^-, respectively. Initial attempts to observe aza[60]fullerenyl radicals by thermal generation yielded erroneous results [4]. Photochemical generation of the radicals appears to be the method of choice. Gruss et al. found that light-induced ESR measurement (LESR) of a solution of $(C_{59}N)_2$ in 1-chloronaphthalene, using 532 nm laser pulses, yielded a spectrum with three equidistant lines of equal intensity, indicative of ^{14}N hyperfine interaction [99]. The hyperfine coupling constant is 3.73 G [100]. The isotropic coupling constant of 10.4 MHz is similar to that of $C_{69}N^{\cdot}$ radical 72 (Fig. 33). The reported values for the g factor of $C_{59}N^{\cdot}$ are 2.0011(1) and 2.0013(2). This value is higher than that of $C_{60}^{\cdot-}$, 1.9991 [101]. The unusually low g value of the C_{60} radical anion was explained in terms of Jahn-Teller distortion that splits the triply degenerate t_{1u} states, thus leading to quenching of angular momentum. Hasharoni et al. give two possible explanations for the positive g shift in the heterofullerene radical: (a) the lower symmetry of $C_{59}N^{\cdot}$ relative to that of $C_{60}^{\cdot-}$ removes the distortion; and (b) the nitrogen heavy atom effect increases spin-orbit coupling in $C_{59}N^{\cdot}$. The symmetry argument is somewhat weakened by the fact that "dissymmetrized" radical anions of C_{60} derivatives (methanofullerenes and fulleroids) show g-values closer to that of C_{60} [102]. Most likely, the major decay channel of the $C_{59}N^{\cdot}$ radicals is redimerization, since a train of more than 10^5 light pulses did not decrease the signal intensity.

LESR measurements on two isomers of $(C_{69}N)_2$, structures 66 and 67, were performed to study the two isomeric $C_{69}N^{\cdot}$ radicals 2-aza[70]fulleren-1-yl 72 and 1-aza[70]fulleren-2-yl 73, respectively [100] (Fig. 33).

The LESR spectrum of dimer 66 also features a reasonably clean three-line spectrum due to ^{14}N hyperfine splitting that has a more positive g-factor, 2.0024(2), and a larger hyperfine coupling, 4.74 G, than those of $C_{59}N^{\cdot}$. This indicates that the unpaired electron and nitrogen wave functions overlap to a greater extent in azafullerenyl 72 and, thus, the electron is more localized in the vicinity of the nitrogen atom.

The LESR spectrum obtained from dimer 67 is more complicated: it shows two ^{14}N hyperfine splittings. The first is identical to that of 72 [g-value 2.0025(2), hfc 4.78 G] and the second stems from a different radical with a g-value of 1.9973(2) and hfc of 0.49 G. This was explained by assuming that the radical 73 actually has two different electron localization sites, where in one of them the unpaired electron is at the pole position as shown for 73 and in the other it is forced to be on the equatorial site because of strain in the molecule [100]. More

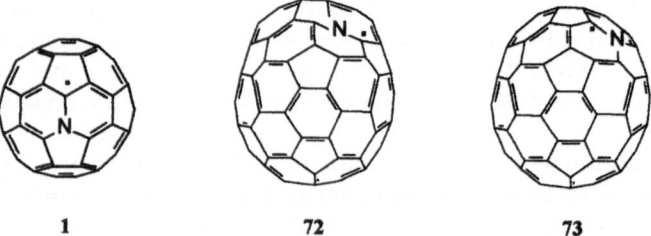

1   **72**   **73**

Fig. 33. Azafullerene radical $C_{59}N^{\cdot}$ (1) and $C_{69}N^{\cdot}$ radicals 2-aza[70]fulleren-1-yl (72) and 1-aza[70]fulleren-2-yl (73)

proof of this phenomenon awaits the detection of LESR signals from preferably all five different aza[70]fullerene homodimers!

4.3
Solid-Phase Properties of Azafullerenes

Solid-Phase Structure of $(C_{59}N)_2$. As was the case for the theoretical calculations mentioned above, the driving force for solving the crystal structure of the dimer was the lack of a ^{13}C NMR signal in the expected region of the spectrum. After several attempts to obtain crystals by sublimation, Prassides' group was able to get a crystalline powder whose XRD spectrum could not easily be interpreted, even after use of a high-resolution synchrotron X-ray diffraction source [103]. The problem was that the dominant peaks were those of a pseudo-cubic solid with lattice dimensions in the same order as those of C_{60}! However, the authors noted the importance of a weak, broad peak at ca 3° that was due to a superlattice which indexed as $(1/2,1/2,1/2)_{cubic}$ within the fundamental cubic cell. This demanded an enlargement of the unit cell to orthorhombic. From the dimensions of this unit cell one could now accommodate the dumbbell-shaped $(C_{59}N)_2$. Further examination revealed that the dumbbells were rotated along their long axis relative to the axis of the lattice, forcing further revision of the structure to monoclinic. With additional help from the theoretical calculations by Andreoni mentioned above, the structure was solved.

In attempts to solve the structure via electron diffraction by TEM, it was discovered during preliminary SEM examination, that the morphology of the heterofullerene dimer was that of spheres of spheres. Giant hollow spheres of ca. 2–8 μm in diameter had formed from smaller (50–100 nm) ones in the process of "crystallization" [104, 105]. Later, Bellavia-Lund discovered the same behavior for $(C_{69}N)_2$ [76]. The $(C_{59}N)_2$ spheres were only poorly crystalline, giving only five lines superimposed on a broad amorphous background. The lines could be indexed to a hexagonal unit cell with a = 9.97 Å and c = 16.18 Å.

The Effect of Pressure on Solid $(C_{59}N)_2$. Synchrotron X-ray powder diffraction experiments on crystalline $(C_{59}N)_2$ up to 22.5 GPa were determined [106]. Upon deriving the pressure-volume equation of state, it was discovered that the material observed at atmospheric pressure remained stable up to the highest hydrostatic compression studied. The bulk modulus (E_o) of 21.5 GPa and its pressure derivative, $dE_o/dP = 4.2$, were only slightly lower than the values observed for C_{60} ($E_o = 18.1$ GPa, $dE_o/dP = 5.7$). By analyzing the pressure dependence of the three lattice constants it was possible to determine various bonding interactions in the solid. For example, the absence of strong anisotropy in compressibility was rationalized as resulting from the nonalignment of the dumbbell's long axis with that of the unit cell axis, mentioned above (Sect. 4.3). However, as the pressure increased, the interdimer distances decreased more than the intradimer ones. By the time a pressure of ca. 6.5 GP had been applied, both distances were observed to become of comparable magnitude, forcing the authors to conclude that a novel solid state structure with almost isotropic bonding between spheres had been produced.

Potassium Doping of $(C_{59}N)_2$. As stated before, the electronic structure of aza[60]fullerene is dramatically different from that of C_{60}, such that the equivalent of K_3C_{60} would be $K_2C_{59}N$. Since the intercluster bond of the dimer was calculated to be very weak, and since addition of a number of electrons to the dimer forces breakage of the intercluster bond (see Sect. 4.4), it was not surprising that treatment of the dimer with excess potassium metal at elevated temperature led to the formation of $K_6C_{59}N$ [107]. Interestingly, the bcc symmetry crystal structure of this salt was found to be identical to that of its C_{60} congener [108,109]. There were subtle differences, however; for example, the unit cell parameter was slightly smaller, giving an intercluster distance of 9.80 Å, compared to 9.86 Å for K_6C_{60}. Because the crystal structures of both potassium salts were so similar and because $C_{59}N$, under stringent conditions (heating in a tungsten crucible to more than 500 °C [110] or electron impact with greater than 70 keV [111]), was known to produce C_{60}, the synchrotron X-ray sample was reported to have been re-oxidized in the presence of pTsOH, to produce a mixture of $C_{59}HN$ and $(C_{59}N)_2$. The modified electronic structure of the azafullerene relative to the all-carbon cage required that in $K_6C_{59}N$ there should be an unfilled band resulting from interaction with the open shell $C_{59}N^{-6(\cdot)}$ in the solid state; in contrast to K_6C_{60}, where the hexa-anion is closed shell due to complete filling of the triply degenerate t_{1u} level, leading to a filled band and the expected insulating state. Hence, in contrast to the all-carbon case, the azafullerene salt was found to be a conductor at room temperature. The partial filling of the band was also reflected in the magnetic susceptibility, which indicated Curie–Weiss behavior [112]. In principle, the most interesting salts would be $K_2C_{59}N$ and $K_1C_{59}N$, where the former is isoelectronic with the superconducting phase of C_{60}. Many attempts by the Sussex group have so far failed to produce a stable phase of either stoichiometry of the K and Rb salts [113]. Some evidence for the formation of "$Rb_{0.7}(C_{59}N)_2$" and "$Rb_{1.8}(C_{59}N)_2$" compositions, with the intercalated Rb residing in the octahedral interstices, has been reported more recently [114]. Preliminary DC magnetization measurements showed a Curie-like paramagnetic component, superimposed on a temperature-independent background.

The Electronic Structure of $(C_{59}N)_2$ and Its Alkali-Metal Salts. The electronic properties of $(C_{59}N)_2$ in thin films were studied using photoemission spectroscopy (PES) and electron energy-loss spectroscopy (EELS). The HOMO of $(C_{59}N)_2$ is clearly different from that of C_{60}, with the electron density strongly concentrated on the N atoms and along the axis of the dimer bond [115,116]. In contrast, the LUMO has mainly C-character. When $(C_{59}N)_2$ was compared to the C_{60}^- -dimer in dimerized Rb_1C_{60}, a lowering of the HOMO and HOMO-1 states of the heterofullerene dimer was observed. This reduces the splitting of the occupied electronic levels in $(C_{59}N)_2$. The N 1s excitation edge of $(C_{59}N)_2$ resembles that of polypyrrole much more than that of polymethineimine. This is in agreement with a [6,6]-closed structure for the [60]azafullerene moieties. The optical gap of the dimer is ~ 1.4 eV, slightly smaller than that of C_{60} (1.8 eV) [117,118]. For $(C_{59}N)_2$, a static dielectric constant $\varepsilon_1(0) \sim 5.6$ was obtained from the energy-loss functions. This value is larger than that of C_{60} [$\varepsilon_1(0) \sim 4$], reflecting the smaller energy gap of the heterofullerene dimer.

The electronic structure of $A_xC_{59}N$ salts (A = K, Rb, Cs) was determined analogously using PES and EELS [119]. Intercalation of the three different metals in $(C_{59}N)_2$ results in very similar PES spectra. The spectrum for very low K content resembles that of the solid solution "α phase" in the KC_{60} system. Upon increasing intercalation, an additional feature at ~1.2 eV binding energy appears; further increase in K content results in the appearance of a broad structure at ~0.5 eV. Upon K saturation, this last feature grows in weight and shifts to 0.7 eV, a situation similar to that observed for K_6C_{60}. No density of states at the Fermi level was observed at any time during the intercalation experiment (analogous to the situation observed for Rb_xC_{70} and K_xC_{70}). All of the PES and EELS spectra are very similar to that of K_6C_{60}. This is still somewhat of a puzzle, if one assumes that the saturated heterofullerene salts have the $A_6C_{59}N$ stoichiometry, since one would expect to find some sign of a singly occupied MO, that is to be present in $C_{59}N^{6-}$.

4.4
Electrochemistry of Azafullerenes

The cyclic voltammogram (CV) of $(C_{59}N)_2$ showed three overlapping pairs of reversible one-electron reductions within the solvent window (E_1 = -997 mV, E_2 = -1071 mV, E_3 = -1424 mV, E_4 = -1485 mV, E_5 = -1979 mV, E_6 = -2089 mV; ferrocene/ferrocenium couple, internal standard) [7]. A combination of linear sweep voltammetry and chronoamperometry established that all overlapping waves were two-electron reductions [120]. There was also an irreversible two-electron oxidation with a peak potential at +886 mV, that is 0.2 V more negative (easier to oxidize) than C_{60} [121]. The appearance of closely spaced pairs of waves in the CV was interpreted in terms of two (identical) weakly interacting electrophores, similar to the dianthrylalkanes [122]. After the third double wave, the process is irreversible, this was interpreted as irreversible cleavage of the dimer bond.

The situation with the electrochemistry of $C_{59}HN$ is somewhat more complicated because the redox waves correspond to electrochemically irreversible steps, a result which was interpreted as arising from the weak C–H bond [66]. The first and second reduction waves occur at more negative potentials (E_1 = -1106 mV, E_2 = -1500 mV; ferrocene/ferrocenium couple, internal standard) than the dimer but very slightly more positive potentials than C_{60} and significantly more positive potentials than any dihydro C_{60} derivative, a reflection of the higher electronegativity of nitrogen vs. carbon. The observed irreversible oxidation wave at +823 mV (vs. ferrocene/ferrocenium couple, internal standard; 570 mV easier to oxidize than C_{60}!) may also be a reflection of the weak C–H bond, or may stem from the ionization of one of the lone pair electrons on the nitrogen atom.

5
Miscellaneous Reports on Other Heterofullerenes

In this section we will briefly summarize the literature with respect to reports on heterofullerenes other than aza[60]fullerene and aza[70]fullerenes. Since we limit ourselves to heterofullerenes in the narrow sense of the definition, non-carbon clusters and most $C_m X_n$ and $C_m X_n Y_p$ clusters are neglected here to leave reports on $C_{59}X$, $C_{58}X_2$, $C_{57}X_3$, $C_{58}XY$, and the corresponding heterofullerenes derived from C_{70} to be reviewed.

Theoretically, even this limitation still leaves us with an enormous number of structures: while for every X there is only one $C_{59}X$ isomer, for hetero[60]fullerenes $C_{58}X_2$ there are already 23 possible isomers (of which some are chiral), not to mention the number of isomers of the hetero[60]fullerenes $C_{57}X_3$! In Fig. 34, the 23 isomers of $C_{58}X_2$ are shown, together with the IUPAC numbering [22].

Borafullerenes. Borafullerenes were the first heterofullerenes to be reported by the Smalley group in 1991 and thereafter [123–126]. Laser vaporization of a graphite/boron nitride composite disk and analysis of the formed clusters by the Fourier transform ion cyclotron resonance (FT-ICR) device yielded mass spectra with "clusters" of peaks corresponding to C_n and $C_{n-x}B_x$ (with n even and > 44). From the deconvoluted mass spectra, the approximate amounts of the bora[60]fullerene ions, relative to C_{60}^+ (= 100%), were calculated to be 95% $C_{59}B^+$, 109% $C_{58}B_2^+$, 82% $C_{57}B_3^+$, 42% $C_{56}B_4^+$, 18% $C_{55}B_5^+$, and 9% $C_{54}B_6^+$. Upon exposure of this set of ions to NH_3 at low pressure, a new series of peaks was observed, corresponding to adducts with formulas $C_{59}B(NH_3)$, $C_{58}B_2(NH_3)_2$, etc, with the same relative abundance. A similar series of bora[70]fullerene ions was observed around the m/z = 840 cluster. The authors speculate that since every boron atom appears to be able to bind an ammonia molecule, the boron atoms are likely to be well separated on the surface of the cage. Further support for the heterofullerene structure of these gas-phase ions was found in the photophysical behavior: exposure to laser pulses yielded spectra showing the typical fullerene "shrink-wrap" peaks from successive loss of C_2 fragments. Later, when a KCl/boron/graphite composite target was laser vaporized in a 1200°C tube furnace, material was collected that showed peaks in FT-ICR mass spectra that were ascribed to $C_{59}B$, $K@C_{59}B$, and $K@C_{58}B_2$ ions. Laser photolysis of these ions again led to the appearance of shrink-wrap peaks [124].

In 1996, Muhr et al. reported on the macroscopic preparation of borafullerenes using the arc-evaporation method on graphite rods doped with either boron nitride, boron carbide, or boron [127]. An extraction and enrichment scheme was used for the heterofullerene content of the soot, involving pyridine extraction and subsequent treatments of the extract with CS_2 (yielding extract 1) and pyridine (extract 2). Negative mode TOF mass spectroscopy of extract 2 showed clear contributions of $C_{59}B^-$ and $C_{69}B^-$ ions to the m/z = 720 and 840 peak clusters. The use of ^{10}B-enriched material confirmed the MS interpretation. Only monoborafullerene ions were observed. XPS spectra of the extract showed a peak at 188.8 eV that was assigned to boron (1s core level) in borafullerene ($C_{59}B$ mainly). Treatment of the soot with THF supposedly led to extraction of the "borafullerenes", and also to their decomposition. These workers further report that the extracted materials appear to be moisture sensitive, leading to the formation of boron oxide or boric acid. Piechota et al. performed a similar production experiment and extracted the soot in the standard manner with toluene

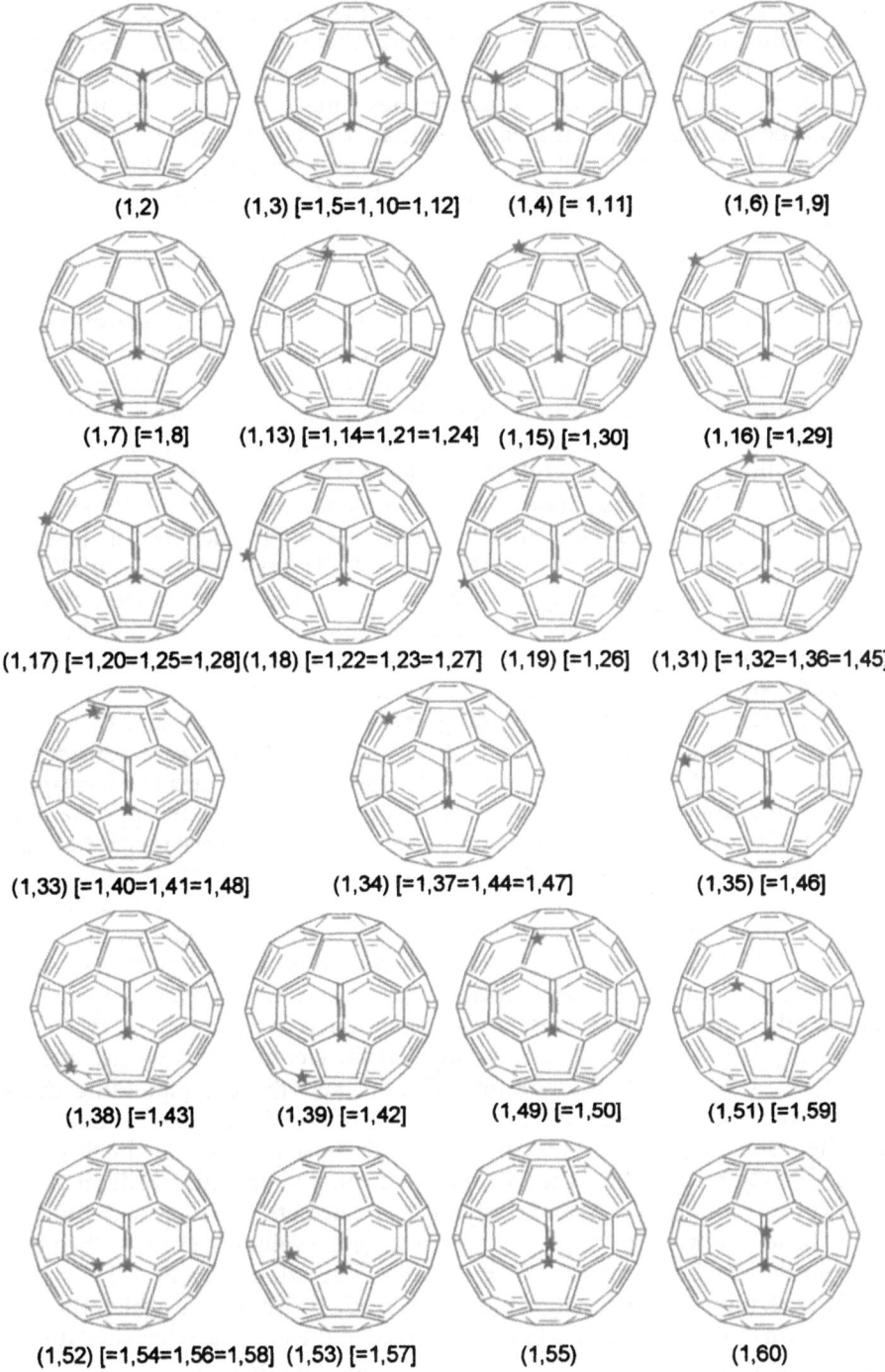

Fig. 34. The 23 possible positions of two heteroatoms in heterofullerenes with formula $C_{58}X_2$ (the positions of the heteroatoms are indicated by *; atom numbering according to IUPAC rules)

and analyzed the extract without any further purification, since too much material seemed to be bound irreversibly to alumina when the latter was used as solid phase for column chromatography. MS analysis of the crude extract did not indicate the presence of "borafullerenes", however [31]. Surprisingly, in none of the two reports was it mentioned that, based on the previously published properties of $C_{59}N$, isolating neutral and unsubstituted monomeric $C_{59}B$ would be very unlikely since it is an open shell molecule. In Chemical Abstracts we found a Chinese report on a gas-phase preparation method that yielded materials showing mass spectra with peaks interpreted as $C_{59}B$, $C_{58}B_2$, and $C_{57}B_3$ ions [128].

Bora[60]fullerene (CAS name for the radical: 2H-1-bora[5,6]fulleren-C_{60}-I_h-2-yl) has been the subject of various types of calculations. Andreoni et al. used the Car–Parrinello method that proved quite accurate in calculating a number of properties of C_{60}. It was found that the B atom remains threefold coordinated and most of the distortion is localized on the B–C bonds. The length of the "6,6" B–C bond increases by 10%, while that of the "5,6" B–C bonds increases by 7.5% relative to C_{60}, according to these calculations. Hence, the bora[60]fullerene cage is expected to be closed. The "impurity state" is strongly localized, as in $C_{59}N$. A radial, inward directed, dipole moment of ~ 1.4 D was calculated [80]. Kurita et al., in a series of papers, reported calculated molecular structures, binding energies, and electronic properties of $C_{59}B$, $C_{58}B_2$, and several other heterofullerenes using a MO method with Harris approximation [129–131]. They envisioned only a 4.2 and 4.7% increase, respectively, in the "6,6" C–B and "5,6" C–B bond lengths in $C_{59}B$. The binding energies for C_{60}, $C_{59}B$, and $C_{59}N$ were found to be almost identical. For $C_{58}B_2$, only two isomers were considered, namely one with the two boron atoms as "nearest-neighbor sites in a pentagon" (i.e. 1,6-bibora[60]fullerene) and one with the boron atoms on opposite sites (i.e. 1,60-dibora[60]fullerene), the latter appearing to have very much the same properties as $C_{59}B$. In the 1,6 isomer, a heavy distortion was found, already evident from the B–B bond length, calculated to be 1.65 Å. However, according to the calculation, this did not significantly affect the binding energy. $C_{59}B$ (and sometimes $C_{58}B_2$) was also considered and compared to $C_{59}N$ with respect to electronic structure, (nonlinear) optical properties, and in metal complexes in several other theoretical studies, already mentioned in Sect. 4.1 [31, 79, 83, 85, 87–89, 92]. Furthermore, Liu et al. calculated the electronic properties of 1,2-, 1,3-, 1,58-, 1,59- and 1,60-dibora[60]fullerene and the corresponding diaza[60]fullerene isomers using the Su–Schrieffer–Heeger (SSH) model and found that the results were not always consistent with those obtained by the SCF-MO method [132]. Xu et al. used the SSH model to compute the third-order polarizability of 1,6- and 1,60-$C_{58}B_2$ and compared the calculated γ-values with those of the N-analogs and $C_{59}X$. 1,60-$C_{58}B_2$ has the largest γ-values [133,134]. Chen and Lin considered all 23 isomers of $C_{58}B_2$ in their calculations on a PC [135].

Xia et al. found in their calculations that in $C_{58}B_2$ isomers in which the two boron atoms were further apart on the surface of the ball, the open shell diradical form becomes more favorable than the closed shell structure [136].

Miyamoto et al. calculated the electronic structure of solid $C_{59}B$ as a metallic fullerene material consisting of single molecules [137]. Taking the experimentally observed consistent dimerizing tendency of the known fullerene mono-radicals and monoradical ions into account, this seems to be a highly hypothetical situation!

Zou et al. calculated the electronic states of C_nB, with n = 59, 179, 239, 419, 539, 719, 779, finding a localized distortion and a "mid-gap state", appearing in the original gap of the pristine fullerenes [138].

In our opinion, by far the most important and challenging future work to be done on borafullerenes is to design and perform a rational synthesis for $C_{59}B$ (as the dimer, a salt or a derivative), most likely starting from C_{60}, as was done for aza[60]fullerene.

Azafullerenes with More Than One Nitrogen Atom. Up to now, no diaza[60]fullerene has been prepared, most likely not even in the gas phase. Since all $C_{58}N_2$ isomers, especially the ones with two adjacent nitrogen atoms, should, in principle, be able to relax to closed shell structures, monomeric pristine $C_{58}N_2$ seems a feasible target. In 1991, it was reported that mass spectroscopy of the toluene extract of soot, obtained by contact-arc vaporization of graphite in the presence of N_2, showed peaks that were tentatively attributed to formulas such as $C_{70}N_2$, $C_{59}N_6$, $C_{59}N_4$, and $C_{59}N_2$. Since no further experimental evidence was given for such structures, this interpretation is pure speculation [23,24]. Glenis et al. reported on a similar experiment, but using pyrrole gas as the source of nitrogen atoms. Elemental analysis of the soot showed almost 1% N-content. The toluene extract was chromatographed on alumina, resulting in a "yellow-orange", a "yellow-green", and a "dark orange" band, subsequently. The material from the second band was analyzed by mass spectroscopy, UV-Vis and fluorescence spectroscopy [26]. Mass spectroscopy revealed peaks at almost every even m/z value between 640 and 790. Especially between 720 and 790 amu, only even masses were observed. The authors conclude from this spectrum that, next to $C_{59}N$, only molecules with even-numbered ratios of C to N atoms are formed (like $C_{58}N_2$, $C_{56}N_4$, etc., etc.). We, however, consider it impossible that *any* heterofullerene would show even mass peaks *only*, unless pure ^{12}C-graphite was used in the experiment! Later, the extract was submitted to HPLC, using a GPC column, to yield ten separated fractions, that were regrettably not submitted to mass spectroscopy but analyzed by absorption and fluorescence spectroscopy only [27].

Diaza[60]fullerenes have been the subject of several theoretical considerations and calculations, however – Karfunkel et al. considered the ten possible isomers of $C_{58}N_2$, with the nitrogen atoms separated by maximally three carbon atoms [139]. Some thermochemical and electronic properties of the closed shell structures were calculated using various methods. It was found that 1,7-diaza[60]fullerene, with the two nitrogen atoms in a 1,3-fashion in the same pentagon, appears to be of remarkable stability (15 kcal mol^{-1} more stable than the next most stable isomer, according to an AM1 calculation). The geometry of all isomers appears to be very much like that of C_{60}, but the predicted vibrational spectra and the HOMO and LUMO values differ significantly. These workers also considered some "truncated" heterofullerenes, based on the C_{60} skeleton. Some of the structures, with formula $C_{54}N_4$, show some resemblance to porphyrins. In several computational studies, efforts were made to compare various properties of $C_{58}N_2$ with those of $C_{58}B_2$ (see above) [79, 83, 92] (23 isomers), [130–134, 136]. The Hirsch group has mentioned an idea for a synthetic route towards $C_{58}N_2$, starting from C_{60} [70]. The remarkable regioselectivity in the bisaddition of azides to C_{60} has prevented a simple "bis-version" of the synthetic routes, described for $C_{59}N$ and $C_{69}N$ compounds. Addition of bisazides in which the two azide moieties are tethered by a (removable) spacer might also be considered as a first step in a route towards certain isomers of $C_{58}N_2$. The synthesis of $C_{58}N_2$ is possibly complicated if the formation of the two N-sites is not simultaneous: in that case, an open shell intermediate, as found in the synthesis of $(C_{59}N)_2$, might give rise to an undesired reaction path.

Bühl has proposed that hexaaza[60]fullerene, $C_{54}N_6$, could be an aromatic molecule, being isoelectronic with C_{60}^{6-} [140]. An ab initio study of six highly symmetrical $C_{54}N_6$ isomers indicated that two isomers with D_{3d} symmetry may have large endohedral chemical shieldings, much more than the other isomers with D_3 symmetry.

Azabora[60]fullerene. Of the many possible isomers of $C_{58}BN$, most (computational) attention has been given to the 1,2-, 1,6- and the 1,60-isomer. There are no reports of formation of $C_{58}BN$ in gas-phase heterofullerene preparation experiments. This may be due to the fact that $C_{58}BN$ has a mass of 720/721 amu ($\sim 1:4$), apart from ^{13}C contributions, which makes it very hard to be detected unambiguously in the MS of mixtures that also contain C_{60}. Piechota et al. do indeed mention a higher 721/720 ratio in their possibly heterofullerene-containing product than in the product from a carbon-only experiment, but found no other proof for $C_{58}BN$ [31]. These workers calculate a HOMO-LUMO gap of 2.2 and 1.8 eV for the 1,2- and 1,6 isomer, respectively, using the so-called Dgauss program, while Liu et al. found similar gaps of ~2 eV for a non-specified "nearest neighbor" BN structure and the 1,60-isomer, using a tight-binding approximation quantum mechanical method [83]. Rustagi et al., without much scientific ado, envision interesting nonlinear optical effects in 1,60-azabora[60]fullerene [92]. Xia et al. found in their calculations that all isomers of $C_{58}BN$ are preferably in the closed shell configuration (in contrast to $C_{58}B_2$). In the 1,2-isomer, B and N have estimated charge densities of +0.0523 and –0.1450, respectively, while in the (presumably) 1,3-isomer and the 1,60-isomer both heteroatoms have negative charge density [136]. Others calculated bond lengths and HOMO-LUMO gaps of the 1,6- and the 1,60-isomer and found a B–N bond of 1.61 Å in the 1,6-isomer and a factor 8.6 difference in the energy gap between the two isomers [131]! Esfarjani et al. performed computations on the 1,2- and 1,60-isomers of $C_{58}BN$ in the solid state [141–143]. These workers discuss the various possibilities for the nature of the B–N bond in the 1,2-isomer, the limits being "covalent" (ylide > $B^- = N^+ <$), "ionic" ("open" structure with > B^+ and $N^- <$), and "no charge transfer" (with a single > B–N < σ-bond). Calculations on the fcc solid-state electronic structure indicated the presence of a single B–N bond with slight charge transfer from N to B. However, since the B–N bond length was fixed *a priori* at 1.40 Å, this computational result, reasonable as it may seem, may still be of limited value, as was recognized by the authors [142].

[60]Heterofullerenes with Heteroatoms Other Than N, B. After unsuccessful attempts by the Smalley group, Clemmer et al. reported the formation of C_nNb^+ (n = 28–50) clusters upon pulsed laser vaporization of a mixed NbC/graphite composite rod [144]. From the difference between even and odd n C_nNb^+ ions in their chemical reactivity towards O_2 and N_2, evidence was found for the odd-C numbered cages to be (quasi)heterofullerenes instead of endohedral complexes. No clusters with a C/Nb ratio of ~60 : 1 were detected, however.

In our opinion, heterofullerenes with divalent heteroatoms, like O and S, are potentially interesting compounds because they could possibly exist as neutral ylide structures with $>{}^-C-O^+<$ and $>{}^-C-S^+<$ bonds, respectively. Alternative structures for $C_{59}O$ and $C_{59}S$ are less ionic truncated quasi-fullerene structures bearing a carbonyl or thiocarbonyl moiety, as shown in Fig. 35.

Kurita et al. calculated a $C_{59}S$ cluster and found a heavily distorted structure with C–S "bond lengths" of 2.01 and 2.16 Å, a rather low binding energy of 5.1 eV/atom, a HOMO-LUMO gap of 0.63 eV, and an atomic charge of +0.5 on the S atom [129]. Later, Glenis et al. tried to prepare thiafullerenes analogously to their previously reported method of contact-arc vaporization of graphite, but in the presence of thiophene or 3-methylthiophene instead of pyrrole [145]. We find it striking that column chromatography of the soot extract (containing 2% S by elemental analysis) yielded three bands with the same colors as in the case of their attempted azafullerene preparation (see above). Many peaks with tempting amu numbers were observed in mass spectroscopy, but there was no proof for any heterofullerene structure.

Of the oxygen-containing heterofullerene family, oxa[60]fullerene, $C_{59}O$, has only been mentioned as possibly present in the form of $C_{59}O^+$ ions in gas-phase experiments by Christian et al. [25,146]. At collision energies between 15 and 75 eV, O^+ ions and gas-phase C_{60} yielded peaks in MS that were interpreted as odd-sized "fullerene oxide" ions of the form $C_{59-2n}O^+$ (n = 0–4), decomposing primarily by C_2 loss, retaining the O atom. The possibility of the structures being $CO@C_{58-2n}$ was slightly favored by the authors, however. Karfunkel et al. reported on a philosophy about truncated heterofullerenes containing O atoms or N and O atoms in such a way that crown-ether-like structures would be present as part of the cage [139].

Purely hypothetical are the mono- and the 23 diphospha[60]fullerene isomers, $C_{59}P$ and $C_{58}P_2$, considered using a PC-based calculation by Chen and Lin [135].

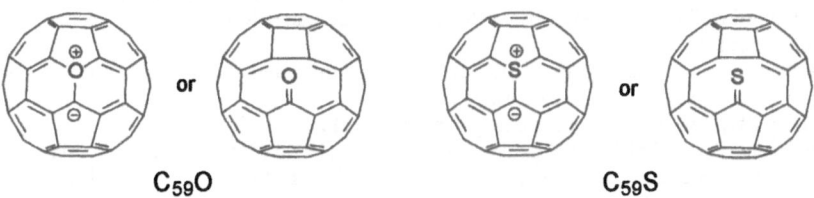

$C_{59}O$ $C_{59}S$

Fig. 35. Possible structures for $C_{59}O$ and $C_{59}S$: ylide heterofullerenes or open-cage (thio)carbonyl quasi-fullerenes

Lastly, we found two reports on silafullerenes [147,148]. In 1993, Jelski et al. computed the structures and relative stability of $C_{59}Si$, 1,2-, 1,6,- and 1,60-$C_{58}Si_2$, and a $C_{48}Si_{12}$ isomer by a AM1 method. These structures were predicted to be stable and to show a steady decrease of the band gaps with the increasing number of Si atoms. These workers did not hesitate to mention the possibility of a polysiloxane type 1,60-disila[60]fullerene polymer. Kimura et al. analyzed the products from pulsed-laser vaporization of silicon-carbon composite ("bulk siliconized carbon") rods by time-of-flight MS. Only small peaks were observed that would match the masses of "SiC_n^+-clusters" ($61 \geq n \geq 56$). Since no further evidence was given, silafullerenes are still best considered as hypothetical structures.

Considering the fact that, at best, scarce proof was given for the existence of only some of the heterofullerenes mentioned in Sect. 5, there are already a surprising number of patents covering, among other things, the preparation and/or uses of such materials [34,149–153]!

6
References

1. Kroto HW, Heath JR, O'Brien SC, Curl RF, Smalley RE (1985) Nature 318:162
2. Krätchmer WA, Lamb LD, Fostiropoulos K, Huffman DR (1990) Nature 347:354
3. Guo T, Jin C, Smalley RE (1991) J Phys Chem 95:4948
4. Hummelen JC, Wudl F (1995) Unpublished results
5. Lamparth I, Nuber B, Schick G, Skiebe A Grösser T, Hirsch A (1995) Angew Chem Int Ed Engl 34:2257
6. Averdung J, Luftmann H, Schlachter I, Mattay J (1995) Tetrahedron 51:6977
7. Hummelen JC, Knight B, Pavlovich J, González R, Wudl F (1995) Science 269:1554
8. Nuber B, Hirsch A (1996) J Chem Soc Chem Commun 1421
9. Tenne R (1995) Adv Mater 7:965
10. Huckzo A (1997) Fullerene Sci Technol 5:1091
11. Huckzo A (1997) Wiad Chem 51:27 (Chem Abstr 127:50671)
12. Ying ZC, Zhu JG, Compton RN, Allard Jr LF, Hettich RL Haufler RE (1997) ACS Symp Ser 679:169 (Chem Abstr 127:365278)
13. Cao B, Zhou X, Gu Z (1997) Huaxue Tongbao 1 (Chem Abstr 128:109852)
14. Prassides K (1997) Curr Opin Solid State Mater Sci 2:433
15. Kuzmany H (1998) Phys Unserer Zeit 29:16
16. Hirsch A (1997) J Phys Chem Solids 58:1729
17. Mattay J, Torres-Garcia G, Averdung J, Wolff C, Schlachter I, Luftmann H, Siedschlag C, Luger P, Ramm M (1997) J Phys Chem Solids 58:1929
18. Averdung J, Torres-Garcia G, Luftmann H, Schlachter I, Mattay J (1996) Fullerene Sci Technol 4:633
19. Wudl F, Bellavia-Lund C, Collins T, González M, Hicks RG, Hummelen JC, Keshavarz-K M, Sastre A, Srdanov G (1996) Proc Robert A Welch Found Conf Chem Res 40th 131–138:87
20. Bellavia-Lund C, Hummelen JC, Keshavarz-K M, González R, Wudl F (1997) J Phys Chem Solids 58:1983
21. Bellavia-Lund C, Keshavarz-K M, González R, Hummelen JC, Hicks R, Wudl F (1997) Phosphorus, Sulfur, and Silicon 120+121:107
22. Godly EW, Taylor R (1997) Pure Appl Chem 69:1411
23. Pradeep T, Vijayakrishnan V, Santra AK, Rao CNR (1991) J Phys Chem 95:10564
24. Rao CNR, Pradeep T, Seshadri R, Govindaraj A (1992) Ind J Chem 31 A+B:F 27
25. Christian JF, Wan Z, Anderson SL (1992) J Phys Chem 96:10597
26. Glenis S, Cooke S, Chen X, Labes MM (1994) Chem Mater 6:1850
27. Glenis S, Cooke S, Chen X, Labes MM (1995) Synth Metals 70:1313

28. Yu R, Zhan M, Cheng D, Yang S, Liu Z, Zheng L (1995) J Phys Chem 99:1818
29. Zhang MX, Yu RQ, Wang YH, Yang SY, Cheng DD, Liu CY, Zheng LS (1995) Gaodeng Xuexiao Huaxue Xuebao 16:1780 (Chem Abstr 124:217583)
30. Xie E, Gong J, Xu C, Chen G (1996) Bandaoti Xuebao 17:177 (Chem Abstr 125:209454)
31. Piechota J, Byszewski P, Jablonski R, Antonova K (1996) Fullerene Sci Technol 4:491
32. Ying ZC, Hettich RL, Compton RN,.Haufler RE (1996) J Phys B:At Mol Opt Phys 29:4935
33. Ying ZC, Zhu JG, Compton RN, Allard LF Jr, Hettich RL Haufler RE (1997) ACS Symp Ser 679:169 (Chem Abstr 127:365278)
34. Yoshida T, Eguchi K, Yoshie K (1992) Eur Pat Appl (Chem Abstr 118:194586)
35. Prato M, Chan LQ, Wudl F, Lucchini V (1993) J Am Chem Soc 115:1148
36. Nuber B, Hampel F, Hirsch A (1995) J Chem Soc Chem Commun 1799
37. Grösser T, Prato M, Lucchini V, Hirsch A, Wudl F (1995) Angew Chem Int Ed Engl 34:1343
38. Prato M, Wudl F, Grösser T, Hirsch A (1995) German Patent DE 95-19507502 (to Hoechst A.-G.)
39. Shiu LL, Chien KM, Liu TY, Lin TI Her GR, Luh TY (1995) J Chem Soc Chem Commun 1159
40. Schick G, Hirsch A, Mauser H, Clark T (1996) Chem Eur J 2:935
41. Shen C, Yu H, Juo CG, Chien KM, Her GR, Luh TY (1997) Chem Eur J 3:744
42. Shen CKF, Chien KM, Juo CG, Her GR, Luh TY (1996) J Org Chem 61:9242
43. Shen CKF, Chien KM, Juo CG, Her GR, Luh TY (1997) J Org Chem 62:4548
44. Banks MR, Cadogan JIG, Gosney I, Hodgson PKG, Langridge-Smith PRR, Millar JRA, Parkinson JA, Rankin DWH, Taylor AT (1995) J Chem Soc Chem Commun 887
45. Schick G, Grösser T, Hirsch A (1995) J Chem Soc Chem Commun 2289
46. Averdung J, Mattay J, Jacobi D, Abraham W (1995) Tetrahedron 51:2543
47. Averdung J, Luftmann H, Mattay J, Claus K.-U, Abraham W (1995) Tetrahedron Lett 36:2957
48. Mattay J, Averdung J, Luftmann H, Schlachter I (1995) German Patent DE 95-19506230 (Chem Abstr 125:221612; 125:247605)
49. Banks MR, Gosney I, Hodgson PKG, Langridge-Smith PRR, Millar JRA, Taylor AT (1995) J Chem Soc Chem Commun 885
50. Averdung J, Wolff C, Mattay J (1996) Tetrahedron Lett 37:4683
51. Averdung J, Mattay J (1996) Tetrahedron 52:5407
52. Ihisda T, Tanaka K, Nogami T (1994) Chem Lett 561
53. Yan M, Cai SX, Keana JF (1994) J Org Chem 59:5951
54. Banks MR, Cadogan JIG, Gosney I, Hodgson PKG, Langridge-Smith PRR, Rankin DWH (1994) J Chem Soc Chem Commun 1365
55. Kuwashima S, Kubota M, Kushida K, Ishida T, Ohashi M, Nogami T (1994) Tetrahedron Lett 35:4371
56. Hirsch A (1994) The chemistry of the fullerenes. Thieme, Stuttgart
57. Hirsch A, Lamparth I, Grösser T, Prato M, Lucchini V, Wudl F (1996) NATO ASI Ser, Ser E 316 [Chemical Physics of Fullerenes 10 (and 5) Years Later, p 267]
58. Mattay J, Siedschlag C, Torres-Garcia G, Ulmer L, Wolff C, Fujitsuka M, Watanabe A, Ito O, Luftmann H (1997) Proc Electrochem Soc 97 14:326
59. Averdung J, Albrecht E, Luftmann H, Schlachter I, Mattay J, Claus K.-U, Jakobi D, Abraham W (1995) Proc Electrochem Soc 95 10:1164
60. Averdung J, Mattay J (1997) Prax Naturwiss Chem 46:28
61. Averdung J, Gerkensmeier T, Ito O, Luftmann H, Luger P, Schlachter I, Siedschlag C, Torres-Garcia G, Mattay J (1997) In: Kuzmany H (ed) Fullerenes Fullerene Nanostruct, Proc Int Wintersch Electron Prop Novel Mater 10th, World Scientific Singapore, p 509
62. Hummelen JC, Prato M, Wudl F (1995) J Am Chem Soc 117:7003
63. O'Brien S, Heath JR, Curl RF, Smalley RE (1988) J Chem Phys 88:220
64. Hummelen JC, Keshavarz-K M, van Dongen JLJ, Janssen RAJ, Meijer EW, Wudl F (1998) J Chem Soc Chem Commun 281
65. Bellavia-Lund C, Keshavarz-K M, Collins T, Wudl F (1997) J Am Chem Soc 119:8101
66. Sastre A, Keshavarz-K M, Wudl F Unpublished results

67. Keshavarz-K M, González R, Hicks RG, Srdanov G, Srdanov VI, Collins TA, Hummelen JC, Bellavia-Lund C, Pavlovich J, Wudl F (1996) Nature 383:147
68. Nuber B, Hirsch A (1998) J Chem Soc Chem Commun 405
69. Bellavia-Lund C, González R, Hummelen JC, Hicks RG, Sastre A, Wudl F (1997) J Am Chem Soc 119:2946
70. Reuther U, Hirsch A (1998) J Chem Soc Chem Comm 1401
71. Birkett PR, Avent AG, Darwish AD, Kroto HW, Taylor R, Walton DRM (1993) J Chem Soc Chem Commun 1230
72. Bellavia-Lund C, Wudl F (1997) J Am Chem Soc 119:943
73. Smith III A, Strongin R, Brard L, Furst G, Romanow W, Owens K, Goldschmidt R (1995) J Am Chem Soc 117:5492 and references within
74. Hassner AJ (1968) J Org Chem 33:2684
75. Hawkins JM, Meyer A, Solow M (1993) J Am Chem Soc 115:7499
76. Bellavia-Lund (1997) Thesis UCSB, Santa Barbara CA (USA)
77. Hasharoni K, Bellavia-Lund C, Keshavarz-K M, Srdanov G, Wudl F (1997) J Am Chem Soc 119:11128
78. De la Cruz P, Wudl F, Bellavia-Lund C (1998) Presented at the 193rd Meeting of the ECS, San Diego, CA
79. Wang S-H, Chen F, Fann Y-C, Kashani M, Malaty M, Jansen SA (1995) J Phys Chem 99:6801
80. Andreoni W, Gygi F, Parrinello M (1992) Chem Phys Lett 190:159
81. Andreoni W, Gygi F, Parrinello M (1992) NATO ASI Ser, Ser C 374:333
82. Kurita N, Kobayashi K, Kumahora H, Tago K, Ozawa K (1992) Chem Phys Lett 198:95
83. Liu J, Gu B, Han R (1992) Solid Sate Commun 84:807
84. Rosen A, Oestling D (1992) Mater Res Soc Symp Proc 270:141
85. Chen F, Singh D, Jansen SA (1993) J Phys Chem 97:10958
86. Breslavskaya NN, D'yachkov PN (1994) Koord Khim 20:803 (Chem Abstr 123:112384)
87. Dong J, Jiang J, Wang ZD, Xing DY (1995) Phys Rev B 51:1977
88. Dong J, Jiang J, Yu J, Wang ZD, Xing DY (1995) Phys Rev B 52:9066
89. Jiang J, Dong J, Xing DY (1997) Solid State Commun 101:537
90. Xu Q, Jiang J, Dong J, Xing DY (1996) Phys Status Solidi B 193:205
91. Jiang J, Dong J, Xu Q, Xing DY (1996) Z Phys D:At Mol Clusters 37:341
92. Rustagi KC, Ramaniah LM, Nair SV (1992) Int J Mod Phys B 6:3941
93. Vancik H, Babic D, Trinajstic N (1995) Fullerene Sci Technol 3:305
94. Piechota J, Byszewski P (1997) Z Phys Chem 200:147
95. Andreoni W, Curioni A (1996) In: Kuzmany H (ed) Fullerenes Fullerene Nanotruct, Proc Int Wintersch Electron Prop Novel Mater, 10th World Scientific Singapore, p 359
96. Bühl M, Curioni A, Andreoni W (1997) Chem Phys Lett 274:231
97. Curioni A, Andreoni W (1998) In: Kuzmany H, Fink J, Mehring M, Roth S (eds) Molecular Nanostructures, World Scientific Singapore, p 81
98. Andreoni W, Curioni A, Holczer K, Prassides K, Keshavarz-K M, Hummelen JC, Wudl F (1996) J Am Chem Soc 118:11335
99. Gruss A, Dinse K-P, Hirsch A, Nuber B, Reuther U (1997) J Am Chem Soc 119:8728
100. Hasharoni K, Bellavia-Lund C, Keshavarz-K M, Srdanov G, Wudl F (1997) J Am Chem Soc 119:11128
101. Allemand P-M, Srdanov G, Koch A, Khemani K, Wudl F, Rubin Y, Diederich M, Alvarez MM, Anz SJ, Whetten RL (1991) J Am Chem Soc 113:2780
102. Janssen RAJ, Hummelen JC, Lee K, Pakbaz K, Sariciftci NS, Heeger AJ, Wudl F (1995) J Chem Phys 103:788
103. Brown CM. Cristofilini L, Kordatos K, Prassides K, Bellavia C, González R, Keshavarz-K, Wudl F, Cheetham AK, Zhang JP, Andreoni W, Curioni A, Fitch AN, Pattison P (1996) Chem Mater 8:2548
104. Prassides K, Keshavarz-K M, Beer E, Bellavia C, González R, Murata Y, Wudl F, Cheetham AK, Zhang JP (1996) Chem Mater 8:2405
105. Prassides K, Wudl F, Andreoni W (1997) Fullerene Sci Technol 5:801

106. Brown CM, Beer E, Bellavia C, Cristofolini L, González R, Hanfland M, Häusermann D, Keshavarz-K M, Kordatos K, Prassides K, Wudl F (1996) J Am Chem Soc 118:8715
107. Prassides K, Keshavarz-KM, Hummelen JC, Andreoni W, Giannozzi P, Beer E, Bellavia C, Cristofilini L, González R, Lappas A, Murata Y, Malecki M, Srdanov V, Wudl F (1996) Science 271:1833
108. Fleming RM, Ramirez AP, Rosseinsky MJ, Murphy DW, Haddon RC, Zahurak SM, Makhija AV (1991) Nature 352:787
109. Tanigaki K, Ebbesen TW, Saito S, Mizuki J, Tsai JS, Kubu Y, Kuroshima S (1991) Nature 352:222
110. Srdanov, V (1996) unpublished results, Santa Barbara
111. Schwarz H (1996) unpublished results, Berlin
112. Thompson JD (1996) unpublished results, Los Alamos National Laboratory
113. Prassides K, Kordatos K (1996–1998) unpublished results, Sussex
114. Kordatos K, Brown CM, Prassides K, Bellavia-Lund C, de la Cruz P, Wudl F, Thompson JD, Fitch AN (1998) In: Kuzmany H, Fink J, Mehring M, Roth S (eds) Molecular Nanostructures World Scientific Singapore, p 73
115. Pichler T, Knupfer M, Golden M, Haffner S, Friedlein R, Fink J, Andreoni W, Curioni A, Keshavarz-K M, Bellavia-Lund C, Sastre A, Hummelen JC, Wudl F (1997) Phys Rev Lett 78:4249
116. Pichler T, Knupfer M, Golden M, Fink J, Winter J, Haluska M, Kuzmany H, Keshavarz-K M, Bellavia-Lund C (1997) Appl Phys A: Mater Sci Process A64:301
117. Pichler T, Knupfer M, Friedlein R, Haffner S, Umlauf B, Golden M, Knauff O, Bauer H-D, Fink J, Keshavarz-K M, Bellavia-Lund C, Sastre A, Hummelen JC, Wudl F (1997) Synth Met 86:2313
118. Haffner S, Pichler T, Knupfer M, Umlauf B, Friedlein R, Golden MS, Fink J, Keshavarz-K M, Bellavia-Lund C, Sastre A, Hummelen JC, Wudl F (1998) Eur Phys J B 1:11
119. Golden MS, Pichler T, Knupfer M, Friedlein R, Haffner S, Fink J, Andreoni W, Curioni A, Keshavarz-K M, Bellavia-Lund C, Sastre A, Hummelen JC, Wudl F (1998) In: Kuzmany H, Fink J, Mehring M, Roth S (eds) Molecular Nanostructures World Scientific Singapore, p 86
120. Malachesky PA (1969) Anal Chem 41:1493
121. Xie FA, Echegoyen L (1993) J Am Chem Soc 115:9818
122. Itaya K, Bard AJ, Szwarc M (1978) Z Phys Chem (Munich) 112:1
123. Guo T, Jin C, Smalley RE (1991) J Phys Chem 95:4948
124. Chai Y, Guo T, Jin C, Haufler RE, Chibante LPF, Fure J, Wang L, Alford JM, Smalley RE (1991) J Phys Chem 95:7564
125. Smalley RE (1991) In: Hammond GS, Kuck VJ (eds) Large carbon clusters. ACS, Washington, DC, p 199
126. Smalley RE (1992) In: Hammond GS, Kuck VJ (eds) Fullerenes: synthesis, properties, and chemistry of large carbon clusters. ACS Symposium Series 481, p 141
127. Muhr H-J, Nesper R, Schnyder B, Kötz R (1996) Chem Phys Lett 249:399
128. Cau B, Zhou X, Shi Z, Jin Z, Gu Z, Xiao H, Wang J (1997) Wuli Huaxue Xuebao 13:204 (Chem Abstr 126:317406)
129. Kurita N, Kobayashi K, Kumahora H, Tago K, Ozawa K (1992) Chem Phys Lett 198:95
130. Kurita N, Kobayashi K, Kumahora H, Tago K (1993) Phys Rev B 48:4850
131. Kurita N, Kobayashi K, Kumahora H, Tago K (1993) Fullerene Sci Technol 1:319
132. Liu M, Wang ZD, Dong J, Xing DY (1995) Z Phys B 97:433
133. Xu Q, Dong J, Jiang J, Xing DY (1996) J Phys B: At Mol Opt Phys 29:1563
134. Xu Q, Dong JM, Jiang J, Xing DY (1996) Acta Phys Sin (overseas edn) 5:175 (Chem Abstr 126:80155)
135. Chen Q, Lin J (1997) Jiegou Huaxue 16:445 (Chem Abstr 128:93396)
136. Xia X, Jelski DA, Bowser JR, George TF (1992) J Am Chem Soc 114:6493
137. Miyamoto Y, Hamada N, Oshiyama A, Saito S (1992) Phys Rev B 46:1749
138. Zou Y, Wang Z, Li W (1997) J Phys Chem Solids 58:1657
139. Karfunkel HR, Dressler T, Hirsch A (1992) J Comp-Aided Mol Design 6:521

140. Bühl M (1995) Chem Phys Lett 242:580
141. Esfarjani K, Ohno K, Kawazoe Y (1994) Phys Rev B 50:17830
142. Esfarjani K, Ohno K, Kawazoe Y (1996) Solid State Commun 97:539
143. Esfarjani K, Ohno K, Kawazoe Y (1996) Surf Rev Lett 3:747
144. Clemmer DE, Hunter JM, Shelimov KB, Jarrold MF (1994) Nature 372:248
145. Glenis S, Cooke S, Chen X, Labes MM (1996) Chem Mater 8:123
146. Christian JF, Wan Z, Anderson SL (1992) Chem Phys Lett 199:373
147. Jelski DA, Bowser JR, James R, Xia X, Xinfu G, Gao J, George TF (1993) J Cluster Sci 4:173
148. Kimura T, Sugai T, Shinohara H (1996) Chem Phys Lett 256:269
149. Oeste FD (1991) German Patents DE 91-4114536 910503 (Chem Abstr 117:91076) and DE 91-4128357 910827 (Chem Abstr 120:22451)
150. Jansen M, Peters G (1991) German Patent DE 91-413959 7911130 (Chem Abstr 119:75799)
151. Yoshida T, Eguchi K, Yoshe K, Kasuya S (1992) Japanese Patent JP 92-318266 321127 (Chem Abstr 121:208539)
152. Onoe J, Ooyama T, Takeuchi K (1992) Japanese Patent JP 92-231173 920831 (Chem Abstr 121:147754)
153. Kurita N, Kobayashi K, Kumado H, Tago K (1992) Japanese Patent JP 92-154720 920615 (Chem Abstr 120:221682)

The Higher Fullerenes: Covalent Chemistry and Chirality

Carlo Thilgen[1] · François Diederich[2]

Laboratorium für Organische Chemie, ETH-Zentrum, Universitätstrasse 16, CH-8092 Zürich, Switzerland

[1] E-mail: thilgen@carb.org.chem.ethz.ch
[2] E-mail: diederich@org.chem.ethz.ch

Many higher fullerenes, including some of their isomers, can be separated by high performance liquid chromatography (HPLC) on a number of stationary phases, a remarkable fact in view of the similarity of the carbon spheroids which differ mainly in shape and electronic properties of their π-systems, in addition to slight variations in size. Except for C_{70}, which is available in preparative amounts from fullerene soot extract without tedious HPLC purification and has been derivatized in many ways, most separations of the larger carbon spheroids are limited to the milligram scale and require a multistep chromatographic purification. Furthermore, taking into account the relatively small amounts of these carbon cages contained in fullerene soot, the availability of pure higher fullerenes has remained the bottleneck in the field of their multifaceted chemistry. Still, a number of pure adducts of C_{76}, C_{78} and C_{84} has now been isolated and characterized, and reactivity as well as regioselectivity principles begin to emerge for the higher fullerenes.

Another fascinating aspect of this research is the chirality of many higher fullerenes and numerous derivatives of chiral as well as achiral parent cages. It can originate from different structural characteristics of the spheroids and has led to the formulation of a new and simple configurational descriptor system. The study of chiral fullerenes and derivatives with a chiral functionalization pattern, initiated by the isolation and characterization of $(\pm)$-D_2-C_{76}, constitutes the central topic of the present review. Following the successful resolution of C_{76} by two different methods, the structural assignment of its enantiomers became possible through comparison of the experimental to calculated circular dichroism spectra.

Keywords: Fullerenes (higher fullerenes), Chirality, Stereo descriptors, Reactivity, Regioselectivity.

Topics in Current Chemistry, Vol. 199
© Springer Verlag Berlin Heidelberg 1999

1
Introduction

The isolation and characterization of D_2-symmetrical C_{76} in 1991 [1] led to important conclusions: It was shown that pure fullerenes with more than 70 carbon atoms could be isolated in macroscopic quantities from soot produced by the Krätschmer-Huffman method [2] and that the vaporization of achiral graphite was able to generate chiral carbon cages. Mass spectrometric evidence for the presence of higher fullerenes – that is, carbon molecules C_n (n $\geq$ 70) – in crude fullerene soot extract had been obtained shortly before, and optimization of the fullerene purification protocol soon afforded milligram samples of material enriched in higher fullerenes [3].

Even though pure or enriched samples of certain higher fullerenes are commercially available nowadays, their prices are prohibitive, in general, for use in preparative work. Thus, any synthesis with C_{76}, C_{78}, or higher fullerenes has to be preceded by a tedious isolation of the starting material from fullerene soot. The availability problem is due in part to the low abundance of the larger carbon spheroids in the fullerene soot [4]. Despite a number of efforts to remedy this situation [5], there has been no major breakthrough in increasing the absolute yield of higher fullerenes in the production process. Furthermore, whereas simple protocols based on chromatography [6–9], crystallization [10], and host-guest-interactions [11, 12] allow a larger scale purification of C_{60} [7, 8] and, to a certain extent, also of C_{70} [9, 13], the isolation of higher fullerenes still relies on multistage chromatography [3, 4]. Pure species are available in 1–100 mg quantities through high performance liquid chromatography (HPLC) purification of fullerene soot extract previously depleted from most of the far more abundant C_{60} and C_{70} [14, 15]. The first isolation of D_2-C_{76} [1], and of C_{78} [16], which could be separated into two isomers of C_{2v}- and D_3-symmetry, was achieved by HPLC on a polymeric C18 reversed phase column. π-Acidic stationary phases, e. g. the tris(2,4-dinitrophenoxy)-based Buckyclutcher I [17], or a brominated polystyrene phase having properties of potent fullerene solvents [18], have proven particularly effective for the preparative scale HPLC purification of higher fullerenes.

Despite a dramatic increase in the number of possible fullerene isomers with expanding cage size [19], an unambiguous structural assignment based on ^{13}C-NMR or X-ray crystallography was possible for D_2-C_{76} [1, 20, 21], C_{2v}-C_{78} [16, 20, 22], D_3-C_{78} [16, 20, 22], C_{2v}'-C_{78} [20, 22], and D_{2d}-C_{84} [23, 24] (Fig. 1). If a rigorous

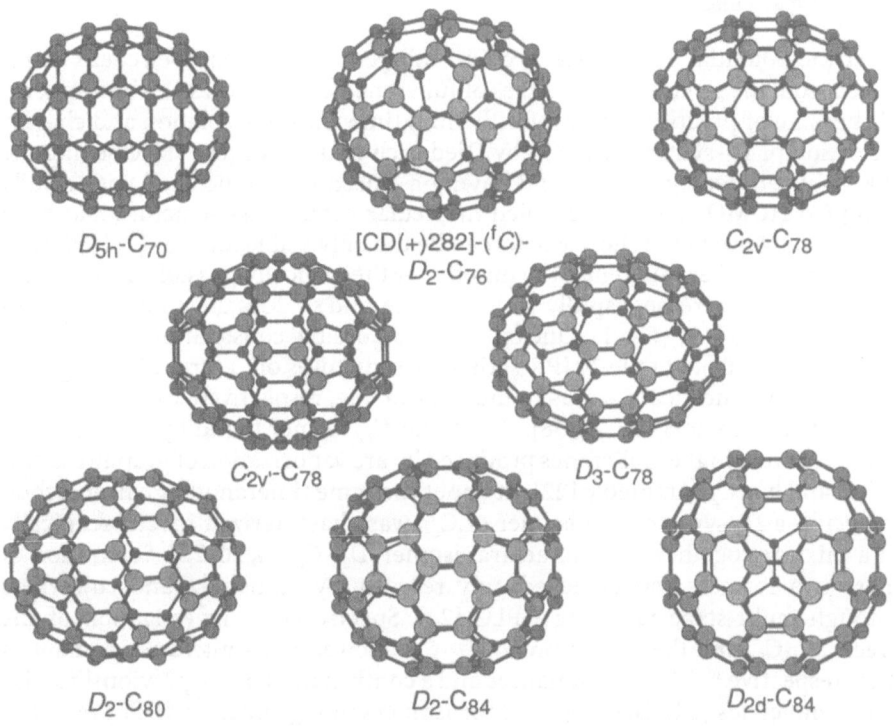

D_{5h}-C_{70} [CD(+)282]-(^{f}C)- C_{2v}-C_{78}
                                D_2-C_{76}

$C_{2v'}$-C_{78} D_3-C_{78}

D_2-C_{80} D_2-C_{84} D_{2d}-C_{84}

Fig. 1. A selection of higher fullerenes for which certain to confident structural assignments were possible. The chiral carbon spheroids are represented by a single enatiomer only; in the case of D_2-C_{76}, the configuration as well as a characteristic CD (circular dichroism) maximium of the shown enantiomer are indicated

experimental isomer assignment is impossible, calculated heats of formation, in conjunction with ^{13}C-NMR spectroscopy, may provide a structural lead, as in the cases of the isolated fullerenes D_2-C_{80} [25], C_2-C_{82} [22], or D_2-C_{84} [4, 20, 22, 24, 26] (Fig. 1).

If the covalent chemistry of higher fullerenes [27], when compared to that of C_{60} [28–32], has been hampered by the scarcity of pure starting material, it nevertheless experienced a considerable development in recent years. Thanks to the increasing number of isolated and characterized derivatives, the reactivity starts to become increasingly understood in terms of regio- or chemoselectivi-ty [27]. Another, most fascinating aspect of many higher fullerenes and higher fullerene derivatives is the particular quality of their chirality which can have various origins [33, 34]. Even though fullerene chirality is not limited to the larger carbon cages when derivatives are included, its study was initiated and developed largely through work on the carbon cages beyond C_{60}.

2
Fullerene Chirality

2.1
Chiral Higher Fullerenes

After the potential occurrence of chiral fullerenes and nanotubes could be in-ferred from the structural principles of fullerenes [35 – 37], these carbon sphero-ids became of practical interest for the first time with the isolation and charac-terization of D_2-symmetrical C_{76} by Diederich and coworkers [1]. Obeying the IPR (isolated pentagon rule) and having a closed electronic shell with a fully occupied HOMO (highest occupied molecular orbital), its structure had been theoretically predicted shortly before to be the only stable form of [76]fullerene [35]. For the larger fullerenes, the number of theoretically possible, IPR-satisfy-ing structures increases rapidly with cage size, and so does the number of chiral representatives [19] (Fig. 1): One out of the 5 possible constitutional isomers of C_{78}, and even 10 out of the 24 IPR-satisfying structures of C_{84} are chiral. The pre-dicted D_3-symmetrical C_{78} was found next to one, respectively two achiral C_{2v}-symmetrical isomers [16, 22, 38]. The major C_{82} isomer found by Kikuchi et al. in a fraction of higher fullerenes produced by arc vaporization of graphite is also chiral and has C_2-symmetry [22]. At about the same time, another chiral carbon molecule, a D_2-symmetrical isomer of C_{84}, was characterized spectroscopically in a mixture together with an achiral isomer, D_{2d}-C_{84} [4, 20, 22]. Separation of these two isomers was achieved only recently by Shinohara and coworkers through multi-stage recycling HPLC [24]. Spectroscopic investigation of the green D_{2d}-C_{84} and the yellow-brown D_2-C_{84} allowed a definitive assignment of their respective ^{13}C-NMR resonances and a confirmation of the previously assig-ned symmetries [24]. Besides, 5 minor isomers of C_{84}, among which there is at least another chiral, D_2-symmetrical isomer, were detected in head and tail HPLC fractions of this fullerene [26]. Five years after the isolation of C_{76} [1] and C_{78} [16], Kappes et al. reported isolation and structural assignment of C_{80} as a D_2-symmetrical structure [25]. It is interesting to note that among the 6 mentioned chiral higher fullerene isomers, four are D_2-symmetrical, whereas D_3- and C_2-symmetry occur only once.

2.2
Configurational Description of Chiral Fullerenes and Fullerene Derivatives with a Chiral Functionalization Pattern

Chirality, in the field of fullerenes, is not limited to the chiral carbon cages themselves. In 1992, the first enantiomerically pure covalent fullerene adducts, C_{60}-sugar conjugates, were prepared [39]. With the explosive development of covalent fullerene chemistry during the last few years [27–32], an increasing number of chiral C_{60} and C_{70} derivatives has been published [27, 33], and the need for an appropriate configurational description became pressing [34].

In many cases, the chirality of fullerene derivatives is neither inherent to the core of the unfunctionalized carbon cage, nor is it due to stereogenic units in the addends, but it can be attributed to an addition pattern lacking reflection symmetry [33, 34].

According to structural criteria of chiral fullerene derivatives, three different types of core functionalization patterns can be distinguished [34] (Fig. 2):

1. If the addition of chiral or achiral addends creates a chiral addition pattern on an achiral fullerene core, irrespective of the addends being identical or different, the functionalization pattern is defined as inherently chiral.

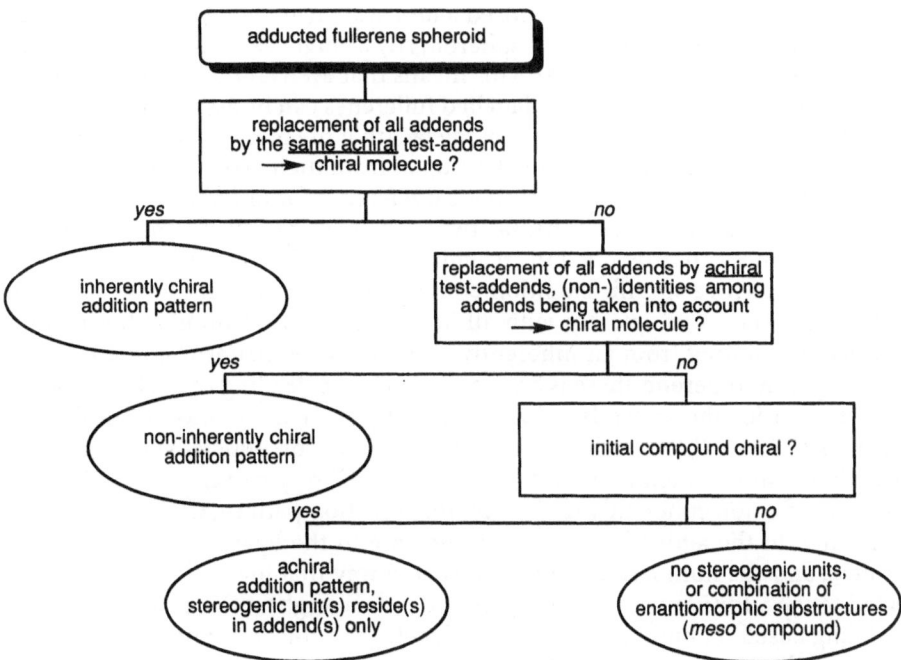

Fig. 2. Flow diagram allowing a facile classification of spheroid chirality in fullerene adducts

2. Addition patterns that are chiral due only to structural differences in the addends (nature, sequence, and steric arrangement of atoms) are termed non-inherently chiral. The analogy to the case of the stereogenic center with a tetrahedral arrangement of ligands is to be noted!
3. Chiral adducts can finally be obtained by functionalizing achiral fullerenes in such a way that chirotopic elements are exclusively located in the addends (no chiral core functionalization pattern).

It is natural that all derivatives of inherently chiral higher fullerenes also have an inherently chiral functionalization pattern. Furthermore, a compound can belong to several of the above classes when different stereogenic elements are superimposed in the same molecule.

The actual type of functionalization pattern can be easily recognized by applying a simple, stepwise substitution test according to the flow diagram of Fig. 2. Checking the structure resulting after each step for chirality quickly reveals the type of addition pattern of the starting spheroid.

In principle, the configurational description of fullerene derivatives with a chiral functionalization pattern could be done by determining the configuration (R or S) of each stereogenic center according to the CIP (Cahn, Ingold, and Prelog) system [40]. This procedure, however, can be overly complicated by the highly branched carbon framework, particularly if it contains a multitude of stereogenic centers. A further complication results from the fact that chiral fullerenes themselves do not have stereogenic centers to be specified by the CIP procedure.

For these reasons, we have proposed a new system allowing the configurational description of chiral fullerene spheroids by a single descriptor, regardless of the functionalization degree [34]. This means that even chiral parent fullerenes, heterofullerenes, and isotopically labelled fullerenes can be assigned an absolute configuration. The system is based on the fact that the numbering schemes proposed for fullerenes, which can be developed from their structures [41, 42], are chiral (helical) and thus constitute an ideal reference for differentiating between enantiomeric carbon cages. Whereas two isometric, mirror-symmetrical numbering schemes can be applied to an achiral parent fullerene, only a unique one is associable with a specific enantiomer of an inherently chiral carbon spheroid [34, 41, 42] and, consequently, with all its derivatives [34]. Similar, for a chiral derivative resulting from an inherently chiral addition pattern "grafted" on an achiral parent fullerene, there is a unique numbering leading to the lowest set of locants [43] for the addends. Depending on whether the motion from C(1) to C(2) to C(3) of this numbering is clockwise (C) or anticlockwise (A), the descriptors are defined as ^fC and ^fA (f = fullerene), respectively (Fig. 3) [34]. In the case of a non-inherently chiral functionalization pattern, CIP priorities are attributed to the addends (in the way they are to the ligands of a stereogenic center), and the numbering is done in such a way that lowest locants are allocated to the addition sites carrying the addends of highest CIP priority.

Being based on numbering, a great advantage of this descriptor system is its easy handling by computers [44]. As different numbering systems have appeared in literature [41, 42], it should be pointed out that the principle of the above

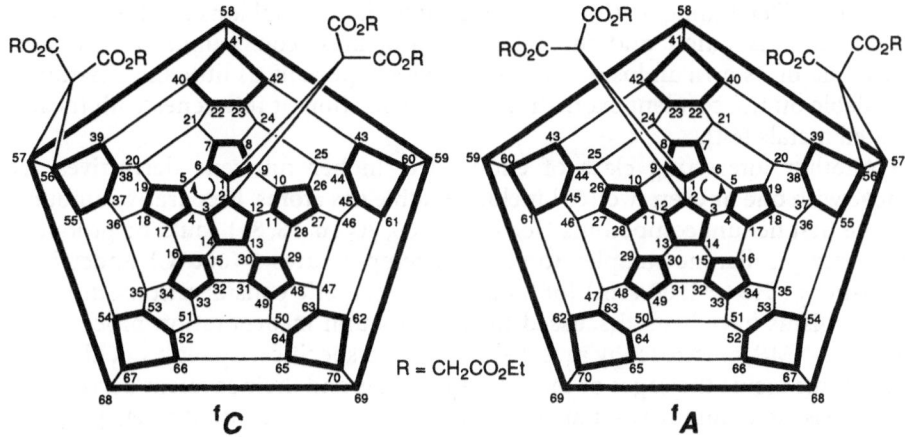

Tetrakis[(ethoxycarbonyl)methyl] 1,2:56,57-bis(methano)[70]fullerene-71,71,72,72-tetracarboxylate

Fig. 3. Schlegel diagrams showing the numbering and configurational descriptors of the enantiomers of bis-adduct (±)-**46** of C_{70} (cf. Sect. 4.2.1 and Fig. 7) with an inherently chiral functionalization pattern

descriptor system stays valid for any helical numbering. Of course, a correct interpretation of a given configurational descriptor requires knowledge of the used numbering system [45].

As to the assignment of absolute configurations [46–49] to isolated fullerene enantiomers [50–52] or fullerene derivatives with a chiral functionalization pattern [53–61] (cf. also the particular case of a chiral core-modified C_{60} derivative [62]), much work remains to be done. Following the initial kinetic resolution of small amounts of the allotropes D_2-C_{76}, D_3-C_{78} and D_2-C_{84} by Hawkins et al [50, 51], Diederich and coworkers recently succeeded in obtaining optically pure C_{76} isomers by functionalization of (±)-C_{76} with an enantiopure chiral addend, separation of the resulting diastereoisomers, and subsequent removal of the functionality [52]. Comparison of the obtained circular dichroism (CD) curves with calculated spectra allowed the assignment of their absolute configuration [46, 49] (cf. Sect. 7).

3
General Aspects of the Functionalization of Higher Fullerenes

The fullerene derivatives at the focus of the present review result from exohedral addition to the larger carbon spheroids [27] with particular emphasis being given to the structural aspects related to chirality. However, a number of other higher fullerene derivatives, including chiral structures as well, is worth being briefly mentioned.

Endohedral inclusion of metal ions, mostly observed with the larger carbon cages, is known since the early days of fullerene chemistry and has suscited strong interest for their electronic properties and potential materials applications [63]. Depending on the cavity size, up to three metal ions can be included. Production, isolation, and characterization of these compounds constitutes a vast field of its own and is not discussed here, especially as little information is available on the covalent exohedral functionalization of fullerenes with incarcerated metals [64].

Another interesting class of endohedral higher fullerene derivatives are those with one or even two [65] included noble gas atoms. Preparative amounts of the first helium compounds He@C_n (n = 60, 70, 76, 78, 84) could be produced by heating fullerene samples to elevated temperatures ($\approx 600\,^{\circ}$C) under high pressure of He (≈ 2700 atm) [66]. Based on energetic considerations, different mechanisms have been discussed for the He-atom incorporation process [67, 68]. ^{3}He-NMR spectroscopy revealed strong and specific magnetic shieldings of ^{3}He atoms inside the cages of different fullerenes and their covalent derivatives, giving rise to a unique resonance for each compound measured so far [69–74]. Endohedral ^{3}He atoms thus constitute a sensitive probe for the presence and magnitude of ring currents in fullerene π-chromophores as a function of the cage structure and the degree as well as the pattern of functionalization [73].

Fullerene derivatization is also possible by structural modification of the core. The simplest case consists in the insertion of a carbon atom or a heteroatom into a C–C bond and is encountered in homo[60]- [28–32, 75] and -[70] fullerenes [27, 75–77]. Only a few deeper modifications, creating small holes in the fullerene shell are known of C_{60} [31, 75] and C_{70} [75, 78]. Looking at structurally modified fullerenes in a broader sense and from the viewpoint not of cage degradation but assembly, it is clear that much more work – mostly aimed at "buckybowls" [79] or the total synthesis of buckminsterfullerene [80] – has been dedicated to this topic.

It should finally be mentioned that non-covalent interactions are at the origin of complexes or clathrates of C_{60} and C_{70} with organic guest molecules like cyclodextrins [11, 81] and calixarenes [12, 82].

3.1
Reactivity and Regioselectivity in Addition Reactions with Higher Fullerenes

Despite gradual differences in their electronic properties, e.g. the electrochemically determined HOMO-LUMO gap [83], C_{60} as well as the higher fullerenes can be considered as electron-deficient polyenes and show a number of similarities in their chemical behavior [27–32], such as the preference for primary addition of nucleophiles, as well as cycloadditions occurring at the bonds adjacent to two six-membered rings (6–6 bonds). An important difference results from the large variety of bond environments encountered in the less symmetrical larger carbon cages when compared to I_h-C_{60}, a situation that can in principle lead to a broad isomeric product distribution. In practice, however, additions to higher fullerenes often show a remarkable regioselectivity [27] (cf. Sects. 4, 5).

Considering the possible modes of addition to fullerene "polyenes", most functionalizations of buckminsterfullerene [28, 29] and higher fullerenes [27] occur as 1,2-additions. If this mode appears obvious in the case of many bridging addends (cycloadditions, addition of α-halocarbanions, carbene addition, formation of transition metal complexes, epoxidation) it is also observed in the hydrogenation and in most additions of nucleophiles followed by quenching with an independent electrophile species. On the other hand, most radical reactions, e.g. halogenations, occur as intrahexagonal 1,4-additions, a mode that in C_{60} leads to the introduction of a double bond into a five-membered ring. This energetically unfavorable situation [84] can be partially compensated by a reduction in steric strain, especially if the addends are bulky. In higher fullerenes, the situation is more complex, and depending on the site, 1,4-addition across a six-membered ring can even cause a reduction in the number of intrapentagonal double bonds which may be present in the parent structure [85, 86].

Fullerenes are highly strained molecules, mostly as a result of σ-strain deriving from the pyramidalization of sp^2-hybridized C-atoms confined within a spheroidal structure. Relief of strain due to rehybridization ($sp^2 \rightarrow sp^3$) of certain C-atoms upon functionalization can be considered as the main driving force in covalent fullerene chemistry [87]. The reactivities of higher fullerene bonds can therefore be estimated by considering the local spheroid curvature, expressed as pyramidalization angles of the corresponding C-atoms [50, 87–89]. In higher fullerenes, local curvature does not only depend on the overall fullerene size, but also on the distribution of the 12 five-membered and the six-membered rings on the fullerene surface. For making *simple* predictions on isomeric product distributions to be expected, we have been considering a qualitative model for the evaluation of local curvature [14]. As the latter results from the incorporation of pentagons into a network of hexagons, a 6–6 bond should be the more reactive, the more it is surrounded by five-membered rings, and the shorter its distance is to the latter. According to this description, the most strained fullerene bonds are located at the center of a pyracylene (= cyclopent[fg]acenaphthylene) substructure (Fig. 4). In C_{70} and C_{76} (cf. Figs. 3, 12) they are found at the poles (type α, surrounded by type β), whereas the flatter equatorial region contains a number of different, less curved bonds.

From a thermodynamic viewpoint, calculations at different levels of theory have shown that 1,2-additions in C_{70} occur preferentially at bonds C(1)–C(2) and C(5)–C(6) [90–92] (for the numbering of C_{70}, cf. Fig. 3). However, isomerization of fullerene derivatives is usually not observed under the employed reaction conditions and with few exceptions, e.g. the formation of transition metal complexes [93] and catalytic hydrogenation [92, 94], most additions to fullerenes seem to be kinetically controlled. Based on this assumption, LUMO coefficients [95, 96] as well as Mulliken charges [90, 97] of the fullerene moiety

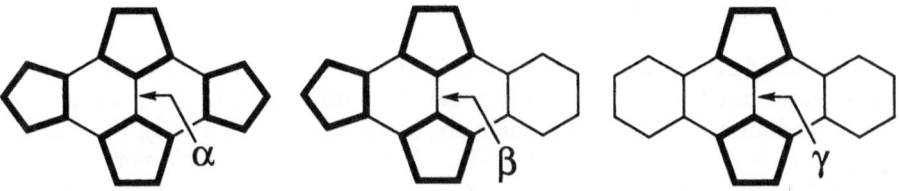

Fig. 4. The three most curved fullerene bond types α, β, and γ in their local environments

have been used in predicting or explaining the regiochemistry of nucleophilic additions and cycloadditions. In multiple functionalizations, possible steric hindrance among addends is another factor to be taken into account [28]. A different approach, finally, uses π-bond orders for evaluating the reactivity of different types of bonds [98–100].

4
The Chemistry of C_{70}

4.1
Mono-Adducts of C_{70}

4.1.1
C(1)–C(2)-Adducts (Type α)

Mono-functionalization of C_{70} affords an isomeric product mixture in many cases. In accordance with the above reactivity principles, the main product usually results from attack of the reagent in the polar region at the most curved bond C(1)–C(2) (type α) followed by addition to the second most curved C(5)–C(6) bond (type β) [27] (Figs. 4, 5).

Provided that the addend [101] does not contain chirotopic units, C(1)–C(2) mono-adducts are C_s-symmetrical. If the addend itself is C_{2v}-symmetrical [101], a single product isomer results. This is the most common type of the known C_{70} derivatives and includes products (Fig. 5) resulting from epoxidation (1) [3, 102, 103], osmylation (2) [89], formation of η^2-Ir- (3) [104] and η^2-Pt-complexes (4) [105], hydrogenation [92, 106], and from many cycloadditions, as well as additions of nucleophiles [27], some of which will be discussed in more detail below. If the addend is C_s-symmetrical, two constitutional isomers (regioisomers) can be formed. These could be observed for example as primary products in cycloaddition reactions with diazoalkanes or azides (cf. Sect. 4.1.3). The C(1)–C(2) addition pattern is per se achiral, and chiral products are possible only when the addend contains chirotopic units such as the stereogenic centers in 5 [54]. Stereogenic centers in the addends can also be generated in the course of the addition reaction, as it was found for the Diels-Alder addition of cyclopenta-1,3-dienes [107], the cyclopropanation by unsymmetrically substituted sulfonium ylids (adduct (±)-6) [108] or the [8+2]cycloaddition of 8-methoxyheptafulvene [109].

4.1.2
C(5)–C(6)-Adducts (Type β)

In the case of the second most favored C(5)-C(6)-adduct (type β) which often accompanies the C(1)–C(2) mono-adduct in product mixtures [27], a C_{2v}-symmetrical addend again affords a single product isomer (e.g. 7-10, Fig. 5), whereas addition of a C_s-symmetrical addend leads to a pair of enantiomers [(±)-11, Fig. 5]. In the latter case, a chiral addition pattern is created on the fullerene spheroid. Being due only to differences in nature, sequence, or steric arrange-

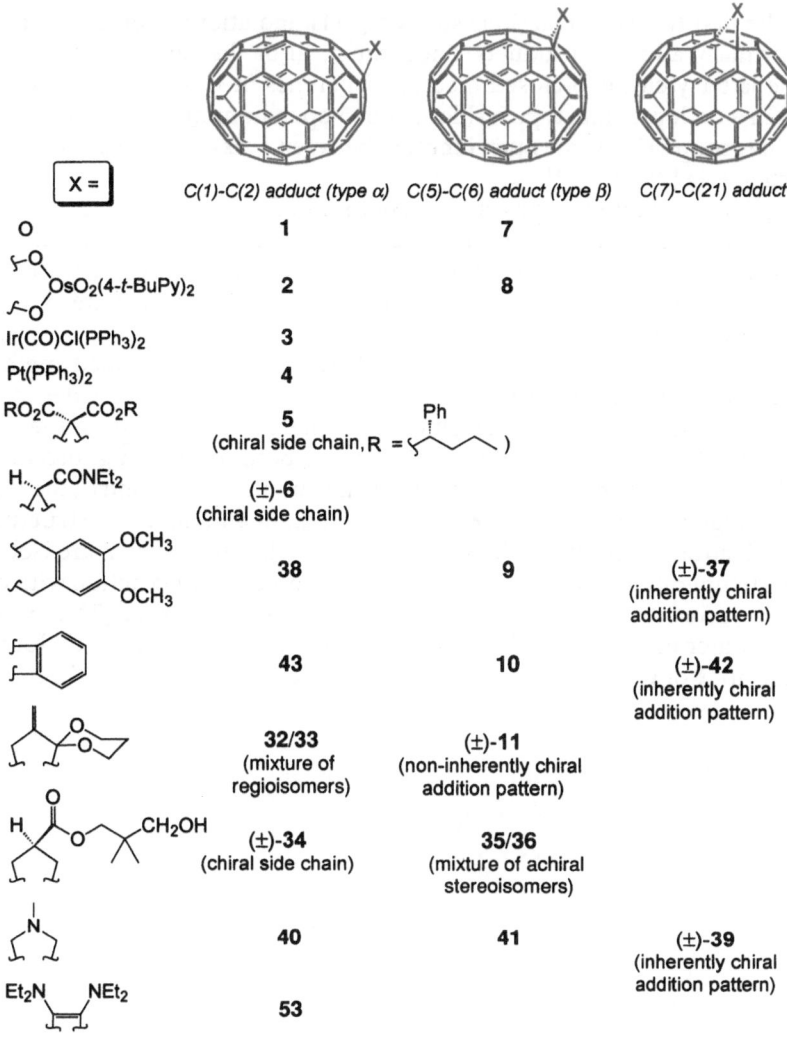

Fig. 5. Three types of mono-adducts of C_{70} resulting from addition across a 6–6 bond, and a selection of synthesized representatives

ment of the atoms bonded to each of the newly generated sp^3-C-atoms of the fullerene core, this addition pattern is termed non-inherently chiral [34].

4.1.3
Product Mixtures Including α- and β-Type Adducts of C_70

Together with the addition of C-nucleophiles, cycloadditions represent the most important reaction for the formation of C–C-bonds with the fullerene framework [28, 29, 110]. [3 + 2]Cycloaddition of diazo alkanes to 6–6 bonds was

among the first fullerene reactions studied [111] and affords isolable pyrazolines as primary addition products. Photochemical or thermal extrusion of N_2
from the latter yields 6–6-closed methanofullerenes in which the 6–6 bond is
bridged in a cyclopropane type fashion, or 6–5 open homofullerenes in which a
methylene group bridges the open junction between a six- and a five-membered
ring (Scheme 1) [28, 29, 112].

Taking into account the 2 possible orientations of the attacking reagent or, in
other words, the C_s-symmetry of the addend, two regioisomeric pyrazolines, 12
and 13, arise from addition at C(1)–C(2), and a racemic mixture of a third constitutional isomer [(±)-14] with a non-inherently chiral addition pattern results
from addition across C(5)-C(6) [76] (Scheme 1).

Irradiation of the pyrazoline mixture 12, 13, and (±)-14 yielded methanofullerenes 15 and 16 (7:1) with a closed transannular 6–6 bond besides minor
amounts of a third constitutional isomer (17), whereas thermolysis of the same
mixture gave 17 and inherently chiral (±)-18 (4:1), both constituting core-modified fullerenes [76] (Scheme 1). With open transannular 6–5 "bonds", the functionalized regions of 17 and (±)-18 resemble methanoannulene sub-structures.
[3 + 2]Cycloaddition of alkyl azides to C_{70} affords triazolines which are isolable
when the temperature does not exceed ca. 50 °C. Similar to the reaction with diazomethane, three constitutionally isomeric adducts were observed [77, 113], and
the regioisomer of type 19 [113] with the substituent of the heterocycle located
above the apex of C_{70}, predominated over its regioisomer of type 20 in two in-

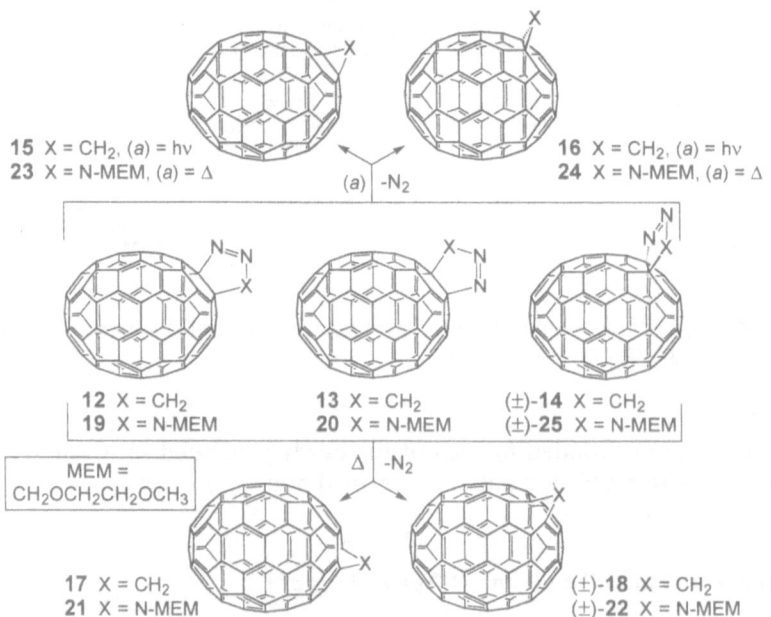

Scheme 1. Isomeric product distribution resulting from [3 + 2]cycloaddition of diazomethane
or a substituted azide to C_{70}, and extrusion-rearrangement sequence to the respective 6–6
closed or 6–5 open structures

dependent studies [77, 113] (Scheme 1). This selectivity was explained by favorable electrostatic interactions between the negatively polarized N-atom of the azide and positive Mulliken charges on the C-atoms in the apical pentagon of C_{70} [77]. Thermal elimination of N_2 from the fullerotriazolines under formation – next to the parent C_{70} – of N-bridged compounds showed a preference for the 6–5 open azahomofullerene structures (types 21 and ($\pm$)-22) as compared to the 6–6 closed aziridine isomers corresponding to 23 and 24 [77, 113] (Scheme 1). In agreement with the lower thermodynamic stability of the β-type triazoline ($\pm$)-25 (minor product of the azide addition), its thermal decomposition occurs at lower temperatures when compared to the α-type adducts [113].

After N-MEM- ([2-(methoxy)ethoxy]methyl) protected azahomo[60]fullerene had been successfully used by Wudl and coworkers for the synthesis of macroscopic amounts of the "dimeric" heterofullerene $(C_{59}N)_2$ [114], they applied the same methodology to azahomo[70]fullerenes 21 and ($\pm$)-22 [75], obtained according to the above general scheme [113] (Scheme 1). [2 + 2]Cyclo-addition of singlet oxygen to the electron-rich enamine type double bonds of 21 and ($\pm$)-22, followed by decomposition of the 1,2-dioxetane intermediates, led to the ketolactam "holey balls" ($\pm$)-26, and a mixture of ($\pm$)-27 (major component) and ($\pm$)-28, respectively (Scheme 2). Acid-induced loss of 2-methoxyethanol from ($\pm$)-26 followed by a rearrangement of the generated carbocation and subsequent extrusion of formaldehyde and carbon monoxide afforded the 2-azonia[70]fullerene cation [75] (not shown). This could be reduced to the aza-fullerenyl radical which dimerized to isomerically pure C_s-symmetrical 2,2'-diaza-1,1'-bi([70]fullerenyl) (29), functionalized across the α-type bond with the N-atom replacing C(2) in each fullerene moiety (Scheme 2). The "dimeric" structure was confirmed by electrospray-MS showing a molecular ion at m/z = 1685 and a strong base peak at m/z = 842 [75]. Starting from ketolactam ($\pm$)-27, the analogous reaction sequence led to the C_s-symmetrical azafullerene "dimer" 1,1'-diaza-2,2'-bi([70]fullerenyl) (30), again functionalized across the α-type bond, but with the N-atom replacing C(1) of both fullerene moieties (Scheme 2). Reaction of the minor ketolactam isomer ($\pm$)-28 resulting from ($\pm$)-22 (Scheme 1), finally, should afford an aza[70]fullerene "dimer" as an interesting mixture of isomers: Each fullerene moiety, unsymmetrically functionalized across the β-type bond C(5)–C(6), has a non-inherently chiral functionalization pattern and can therefore exist in two enantiomeric forms; consequently, three possible "dimer" combinations are expected to lead to a pair of C_2-symmetrical enantiomers as well as an achiral, C_s-symmetrical *meso*-form. This particular case of fullerene chirality was also proposed [33, 34] for a "dimeric" 1,4-adduct of C_{60} described in literature [115].

In a different approach to aza[70]fullerenes by Hirsch and coworkers, twofold azide addition yielded C_s-symmetrical 31 which results from bridging two adjacent 6–5 open "bonds" and is a suitable starting material for the macroscopic synthesis of the C_s-symmetrical 2,2'-diaza-1,1'-bi([70]fullerenyl) 29 [116] (Scheme 2) formed together with the "monomeric" 1-[2-(methoxy)ethoxy]-1H-2-aza[70]fullerene [117] (not shown). The latter probably results from trapping of $C_{69}N^+$ intermediates by the acetal cleavage product 2-methoxyethanol. Aza-fullerene formation occurs through elimination of one C-atom (presumably

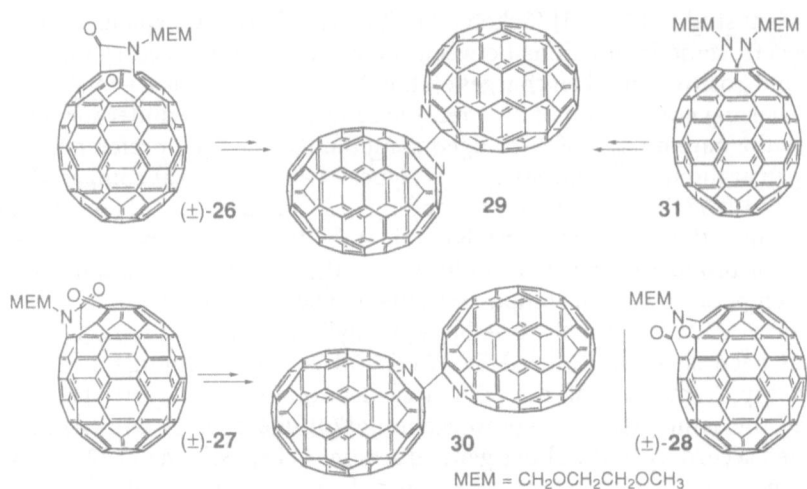

MEM = CH$_2$OCH$_2$CH$_2$OCH$_3$

Scheme 2. Isolated "dimeric" aza[70]fullerene isomers, and aza[70]fullerene precursors

C(1)) as an isonitrile or carbodiimide species from the carbon cage of **31** (Scheme 2) and this requires a facile deprotection of its *N*-bridges [117].

The same type of isomerism encountered in the addition of diazoalkanes and azides was found by Meier et al. [118], and Irngartinger et al. [119] in the synthesis of fulleroisoxazolines by reaction of variously substituted isonitrile oxides with C$_{70}$. Whereas the α-type bond C(1)–C(2) showed the higher reactivity, there was essentially no preference for the orientation of the reagent upon addition to this bond. Furthermore, ^{1}H-NMR spectroscopic investigation of the two regioisomeric C(1)–C(2) adducts of methyl isonitrile oxide showed that the apical five-membered ring has a stronger deshielding effect than the lateral one [118].

An interesting stereochemical course was taken in a reaction sequence starting with the addition of a dipolar trimethylenemethane species generated by thermolytic ring opening of a methylenecyclopropane derivative [120]. In analogy to the addition of diazoalkanes, primary addition of the C_s-symmetrical addend, presumably involving a SET (single electron transfer) from the trimethylenemethane to the fullerene, occurs across the α- and β-type bonds of C$_{70}$ giving rise to the achiral regioisomers **32** and **33** as well as a to a third constitutional isomer appearing as a pair of enantiomers [(±)-**11**] with a non-inherently chiral addition pattern (Fig. 5). Subsequent rearrangement of these products, taking place via an intermediate with a C_{2v}-symmetrical addend, and hydrolysis finally yielded a pair of enantiomers with a stereogenic center in the side chain [(±)-**34**] as well as a pair of achiral *cis/trans* isomers (**35** and **36**; *cis* and *trans* with respect to the faces of the pentagon fused to the β-type bond of C$_{70}$). In brief, there was a formal transformation of a pair of regioisomers (constitutional isomers) (C(1)–C(2)-adducts **32** and **33**) into a pair of enantiomers [(±)-**34**], and of a constitutionally different pair of enantiomers (C(5)–C(6) adduct (±)-**11**) into a pair of achiral stereoisomers (**35** and **36**) [120].

4.1.4
C(7)–C(21)-Adducts

All C(7)–C(21) adducts of C_{70} are C_1-symmetrical and interesting from a stereo-chemical viewpoint because they are the only 6–6 bond mono-adducts of this fullerene with an inherenly chiral functionalization pattern [34]. The bond in question, however, is not among the types α–γ (Fig. 4) and has a relatively low local curvature [14]. Furthermore, its reactivity is probably reduced by its location in a six-membered ring with a relatively high benzenoid character [121] (cf. Sect. 4.1.5 and Fig. 6). Despite its high frequency (20 bonds vs. 10 bonds of type α and 10 bonds of type β in C_{70}), the low reactivity may therefore account for the few observations of C(7)–C(21) adducts.

The first case was detected by Diederich and coworkers in the context of a regioselectivity investigation of the Diels-Alder addition of 4,5-dimethoxy-o-quinodimethane to C_{70} and C_{76} [14]. It was isolated as a minor product [($\pm$)-37] next to the more abundant adducts of type α (38) and type β (9) (Fig. 5). Isomer assignments relied on the following observation: In the Diels-Alder adducts, boat-to-boat interconversion of the newly formed cyclohexene ring becomes fast on the ^{1}H-NMR time scale around 80°C, and the average symmetry of the adducts changes upon transition from below to above coalescence temperature (T_c). A unique symmetry combination for $T < T_c$ and $T > T_c$, together with ^{13}C-NMR data, allowed the structural assignment of all three isolated constitutional isomers. The structures of 38 and 9 were confirmed by X-ray crystallography [122].

[3 + 2]Cycloaddition of N-methylazomethine ylid to C_{70} yielded [70]fullero-pyrrolidine ($\pm$)-39 with the heterocycle fused across the fullerene bond C(7)–C(21) as a minor product besides the α- and β-type adducts 40 and 41 [123] (Fig. 5). The isomeric assignment was based on ^{1}H-NMR spectroscopy, the C_1-symmetrical C(7)–C(21) adduct being distinguished by two different methylene groups, each carrying two non-equivalent H-atoms.

The inherently chiral C(7)–C(21) addition pattern was also observed in one [($\pm$)-42] of the minor mono-adduct isomers resulting from [2 + 2]cycloaddition of benzyne to C_{70} [86, 121]. It was accompanied by the common α- (43, major product) and β-type (10) adducts (Fig. 5) as well as by another, unusual isomer (cf. Sect. 4.1.5).

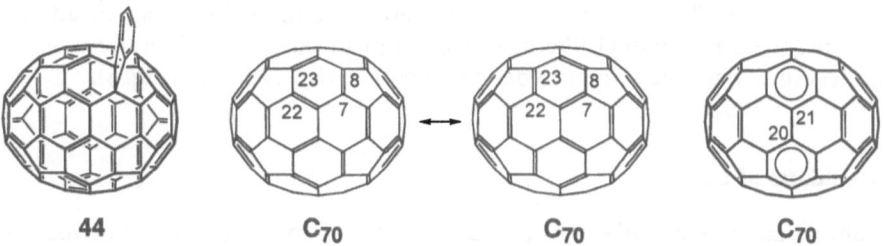

Fig. 6. One of the four isomers resulting from mono-addition of benzyne to C_{70} (C(7)–C(8)-adduct), and important resonance structures of C_{70} illustrating the double bond character of the C(7)–C(8)-bond as well as the benzenoid character of the equatorial hexagons

4.1.5
Other Addition Modes of C$_{70}$

In a first report on the mono-addition of benzyne to C$_{70}$ [86], the fourth product isomer (34%) besides **43, 10** and (±)-**42** (Fig. 5) was suggested to result from an unusual transequatorial 1,4-addition mode. The result is a C$_2$-symmetrical molecule with an inherently chiral functionalization pattern in which the positions C(7) and C(23) (Fig. 6) are bridged. This equator-spanning addition mode was also proposed to occur in C$_{70}$Cl$_{10}$ [85, 124] (cf. Fig. 11) as well as in multiple benzyne adducts of C$_{70}$ [86, 125].

In a recent report by Meier et al. on the mono-addition of benzyne [121], however, the isolated fourth product (13%) was identified as a different isomer, resulting from addition across the 6–5 bond C(7)–C(8) (**44**, Fig. 6). The C$_s$-symmetrical molecule represents the first example of a direct addition to a 6–5 ring fusion and of a 6–5 adduct in which the fusion bond remains intact. This addition mode was explained by the relatively high double bond character of the 6–5 bond C(7)–C(8) when compared to the other 6–5 bonds of C$_{70}$ (C(2)–C(3), C(1)–C(6), C(6)–C(7)) (cf. Fig. 3) and is made plausible by comparing Kekulé-structures of [60]- and [70]fullerene [121]. In C$_{60}$, all double bonds can be drawn exocyclic to five-membered rings, thus resulting in [5]radialene and cyclohexatriene substructures [28, 29, 31]. Addition across a 6–5 junction is accompanied by a strain-inducing [84] shift of two double bonds into pentagons [112]. The parent C$_{70}$ structure, on the other hand, must accommodate at least five double bonds in pentagons; they can be located at C(7)–C(8) and the symmetry-related positions or, in an alternative, iso-energetic Kekulé-structure, at C(22)–C(23) and the symmetry-related positions (Fig. 6). Adduct formation can reduce the number of intrapentagonal double bonds, notably when reaction occurs across the 6–5 junction C(7)–C(8) [121] or in a transequatorial 1,4-mode across C(7) and C(23) [85, 86]. The representation of C$_{70}$ as a hybrid of the mentioned two important resonance structures, as well as the comparable bond lengths of C(7)–C(8) and C(7)–C(21) [122, 126] underline the benzenoid character of the equatorial hexagons, the reactivity of which is ascribed to curvature [121] (Fig. 6).

Addition to the fourth and least curved 6–6 bond C(20)–C(21) lying on the equator of the C$_{70}$ spheroid (Fig. 6) seems unlikely, although it was suggested for the product of the photochemically induced 1,1,2,2-tetramesityl-1,2-disilirane addition [127]. This bond type is completely surrounded by hexagons and can be considered as biphenyl-like, as suggested by the above considerations as well as structural data showing it to be the longest in C$_{70}$ [122, 126] (Fig. 6).

4.2
Bis-Adducts of C$_{70}$

Multiple adducts of fullerenes [27, 28, 29, 31, 112] provide the opportunity of studying the changes in chemical reactivity and physical properties which occur when the conjugated fullerene chromophore is reduced as a result of increasing functionalization [128, 129]. Moreover, the development of methods for selective multiple addition provides access to an unprecedented variety of three-dimensional building blocks for organic chemistry, which nicely complement the present repertoire of two-dimensional acetylenic, olefinic, and benzenoid components for the construction of tailor-made functional molecules and polymers.

With eight distinct bonds (four 6–6 and four 6–5 bonds) potentially available for functionalization, the chemistry of C_{70} may appear complex in comparison to that of C_{60} which comprises only two bond types. In order to avoid overly complex product mixtures, multiple functionalization of C_{70} is best explored with a reaction that minimizes the number of isomers in the first addition step.

A reaction fulfilling this requirement is the *Bingel* addition [130], a methanofullerene synthesis in which the attack of 2-bromomalonate carbanions on the fullerene is followed by intramolecular halide displacement. In the mono-functionalization of C_{70}, it shows a high preference for the C(1)-C(2) bond [54, 130]. This can be rationalized by the preference of the sp^3-hybridized carbanionic center, which develops in the transition state in α-position of the site of nucleophilic attack [28], for a region of high curvature, thus releasing a large amount of strain energy.

4.2.1
Malonate Adducts

The X-ray crystal structures of Diels-Alder mono-adducts **38** and **9** (Fig. 5) had revealed that the electronic configuration at the unfunctionalized pole of C_{70} closely resembles that of the parent fullerene [122]. As a consequence, second Bingel addition to C_{70} [54, 131] should take place at one of the five α-type bonds C(41)–C(58), C(56)–C(57), C(59)–C(60), C(67)–C(68), or C(69)–C(70) at the unfunctionalized pole (Fig. 7; for the numbering of C_{70}, cf. Fig. 3). Next to the three-dimensional representations, the resulting situation is shown schematically in a *Newman*-type projection along the C_5-axis of C_{70} (Fig. 7). Of the two concentric five-membered rings, the inner one corresponds to the polar pentagon closest to the viewer, and the attached vertical line represents the bond C(1)–C(2) where the first addition occurred. The involved α-type bonds at the distal pole, preferred sites of second addition, depart radially from the outer pentagon.

In the case of achiral, C_{2v}-symmetrical malonate addends, three constitutionally isomeric bis(methano)fullerenes are formed [54, 131]: An achiral one (C_{2v}-symmetrical **45**), and two chiral ones (C_2-symmetrical (±)-**46** and (±)-**47**) which are obtained as pairs of enantiomers with an inherently chiral addition pattern [54] (Fig. 7). The tree constitutional isomers are formed in the approximate ratio 2.8:6.2:1 which differs substantially from the statistical 1:2:2 ratio and shows a marked preference for the second addition taking place at bonds C(56)–C(57)/C(59)–C(60) as compared to C(67)–C(68)/C(69)–C(70) (for the numbering of C_{70}, cf. Fig. 3). Electronic factors such as the coefficients of the LUMO of the mono-adduct, to which the electron density of the incoming nucleophile is transferred [95], may explain the observed selectivity among distant bonds of (nearly) identical local curvature.

Twofold addition of bis[(S)-1-phenylbutyl] 2-bromomalonate to C_{70} leads to a superimposition of the chirality of the addend and, in the occurrence, that of

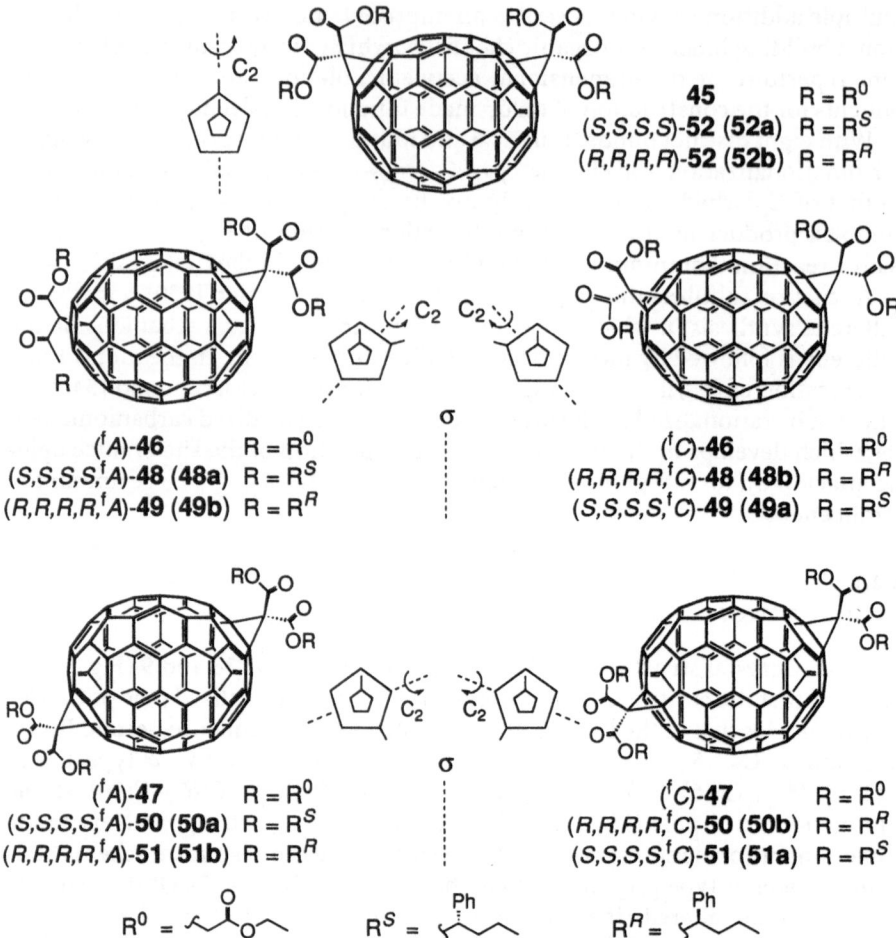

Fig. 7. Combinations of an achiral and two inherently chiral C$_{70}$ addition patterns with achiral and chiral addends, leading to a large variety of constitutional isomers, diastereoisomers, and enantiomers

the core functionalization pattern, thus affording a total of five optically active isomers, two constitutionally isomeric pairs of C$_2$-symmetrical diastereoisomers (**48a/49a** and **50a/51a**) as well as a third constitutional C$_2$-symmetrical isomer (**52a**) [54] (Fig. 7). As the mixture of these five chiral C$_{70}$ derivatives does not include any pair of enantiomers, they could all be separated by HPLC on silica gel, and the same accounts for the products **48b–52b** resulting from addition of the respective (*R,R*)-configured esters [56]. As a consequence, ten pure optically active compounds were obtained; they comprise five pairs of enantiomers, four with an inherently chiral fullerene functionalization pattern and chiral side chains (**48a/b, 49a/b, 50a/b, 51a/b**), and two with stereogenic units in the addends only (**52a/b**) (Fig. 7).

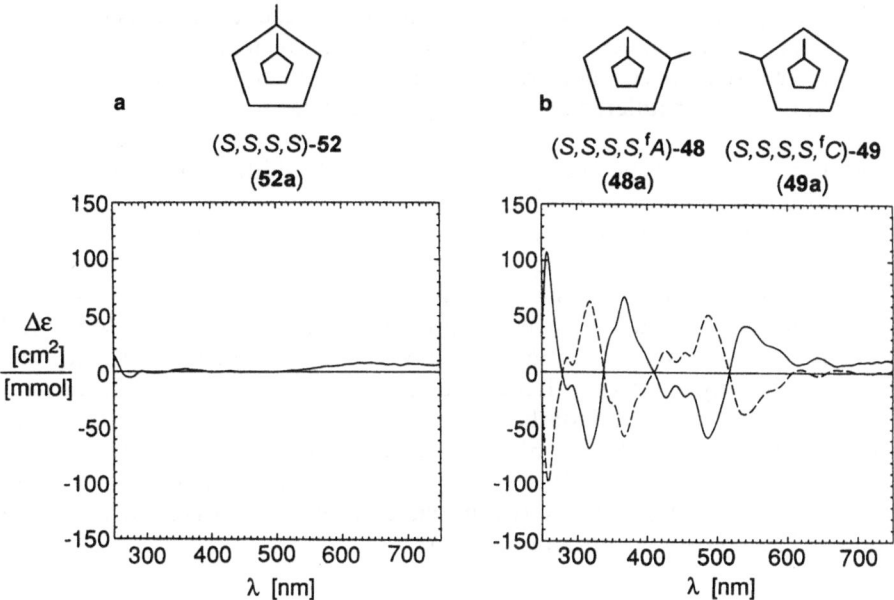

Fig.8. CD spectra of **a 52a** with chiral side chains, but an achiral fullerene addition pattern, and **b** the diastereoisomeric **48a** and **49a** having enantiomeric, inherently chiral fullerene functionalization patterns in addition to the (S)-configured side chains

Inherently chiral addition patterns give rise to chiral π-systems manifesting themselves by very strong Cotton effects in the circular dichroism (CD) spectra [33, 52–61] (Figs. 8, 17) (cf. also the particular case of a chiral core-modified C_{60} derivative exhibiting relatively low $\Delta\varepsilon$-values in its CD spectrum [62]), generally in contrast to sole perturbation of achiral fullerene chromophores by chiral addends [33, 39, 48, 54, 108, 132–139]. Furthermore, and similar to the ^{13}C-NMR and UV/Vis spectra, matching core functionalization patterns lead to (nearly) identical CD curves, irrespective of the side chain configuration. This is nicely illustrated by the spectra of diastereoisomeric bis-adducts (S,S,S,S,^{f}A)-**48** and (S,S,S,S,^{f}C)-**49** which have enantiomeric addition patterns and display almost mirror-image shapes since the chiroptical contributions of the chiral addition pattern of the fullerene strongly dominate those of the chiral addends [54] (Fig. 8). A characteristic band around 460 nm was observed in the CD spectra of all the above C_{70} adducts with an inherently chiral functionalization pattern; it may be helpful in assigning absolute configurations [56].

4.2.2
[2 + 2]Cycloaddition of Alkynes

Zhang and Foote reported [2+2]cycloaddition of the electron-rich bis(diethylamino)ethyne and of 1-alkylthio-2-(diethylamino)ethynes to C_{60} and C_{70} [140]. The resulting adducts include both a photosensitizer (dihydrofullerene) [141]

and an easily photooxydizable group (enamine or thioenol ether) in the same molecule, thus making the cyclobutene bond prone to cleavage by singlet oxygen. Such self-sensitized photooxygenation was an important step in the preparation of the first fullerene[60]- and -[70]-1,2-dicarboxylic acid anhydrides [140].

Next to a single α-type mono-adduct (53, Fig. 5), addition of the diamine to C_{70} yielded two isomeric bis-adducts, tentatively assigned as (±)-54 and (±)-55, in a 9:1 ratio (Fig. 9). The regioselectivity observed in the formation of these C_2-symmetrical bis-adducts with an inherently chiral functionalization pattern is even more pronounced than that of the double Bingel reaction [54, 131], and the C_{2v}-symmetrical bis-adduct corresponding to 45 [54] (Fig. 7) was not observed at all [140]. Furthermore, addition of 1-alkylthio-2-(diethylamino)ethynes was found to be highly selective with regard to the relative orientation of the reagent and the attacked α-type bond; this was verified for the C_s-symmetrical mono-adduct (analogue of 53, Fig. 5) as well as for the major C_2-symmetrical bis-adduct (analogue of (±)-54, Fig. 9). However, it could not be demonstrated unequivocally which one of the two possible orientations was induced by the highly polarized α-type bonds of C_{70} [140].

4.2.3
Transition Metal Complexes of C_{70}

Inherently chiral bis-adduct functionalization patterns were also found in transition metal complexes of C_{70}. These derivatives are characterized by reversible,

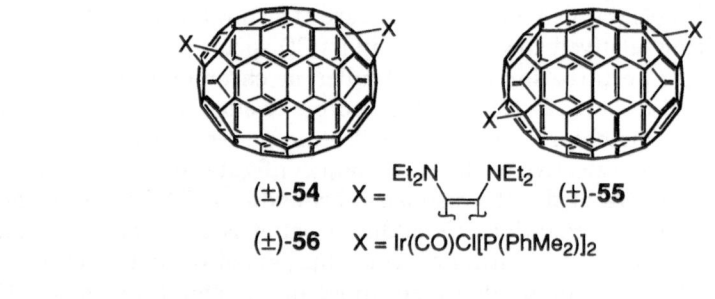

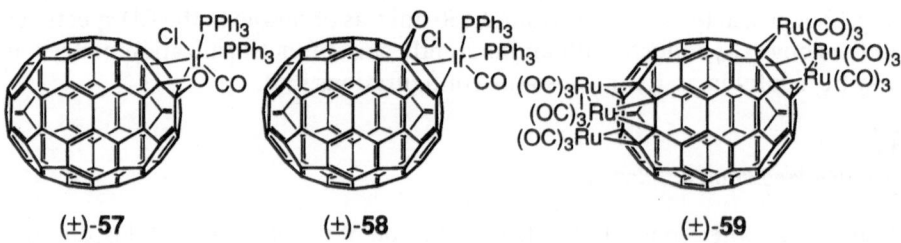

Fig. 9. Bis-adducts of C_{70} with chiral functionalization patterns, resulting from [2+2]cycloaddition of alkynes and from transition metal complex formation

thermodynamically controlled association and a pronounced tendency towards formation of crystals suitable for X-ray analysis [93]. High selectivity for η^2-complexation by Ir and Pt at α-type bonds is found in mono-adducts (3 and 4, Fig. 5) [104,105] as well as in multiple adducts [142,143] of C_{70}. The only crystalline bis-adduct obtained from reaction of C_{70} with [Ir(CO)Cl(PPhMe$_2$)$_2$] was a C_2-symmetrical C(1)–C(2):C(56)–C(57) adduct [($\pm$)-56, Fig. 9]. This addition pattern selectivity corresponds qualitatively to that observed in the Bingel reaction [54, 131] and the [2+2]cycloaddition of alkynes [140]. It has to be stressed though, that in the case of complex formation, crystal packing effects may play an additional role in determining the regioselectivity.

Chiral functionalization patterns also occur in the iridium complexes [144] cocrystallized from a solution containing the two $C_{70}O$ isomers 1 and 7 (Fig. 5) [102,103]. In both complexes, the transition metal and the epoxy bridge are located not only within the same hemisphere of the fullerene, but they are bound to 6–6 junctions within the same six-membered ring adjacent to the polar pentagon [($\pm$)-57 and ($\pm$)-58, Fig. 9]. For C_{60}, this relative arrangement of addends (cis-1) was shown to be among the most favored, provided there is no steric hindrance between the addends [96]. In both transition metal complexes of $C_{70}O$, iridium is bound to an α-type bond; this leads to a non-inherently chiral functionalization pattern in ($\pm$)-57 where the oxygen atom is bound to a neighboring α-type bond and to an inherently chiral functionalization pattern in ($\pm$)-58 with the epoxy bridge located at an adjacent β-type bond [144] (Fig. 9). The latter addition pattern was also found in the C_1-symmetrical tetrahydro[70]fullerene ($\pm$)-1,2,5,6-$C_{70}H_4$, characterized besides the C_s-symmetrical 1,2,3,4-$C_{70}H_4$ in the product mixture resulting from diimide reduction of C_{70} [106].

A singular chiral adduct was formed between C_{70} and two Ru$_3$(CO)$_9$ units which are known for the complexation of arenes. From the corresponding "mono"-adduct it was known that the trinuclear ruthenium moieties add preferentially to the hexagons of highest local curvature [143]. Assuming addition of two Ru$_3$(CO)$_9$ units at opposite poles, three constitutional isomers of [{Ru$_3$(CO)$_9$}$_2$(μ_3-η^2,η^2,η^2-C_{70})] are possible in analogy to the addition of achiral divalent addends to α-type bonds (cf. Sect. 4.2.1) [54, 131]: One of them has C_{2v}-symmetry and two have C_2-symmetry. Of the three formed isomers, the major one afforded crystals suitable for X-ray analysis; it has an inherently chiral addition pattern and corresponds to structure ($\pm$)-59 [143] (Fig. 9).

4.3
Tris- and Tetrakis-Adducts of C_{70}

The two major Bingel bis-adducts with achiral malonate addends, 45 and ($\pm$)-46 (Fig. 7), were further reacted with diethyl bromomalonate to explore the regioselectivity in the formation of higher adducts. Whereas an achiral tetrakis-adduct (C_{2v}-symmetrical 60) was obtained from C_{2v}-symmetrical bis-adduct 45, the chiral, C_2-symmetrical bis-adduct ($\pm$)-46 afforded the C_2-symmetrical ($\pm$)-61 with an inherently chiral functionalization pattern [54] (Fig. 10). Once C_{70} bis-adducts are formed, further addition must take place in a hemisphere

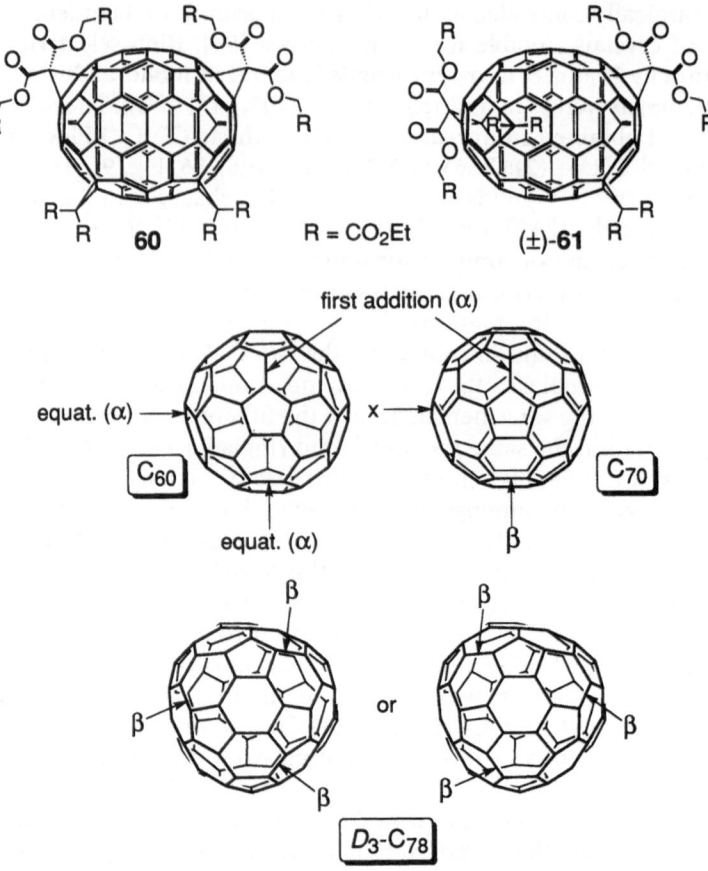

Fig. 10. Two Bingel type tetrakis-adducts of C_{70}, and a comparison of the intrahemispheral relative arrangement of their addends to that of equatorial bis-adducts of C_{60}, as well as of a C_3-symmetrical tris-adduct of D_3-C_{78} (2 isomers are possible, cf. Fig. 14)

that is already functionalized. For steric and probably also electronic reasons, this does not occur at another α-type bond. Instead, a functionalization pattern similar to that of a second bromomalonate addition to C_{60}, preferentially occurring at an equatorial bond, is observed. Given the two alternatives corresponding to this relative arrangement of addends in a C_{70} hemisphere, the next addition takes place at a β-type bond rather than in the less curved equatorial region (6–6 bond of type C(7)–C(21), marked by x in Fig. 10), thus leading to the respective tris-adducts which were characterized next to tetrakis-adducts **60** and (±)-**61** [54]. A comparable relative arrangement of three malonate addends, corresponding to the e,e,e-pattern in C_{60} [145], was found in a tris-adduct of D_3-C_{78} (Figs. 10, 14) [15]. It appears that – despite the larger number of theoretically possible isomers – C_{70}, and possibly other higher fullerenes [15], show a higher selectivity in multiple nucleophilic additions [54, 90] than C_{60} [96, 145].

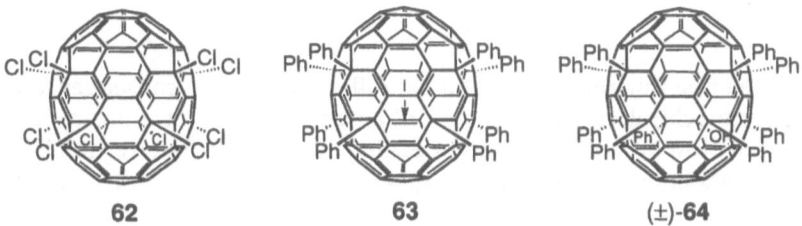

62   **63**   **(±)-64**

Fig. 11. $C_{70}Cl_{10}$ (**62**) can be further derivatized, e.g. to $C_{70}Ph_8$ (**63**) with an intrapentagonal double bond (marked by an arrow) or to $C_{70}Ph_9OH$ [(±)-**64**] displaying a non-inherently chiral functionalization pattern

4.4
C_1-Symmetrical Higher Adduct of C_{70}

The relatively easy access to the isomerically pure, C_s-symmetrical $C_{70}Cl_{10}$ (**62**) with its remarkable intrahexagonal 1,4-addition patterns [124] allowed Taylor and coworkers to investigate the chemistry of this most interesting higher fullerene halide [78, 146, 147] (Fig. 11). In analogy to $C_{60}Cl_6$ [148], $C_{70}Cl_{10}$ undergoes Friedel-Crafts alkylation with benzene/$FeCl_3$ to yield $C_{70}Ph_{10}$ via the octakis-adduct intermediate $C_{70}Ph_8$ (**63**) with a reactive [149] intrapentagonal double bond [C(7)–C(8), marked with an arrow in **63**, Fig. 11] [146]. A by-product of this reaction is fullerenol $C_{70}Ph_9OH$ [(±)-**64**], the formation of which may occur by nucleophilic substitution of another intermediate of the Friedel-Crafts reaction, $C_{70}Ph_9Cl$, by water [147]. The non-inherently chiral functionalization pattern of this first well-defined monohydroxyfullerene molecule can be ascribed to the formal addition of two different monovalent residues (OH and Ph) across the 6–5 bond C(7)–C(8).

5
Covalent Chemistry of the Fullerenes Beyond C_{70}

In contrast to the considerable progress made over the last few years in the production and purification of C_{60} [7, 8] and of C_{70} [9, 13], the availability of the larger carbon cages has remained unsatisfactory. As a consequence, much of the chemistry of these fullerenes waits to be explored, and only a handful of pure derivatives has been isolated and characterized.

Apart from these, a number of reactions, e.g. ozonation [150], hydrogenation [151], fluorination [152], and the formation of methylene adducts by reaction with THF [153], have led to various mixtures of higher fullerene derivatives. Finally, as the methods for production and purification of endohedral fullerene derivatives are being improved, the exohedral functionalization of C_{60} [70, 73, 74] and C_{70} [73] cages containing noble gas atoms, and of endohedral metallofullerenes [64, 154] emerge as new fields of fullerene chemistry.

5.1
Adducts of D_2-C_{76}

D_2-C_{76} is the first inherently chiral higher fullerene, and as a consequence, all its derivatives are chiral, except for as yet unknown *meso* type intramolecular combinations of both of its optical antipodes. Regardless of the nature, number, and position of addends, all adducted C_{76} spheroids have an inherently chiral functionalization pattern, and the according stereo-descriptor is determined by the configuration of the parent fullerene enantiomer [34]. The presence of 15 types of 6–6 bonds in D_2-C_{76} can give rise to a large number of mono-adduct isomers, potential adducts across 6–5 bonds not even considered. Provided the addend is C_2- or C_{2v}-symmetrical, twelve of the fifteen 6–6 bond adducts have C_1-symmetry, and the remaining three have C_2-symmetry. The latter result from addition across the bonds intersected by the C_2-axes of D_2-C_{76}; they are only half as abundant as the other 6–6 bonds. If one focuses on the most curved 6–6 bonds (types α and β; type γ not present) as the potentially most reactive sites, they are concentrated at each pole within four adjacent six-membered rings forming a chrysene-type arrangement centered at the intersection of the spheroid with one of the C_2-axes (Fig. 12). Whereas 3 α-type bonds make up the 3 fusions between adjacent hexagons of this chrysene sub-structure, 6 bonds of type β are distributed over its periphery.

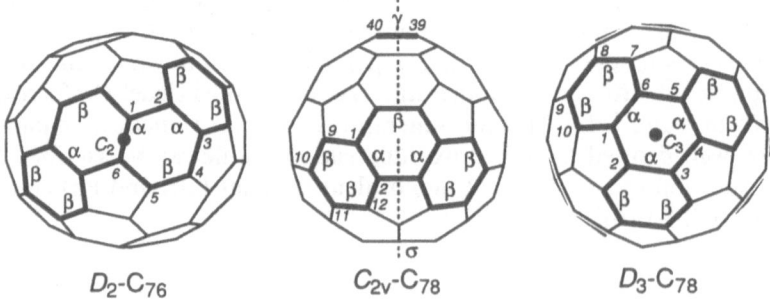

Fig. 12. Caps of D_2-C_{76}, C_{2v}-C_{78} and D_3-C_{78} showing the distribution of the most curved bonds (types α and β, as well as one γ-type bond of C_{2v}-C_{78}) within a hemisphere of these fullerenes

In our program aimed at the investigation of the regioselectivity of addition reactions with higher fullerenes, we found that at least 6 out of the 15 constitutionally isomeric 6–6 adducts were formed in the addition of 3,4-dimethoxy-*o*-quinodimethane to racemic D_2-C_{76} [14]. ^{1}H-NMR spectroscopy, together with the consideration of bond reactivities modulated by local curvature, led to the tentative assignment of the C_1-symmetrical C(2)–C(3)-adduct structure (±)-**65** (Fig. 13, cf. also Fig. 12) resulting from attack at an α-type bond to the major product which was isolated in pure form. Based on similar reasoning, the structure of the apical C(1)–C(6)-adduct (±)-**66** formed by attack at the other α-type bond, was assigned to a C_2-symmetrical mono-adduct, which was isolated as a

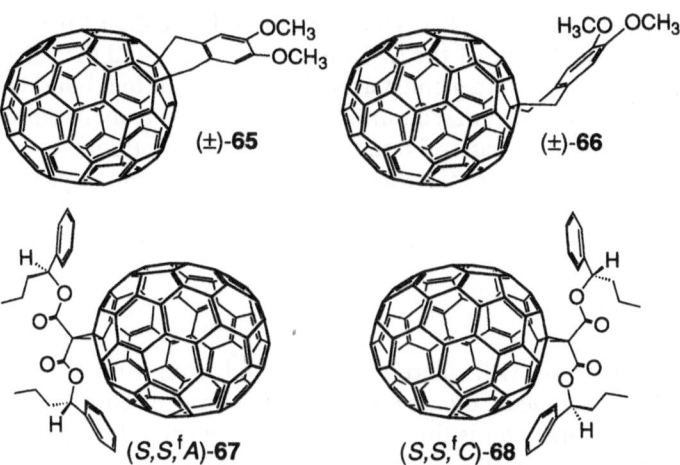

Fig. 13. C$_{76}$ derivatives for which confident structural assignments were possible: Two constitutionally isomeric Diels-Alder adducts [($\pm$)-**65** and ($\pm$)-**66**], and a diastereoisomeric pair of Bingel adducts (**67** and **68**)

mixture together with two C_1-symmetrical isomers [14] (Fig. 13, cf. also Fig. 12). The C(2)–C(3) and C(1)–C(6) addition patterns of ($\pm$)-**65** and ($\pm$)-**66** involve the most pyramidalized C-atoms of C$_{76}$ (Fig. 12) and correspond to those proposed by Hawkins et al. for the main products of the osmylation [50]. With regard to photophysical applications, it is worth mentioning that – similar to the parent C$_{76}$ [1] – the onset of the electronic absorption of ($\pm$)-**65** is located around 880 nm, whereas C$_{70}$ and its derivatives show no significant absorption above 700 nm [14].

In further investigations, we subjected C$_{76}$ to the Bingel reaction which had proven to be more selective than the Diels-Alder addition for the bonds of highest local curvature in C$_{70}$ [54]. In a nucleophilic cyclopropanation of racemic C$_{76}$ with an (S,S)-configured 2-bromomalonate, superimposition of the chirality of the addend and that of the inherently chiral fullerene yielded mainly three constitutionally isomeric pairs of diastereoisomeric mono-adducts, two of C_1-, and one (**67/68**, Fig. 13) of C_2-symmetry (approximate ratio: 8:7:1) [55]. All six compounds were isolated by HPLC to afford the first optically pure adducts of an inherently chiral fullerene. UV/Vis, CD, and ^{13}C-NMR spectra showed distinct similarities for isoconstitutional stereoisomers, thus allowing facile identification of the three pairs of diastereoisomers. The CD spectra of these derivatives of an inherently chiral fullerene display very pronounced Cotton effects with the maximum $\Delta \varepsilon$ values ($\sim$ 250 cm^2 mmol^{-1}) [55] being comparable to those of the optically pure parent C$_{76}$ [52] (cf. Fig. 17) and twice as high as those measured in the case of inherently chiral addition patterns resulting from functionalization of the achiral C$_{70}$ [54] (Fig. 8).

According to the local curvature model used in predicting bond reactivities, the minor C_2-symmetrical diastereoisomeric pair (**67/68**) most probably results from malonate addition across the apical α-type bond C(1) – C(6) intersected by a C_2 axis of D_2-C_{76} (Figs. 12, 13). This leaves another α-bond (C(2)–C(3)) for the addition leading to one of the C_1-symmetrical pairs, and the third, C_1-symmetrical constitutional isomer must arise from addition to a bond located in a less curved region. The fact that a minor product results from α-bond addition, whereas reaction at a less curved bond leads to a major adduct, is indicative of other factors, such as electronic effects, contributing to the regioselectivity in C_{76}, and possibly other higher fullerenes [55].

5.2
Adducts of C_{2v}- and D_3-C_{78}

Our synthetic work on [78]fullerene [15] started from a mixture of C_{2v}- and D_3-C_{78} [16]. Whereas the former is one out of two known achiral C_{2v}-symmetrical isomers of C_{78}, the latter is the second inherently chiral homologue of buckminsterfullerene [16, 20, 22]. C_{2v}-C_{78} is the first fullerene containing γ-type bonds, one of which is situated at the intersection of the carbon cage with the C_2-axis and the two mirror planes (Fig. 12). Among the total of 18 different 6–6 bonds, those of types α and β are located at the fusion bonds and the periphery, respectively, of two phenanthrene subunits bisected by one of the mirror planes (Fig. 12). Similar to D_2-C_{76}, all adducted spheroids of D_3-C_{78} display an inherently chiral functionalization pattern having the same configuration as the parent fullerene enantiomer [34]. The most reactive of the 10 types of 6–6 bonds of D_3-C_{78} are located within a triphenylene substructure centered at the intersection of the spheroid with the C_3-axis. Again, the bonds of type α constitute the hexagon-hexagon fusions of this substructure, whereas those of type β are spread over the periphery, and γ-type bonds are absent in this isomer.

Nucleophilic cyclopropanation of a 3:1 isomeric mixture of C_{2v}- and D_3-C_{78} with ca. 2 equivalents of diethyl 2-bromomalonate showed a high reactivity of these fullerene isomers since at least eight tris-adducts were formed as main products besides two C_1-symmetrical bis-adducts [15]. Most of the derivatives could be isolated in pure state by HPLC.

In the case of the tris-adducts, three isomers displayed a higher symmetry than C_1 and could therefore be unambiguously assigned to the parent C_{78} isomer. Based on a comparison of the UV and ^{13}C-NMR spectra obtained for a C_2-symmetrical tris-adduct of C_{2v}-C_{78} and those of the bis-adducts, three possible structures [($\pm$)-69] were proposed for one of the C_1-symmetrical bis-adducts, identified as direct precursor to the corresponding tris-adduct ($\pm$)-70 (3 possible structures, Fig. 14). Both ($\pm$)-69 and ($\pm$)-70 have an inherently chiral functionalization pattern. For reasons of symmetry, a common addend position in these two adducts has to be the γ-type bond C(39)–C(40) located at the intersection of the parent C_{2v}-C_{78} and the two mirror planes (Figs. 12, 14). For the second addend, one out of three positions within a six-membered ring containing one bond of type α and two of type β is most likely [15] (Figs. 12, 14). π-Bond order considerations [100], taking into account steric effects, predict C(9)-C(10) (Figs. 12, 14) and the symmetry-related C(69)–C(70) to be the most likely positions for the second and third addends in ($\pm$)-70.

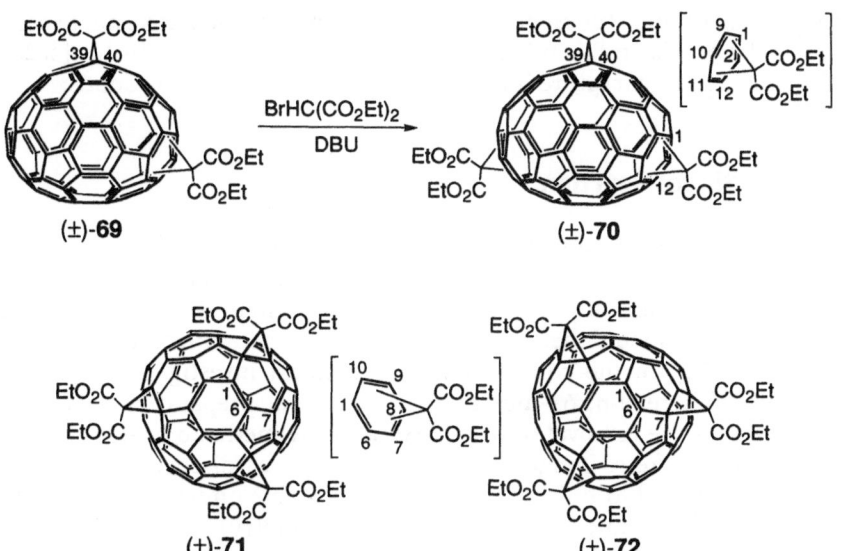

Fig. 14. Structures proposed for a tris-adduct of C_{2v}-C_{78} [($\pm$)-**70**] and its bis-adduct precursor [($\pm$)-**69**] (3 possible isomers; insert shows the numbering for one of the concerned six-membered rings), and for a tris-adduct of D_3-C_{78} (2 possible isomers, ($\pm$)-**71** and ($\pm$)-**72**; insert shows the numbering for one of the concerned six-membered rings)

Two constitutionally isomeric structures, ($\pm$)-**71** or ($\pm$)-**72**, were finally proposed for a C_3-symmetrical pure tris-adduct of D_3-C_{78} (Fig. 14); each of them results from addition to 3 equivalent β-type bonds (Figs. 12, 14) and shows the same relative, pairwise arrangement of addends as that observed within a hemisphere of tris- and tetrakis-adducts of C_{70} (cf. Sect. 4.3 and Fig. 10) [15]. According to π-bond order considerations [100], ($\pm$)-**72** should be formed preferentially.

5.3
Iridium Complex of D_{2d}-C_{84}

The high selectivity in the formation of transition metal complexes (cf. Sect. 4.2.3) was also seen in the exclusive crystallization of $[\eta^2$-$(D_{2d}$-$C_{84})]$ Ir(CO)Cl(PPh$_3$)$_2$ from a solution containing both D_{2d}- and D_2-C_{84}. Furthermore, transition metal binding occurring selectively at a bond intersected by one of the C_2-axes of D_{2d}-C_{84} [C(32)-C(53) = central bond in the D_{2d}-C_{84} structure of Fig. 1], this adduct represents the first isomerically pure derivative of [84]fullerene [23]. It is interesting, in this context, to note that Hückel calculations showed D_{2d}-C_{84} to have the most localized π-bonding among the homologues C_{60}, C_{70}, C_{76}, C_{78} and C_{84}. This suggests a high reactivity in addition reactions, particularly for the γ-type bond that takes part in the above-mentioned η^2-complexation by Ir and was shown to have the highest π-density [100]. Remarkably, among the mentioned fullerenes, D_{2d}-C_{84} is the first representative not to contain α- or β-type bonds. In this fullerene, as well as in the isomeric D_2-C_{84} [4, 20, 22, 24, 26], the highest curvature is expressed in a total of 10 γ-type bonds.

6
The Retro-Bingel Reaction

Even though a mixture of two isomers was used as the starting material in our synthetic work on C_{78} [15] (cf. Sect. 5.2), most of the adducts of C_{2v}- and D_3-C_{78} were chromatographically isolated as (racemic) single isomers. A reversible functionalization should therefore be useful in the separation of higher fullerene isomers [50–52]. In the course of electrochemical investigations on diethyl methano[60]fullerene-61,61-dicarboxylate, a general, preparative electrochemical method for the removal of bis(alkoxycarbonyl)methano bridges from methanofullerenes ("retro-Bingel reaction", Fig. 15) was discovered [52]. Cyclopropane rings fused to fullerenes are generally stable under conditions of CV (cyclic voltammetry) and SSV (steady state voltammetry) [128, 129, 155], as well as against thermal [156] or wet-chemical removal. For a few methanofullerene

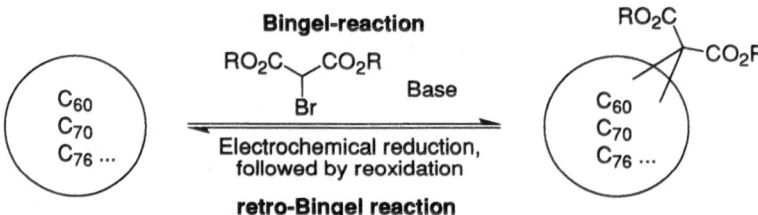

Fig. 15. The electrochemical retro-Bingel reaction

derivatives, however, decomposition was noted during electrochemical investigations [129, 157, 158]. The first two reduction steps of diethyl 1,2-methano[60]fullerene-61,61-dicarboxylate are perfectly reversible on the CV time scale [52, 128, 157]. Furthermore, CPE (controlled potential electrolysis) reduction led to the stable mono-anionic state and corresponds to a charge transfer of one electron per molecule [52]. CPE to the dianion, on the other hand, was surprising in that it corresponded to a transfer of an additional three electrons per molecule instead of only one [52]. Clear changes in the CV indicated that some chemical reaction had taken place, and reoxidation of the solution, followed by chromatography yielded pure C_{60} (> 80%). The generality of the retro-Bingel reaction, demonstrating the introduction of a fused cyclopropane ring as a new protecting group in fullerene chemistry, was proven by its successful extension to Bingel-type e- and trans-3-bis-adducts of C_{60}, to a 1,2-methano[70]fullerene-71,71-dicarboxylate, as well as to methanofullerenedicarboxylates of C_{76} (73 and 74, Fig. 16) [52].

7
Resolution of Chiral Higher Fullerenes and Assignment of Absolute Configurations to the Enantiomers of D_2-C_{76}

Chiral higher fullerenes are among the rare examples of chiral element modifications (allotropes). Isolation of the first representative, D_2-C_{76}, as a racemic mixture [1] suscited strong interest in the separation of its enantiomers and the investigation of their chiroptical properties. This goal was first achieved in 1993 by Hawkins et al. who realized a small scale kinetic resolution by asymmetrical Sharpless osmylation of racemic C_{76}, using reagents with enantiomerically pure ligands derived from a cinchona alkaloid [50] (Fig. 16, top). Uncomplete reaction of the racemic mixture with the osmium reagent led to an enantiomeric enrichment of the unreacted C_{76}, whereas the formed adduct was enriched in the other

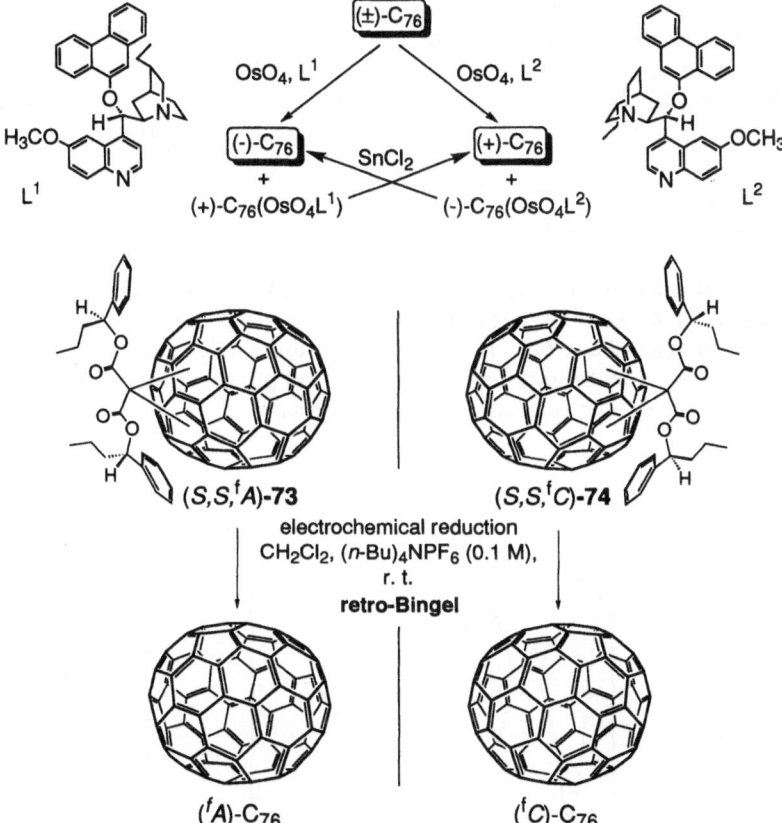

Fig. 16. Resolution of inherently chiral fullerenes. *Top:* Kinetic resolution based on the differential reactivity of an optically pure osmium complex towards the enantiomers of D_2-C_{76}. Hawkins et al. utilized this method also for the resolution of D_3-C_{78} and D_2-C_{84}. *Bottom:* Separation of the enantiomers of D_2-C_{76} by application of the retro-Bingel reaction to each of two optically pure, diastereoisomeric C_{76} derivatives having enantiomeric carbon cores

enantiomer. The latter could be regenerated by reduction of the osmate with SnCl$_2$. Use of a "pseudoenantiomeric" alkaloid-derived ligand in the osmylation reaction afforded an enrichment of the opposite optical C$_{76}$ antipodes in the starting material and the product, respectively. Being able, thus, to obtain each C$_{76}$ enantiomer in an enriched form, Hawkins was able to record the first CD spectra of these chiral carbon molecules [50]. Repetition of the experiment with C$_{78}$ and C$_{84}$ afforded the respective information on the optical antipodes of D_3-C$_{78}$ and D_2-C$_{84}$ [51]. Furthermore, no sign of racemization was shown by the enantiomers of D_2-C$_{76}$ or D_2-C$_{84}$ when heated (600/700 °C) or irradiated (193 nm). This experiment indicates an activation barrier of at least 83 kcal mol^{-1} for Stone-Wales pyracylene rearrangements [159], two of which are necessary for the interconversion of the two D_2-C$_{84}$ enantiomers via the achiral intermediate D_{2d}-C$_{84}$ [51].

A comparison of the CD spectra reported by Hawkins and Meyer for the C$_{76}$ enantiomers ($\Delta\varepsilon$-values of up to 32 cm^2 mmol^{-1} ($\lambda = 405$ nm)) [50] to those of a number of optically pure, covalent C$_{76}$ derivatives prepared by Diederich and coworkers ($\Delta\varepsilon$-values of up to 250 cm^2 mmol^{-1}) [55] (cf. Sect. 5.1), had revealed a large difference in the magnitude of the Cotton effects. In order to reinvestigate the chiroptical properties of the enantiopure allotrope, the retro-Bingel reaction (Fig. 15) was applied [52] (Fig. 16, bottom) to each of two optically pure, diastereoisomeric mono-adducts of C$_{76}$ (**73** and **74**) having enantiomeric carbon cores [55]. The CD spectra of the individually generated optical antipodes of D_2-C$_{76}$ (Fig. 17) displayed the expected mirror image shapes with band positions in full agreement with those of Hawkins and Meyer [50]. In contrast, however, the Cotton effects reach up to ca. 210 cm^2 mmol^{-1} ($\lambda = 406$ nm) and are therefore much more in agreement with the values measured for optically pure C$_{76}$ derivatives such as **67** and **68** (Fig. 13) [55]. The noticeably lower values reported by Hawkins et al. may be ascribed to a low optical purity of their samples or inaccuracies in the determination of concentrations.

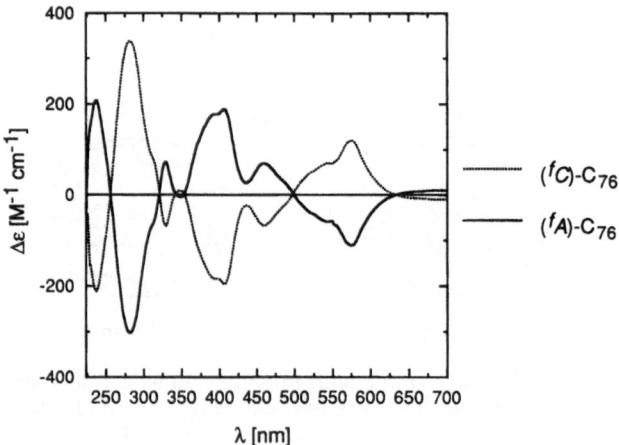

Fig. 17. Circular dichroism spectra of the optically pure enantiomers of D_2-C$_{76}$ in CH$_2$Cl$_2$

Theoretical calculation of the CD spectra [46, 49] by the π-electron SCF-CI-DV MO (self consistent field – configuration interaction – dipole velocity molecular orbital) method [49, 160, 161] and comparison with the experimental data [50, 52] allowed the assignment of absolute configurations to the separated enantiomers of D_2-C_{76}, which can now be addressed as [CD(-)282]-(^{f}A)-D_2-C_{76} and [CD(+)282]-(^{f}C)-D_2-C_{76} [49] (Figs. 1, 16, 17).

8
Concluding Remarks

Despite the laborious purification process necessary for obtaining higher fullerenes in quantities sufficient for derivatization, the chemistry of these carbon spheroids has made considerable progress over the last few years. This is particularly true for the functionalization of C_{70} where a relatively large body of experimental data has led to first reactivity and regioselectivity principles being recognized. These include a tendency for additions at 6–6 bonds in the polar regions of highest local curvature and for bis-additions to take place in opposite hemispheres, unless addends with a particularly low steric requirement are involved. Furthermore, it appears that next to local curvature and steric effects, as yet poorly understood electronic factors govern mono- and multiple additions to C_{70} and to the larger spheroids. Further elucidation of the electronic interrelationships among the double bonds in fullerene π-systems has to be the object of further experimental and theoretical investigations. Compared to the chemistry of C_{60}, an acute problem with the less symmetrical higher fullerenes is the difficulty in making structural assignments for their derivatives which can occur in a host of isomeric, often C_1-symmetrical forms. Furthermore, as demonstrated by the mono-addition of benzyne to C_{70}, higher fullerenes display addition modes that are unknown for buckminsterfullerene itself. These difficulties can be partially compensated in multiple functionalizations of higher fullerenes which may show a regioselectivity exceeding that observed for C_{60}.

The synthetic methodology in fullerene chemistry was recently enriched by the electrochemical retro-Bingel reaction which allows the use of easily introduced malonate addends as protecting groups. This should facilitate the access to derivatives with new addition patterns for higher fullerenes as well as for C_{60}, thus making further tailor-made building blocks available for materials and biological applications.

Already, the retro-Bingel reaction has been of great use in the resolution of the inherently chiral D_2-C_{76}, the enantiomers of which could be assigned by comparison of their experimental to calculated spectra. Hopefully, this will trigger further experimental as well as theoretical studies in the fascinating field of fullerene chirality and ultimately lead to a better understanding of inherently chiral chromophores and their chiroptical properties.

Acknowledgements: Our work was supported by the *Swiss National Science Foundation*. We thank *Hoechst AG* for samples of pure C_{70} and fullerene-soot extract enriched in higher fullerenes.

9
References

1. Ettl R, Chao I, Diederich F, Whetten RL (1991) Nature 353:149
2. Krätschmer W, Lamb LD, Fostiropoulos K, Huffman DR (1990) Nature 347:354
3. Diederich F, Ettl R, Rubin Y, Whetten RL, Beck R, Alvarez M, Anz S, Sensharma D, Wudl F, Khemani KC, Koch A (1991) Science 252:548
4. a) Diederich F, Whetten RL (1992) Acc Chem Res 25:119 b) Thilgen C, Diederich F, Whetten RL (1993) The Higher Fullerenes. In: Billups WE, Ciufolini MA (eds) Buckminsterfullerenes. VCH, New York, p 59
5. a) Holmes Parker D, Wurz P, Chatterjee K, Lykke KR, Hunt JE, Pellin MJ, Hemminger JC, Gruen DM, Stock LM (1991) J Am Chem Soc 113:7499 b) Bunshah RF, Jou S, Prakash S, Doerr HJ, Isaacs L, Wehrsig A, Yeretzian C, Cynn H, Diederich F (1992) J Phys Chem 96:6866 c) Peters G, Jansen M (1992) Angew Chem 104:240; Angew Chem Int Ed Engl 31:223 d) Tohji K, Paul A, Moro L, Malhotra R, Lorents DC, Ruoff RS (1995) J Phys Chem 99:17785 e) Kimura T, Sugai T, Shinohara H, Goto T, Tohji K, Matsuoka I (1995) Chem Phys Lett 246:571 f) Pietzak B, Wolf C, Mockel HJ, Weidinger A (1997) Chem Phys Lett 265:200
6. Theobald J, Perrut M, Weber JV, Millon E, Muller JF (1995) Separ Sci Technol 30:2783
7. Scrivens WA, Bedworth PV, Tour JM (1992) J Am Chem Soc 114:7917
8. Isaacs L, Wehrsig A, Diederich F (1993) Helv Chim Acta 76:1231
9. Scrivens WA, Cassell AM, North BL, Tour JM (1994) J Am Chem Soc 116:6939
10. a) Coustel N, Bernier P, Aznar R, Zahab A, Lambert J-M, Lyard P (1992) J Chem Soc Chem Commun: 1402 b) Doome RJ, Fonseca A, Richter H, Nagy JB, Thiry PA, Lucas AA (1997) J Phys Chem Solids 58:1839 c) Zhou X, Gu Z, Wu Y, Sun Y, Jin Z, Xiong Y, Sun B, Wu Y, Fu H, Wang J (1994) Carbon 32:935
11. Cabrera K, Wieland G, Schäfer M (1993) J Chromatogr 644:396
12. a) Atwood JL, Koutsantonis GA, Raston CL (1994) Nature 368:229 b) Suzuki T, Nakashima K, Shinkai S (1994) Chem Lett:699 c) Suzuki T, Nakashima K, Shinkai S (1995) Tetrahedron Lett 36:249
13. Darwish AD, Kroto HW, Taylor R, Walton DRM (1994) J Chem Soc Chem Commun:15
14. Herrmann A, Diederich F, Thilgen C, ter Meer H-U, Müller WH (1994) Helv Chim Acta 77:1689
15. Herrmann A, Diederich F (1997) J Chem Soc Perkin Trans 2:1679
16. Diederich F, Whetten RL, Thilgen C, Ettl R, Chao I, Alvarez MM (1991) Science 254:1768
17. Welch CJ, Pirkle WH (1992) J Chromatogr 609:89
18. Scrivens WA, Rawlett AM, Tour JM (1997) J Org Chem 62:2310
19. Fowler PW, Manolopoulos DE (1995) An atlas of fullerenes. Clarendon Press, Oxford
20. Taylor R, Langley GJ, Avent AG, Dennis TJS, Kroto HW, Walton DRM (1993) J Chem Soc Perkin Trans 2:1029
21. Michel RH, Kappes MM, Adelmann P, Roth G (1994) Angew Chem 106:1742; Angew Chem Int Ed Engl 33:1651
22. Kikuchi K, Nakahara N, Wakabayashi T, Suzuki S, Shiromaru H, Miyake Y, Saito K, Ikemoto I, Kainosho M, Achiba Y (1992) Nature 357:142
23. Balch AL, Ginwalla AS, Lee JW, Noll BC, Olmstead MM (1994) J Am Chem Soc 116:2227
24. Dennis TJS, Kai T, Tomiyama T, Shinohara H (1998) Chem Commun:619
25. Hennrich FH, Michel RH, Fischer A, Richard-Schneider S, Gilb S, Kappes MM, Fuchs D, Bürk M, Kobayashi K, Nagase S (1996) Angew Chem 108:1839; Angew Chem Int Ed Engl 35:1732
26. Avent AG, Dubois D, Pénicaud A, Taylor R (1997) J Chem Soc Perkin Trans 2:1907
27. Thilgen C, Herrmann A, Diederich F (1997) Angew Chem 109:2362; Angew Chem Int Ed Engl 36:2269, and references cited therein
28. Hirsch A (1994) The chemistry of the fullerenes. Thieme Verlag, Stuttgart, and references cited therein
29. a) Diederich F, Thilgen C (1996) Science 271:317, and references cited therein b) Diederich F (1997) Pure Appl Chem 69:395, and references cited therein

30. Smith III AB (ed) (1996) Fullerene chemistry, Tetrahedron symposia-in-print number 60. Elsevier Science Ltd, Oxford
31. Hirsch A (1997) J Phys Chem Solids 58:1729
32. Kadish KM, Ruoff, RS (eds.) (1994–1997) Proceedings of the symposium on fullerenes and related materials, volumes 1–5. The Electrochemiocal Society Inc, Pennington NJ
33. Diederich F, Thilgen C, Herrmann A (1996) Nachr Chem Tech Lab 44:9, and references cited therein
34. Thilgen C, Herrmann A, Diederich F (1997) Helv Chim Acta 80:183
35. Manolopoulos DE (1991) J Chem Soc Faraday Trans 87:2861
36. Fowler PW, Batten RC, Manolopoulos DE (1991) J Chem Soc Faraday Trans 87:3103
37. Manolopoulos D (1993) Scientist 7:16
38. Taylor R, Langley GJ, Dennis TJS, Kroto HW, Walton DRM (1992) J Chem Soc Chem Commun: 1043
39. a) Vasella A, Uhlmann P, Waldraff CAA, Diederich F, Thilgen C (1992) Angew Chem 104:1383; Angew Chem Int Ed Engl 31:1388 b) Uhlmann P, Harth E, Naughton AB, Vasella A (1994) Helv Chim Acta 77:2335
40. a) Cahn RS, Ingold SC, Prelog V (1966) Angew Chem 78:413; Angew Chem Int Ed Engl 5:385; errata ibid p 511 b) Prelog V, Helmchen G (1982) Angew Chem 94:614; Angew Chem Int Ed Engl 21:567
41. Wudl F, Smalley RE, Smith AB, Taylor R, Wasserman E, Godly EW (1997) Pure Appl Chem 69:1412
42. Goodson AL, Gladys CL, Worst DE (1995) J Chem Inform Comput Sci 35:969
43. Panico R, Powell WH, Richer J-C (ed) (1993) A guide to IUPAC nomenclature of organic compounds. Blackwell Scientific Publications, Oxford, Sect. R-0.2.4.2
44. Calculation of the configuration of C_{60} and C_{70} derivatives with a chiral functionalization pattern can be easily done via the following interactive world wide web page: http://www.diederich.chem.ethz.ch/~mac.chirafull
45. The numbering used in this review is that introduced by Taylor R in refs 100 and 124, and described in extenso in ref. 41
46. Orlandi G, Poggi G, Zerbetto F (1994) Chem Phys Lett 224:113
47. Fanti M, Orlandi G, Poggi G, Zerbetto F (1997) Chem Phys 223:159
48. Wilson SR, Lu Q, Cao JR, Wu Y, Welch CJ, Schuster DI (1996) Tetrahedron 52:5131
49. Goto H, Harada N, Crassous J, Diederich F (1998) J Chem Soc Perkin 2:1719
50. Hawkins JM, Meyer A (1993) Science 260:1918
51. Hawkins JM, Nambu M, Meyer A (1994) J Am Chem Soc 116:7642
52. Kessinger R, Crassous J, Herrmann A, Rüttimann M, Echegoyen L, Diederich F (1998) Angew Chem 110:2022; Angew Chem Int Ed Engl 37:1919
53. Hawkins JM, Meyer A, Nambu M (1993) J Am Chem Soc 115:9844
54. Herrmann A, Rüttimann M, Thilgen C, Diederich F (1995) Helv Chim Acta 78:1673
55. Herrmann A, Diederich F (1996) Helv Chim Acta 79:1741
56. Herrmann A, Gibtner T, Diederich F, unpublished results
57. Nierengarten J-F, Gramlich V, Cardullo F, Diederich F (1996) Angew Chem 108:2242; Angew Chem Int Ed Engl 35:2101
58. Nakamura E, Isobe H, Tokuyama H, Sawamura M (1996) Chem Commun: 747
59. Gross B, Schurig V, Lamparth I, Herzog A, Djojo F, Hirsch A (1997) Chem Commun:1117
60. Gross B, Schurig V, Lamparth I, Hirsch A (1997) J Chromatogr A 791:65
61. Dojo F, Hirsch A (1998) Chem Eur J 4:344
62. Hummelen JC, Keshavarzk-K M, Vandongen JLJ, Janssen RAJ, Meijer EW, Wudl F (1998) Chem Commun: 281
63. a) Heath JR, O'Brien SC, Zhang Q, Liu Y, Curl RF, Kroto HW, Tittel FK, Smalley RE (1985) J Am Chem Soc 107:7779 b) Chai Y, Guo T, Jin C, Haufler RE, Chibante LPF, Fure J, Wang L, Alford JM, Smalley RE (1991) J Phys Chem 95:7564 c) Gillan EG, Yeretzian C, Min KS, Alvarez MM, Whetten RL, Kaner RB (1992) J Phys Chem 96:6869 d) Bethune DS, Johnson RD, Salem JR, de Vries MS, Yannoni CS (1993) Nature 366:123 e) Edelmann FT (1995) Angew Chem 107:1071; Angew Chem Int Ed Engl 34:981 f) Nagase S, Kobayashi K,

Akasaka T (1996) Bull Chem Soc Jpn 69:2131 g) Akasaka T, Nagase S, Kobayashi K, Walchli M, Yamamoto K, Funasaka H, Kako M, Hoshino T, Erata T (1997) Angew Chem 109:1716; Angew Chem Int Ed Engl 36:1643 h) Nakane T, Xu ZD, Yamamoto E, Sugai T, Tomiyama T, Shinohara H (1997) Fullerene Sci Technol 5:829 i) Sun DY, Liu ZY, Guo XH, Xu WG, Liu SY (1997) J Phys Chem B 101:3927

64. a) Akasaka T, Kato T, Kobayashi K, Nagase S, Yamamoto K, Funasaka H, Takahashi T (1995) Nature 374:600 b) Akasaka T, Nagase S, Kobayashi K, Suzuki T, Kato T, Yamamoto K, Funasaka H, Takahashi T (1995) J Chem Soc Chem Commun: 1343 c) Akasaka T, Nagase S, Kobayashi K, Suzuki T, Kato T, Kikuchi K, Achiba Y, Yamamoto K, Funasaka H, Takahashi T (1995) Angew Chem 107:2303; Angew Chem Int Ed Engl 34:2139 d) Sun DY, Liu ZY, Liu ZQ, Guo XH, Hao CY, Xu WG, Liu SY (1997) Fullerene Sci Technol 5:1461

65. Giblin DE, Gross ML, Saunders M, Jiménez-Vázquez H, Cross RJ (1997) J Am Chem Soc 119:9883

66. a) Saunders M, Jiménez-Vázquez HA, Cross RJ, Poreda RJ (1993) Science 259:1428 b) Saunders M, Jiménez-Vázquez HA, Cross RJ, Mroczkowski S, Gross ML, Giblin DE, Poreda RJ (1994) J Am Chem Soc 116:2193

67. Murry RL, Scuseria GE (1994) Science 263:791

68. a) Patchkovskii S, Thiel W (1996) J Am Chem Soc 118:7164 b) Patchkovskii S, Thiel W (1998) J Am Chem Soc 120:556

69. Saunders M, Jiménez-Vázquez HA, Cross RJ, Mroczkowski S, Freedberg DI, Anet FAL (1994) Nature 367:256

70. Saunders M, Cross RJ, Jiménez-Vázquez HA, Shimshi R, Khong A (1996) Science 271:1693

71. Saunders M, Jiménez-Vázquez HA, Cross RJ, Billups WE, Gesenberg C, Gonzalez A, Luo W, Haddon RC, Diederich F, Herrmann A (1995) J Am Chem Soc 117:9305

72. Cross RJ, Jiménez-Vázquez HA, Lu Q, Saunders M, Schuster DI, Wilson SR, Zhao H (1996) J Am Chem Soc 118:11454

73. Rüttimann M, Haldimann RF, Isaacs L, Diederich F, Khong A, Jiménez-Vázquez H, Cross RJ, Saunders M (1997) Chem Eur J 3:1071

74. Jensen AW, Khong A, Saunders M, Wilson SR, Schuster DI (1997) J Am Chem Soc 119:7303

75. Bellavia-Lund C, Hummelen J-C, Keshavarz-K M, González R, Wudl F (1997) J Phys Chem Solids 58:1983

76. Smith III AB, Strongin RM, Brard L, Furst GT, Romanow WJ, Owens KG, Goldschmidt RJ, King RC (1995) J Am Chem Soc 117:5492

77. Nuber B, Hirsch A (1996) Fullerene Sci Technol 4:715

78. Birkett PR, Avent AG, Darwish AD, Kroto HW, Taylor R, Walton DRM (1995) J Chem Soc Chem Commun: 1869

79. Rabideau PW, Sygula A (1996) Acc Chem Res 29:235

80. a) Rubin Y (1997) Chem Eur J 3:1009 b) Rubin Y (1998) Chimia 52:118

81. a) Andersson T, Nilsson K, Sundahl M, Westman G, Wennerström O (1992) J Chem Soc Chem Commun: 604 b) Yoshida ZI, Takekuma H, Takekuma SI, Matsubara Y (1994) Angew Chem 106:1658; Angew Chem Int Ed Engl 33:1597 c) Braun T (1997) Fullerene Sci Technol 5:615, and references cited therein d) Sui GD, Zhang DD, Yang Y (1997) Supramol Chem 8:379

82. a) Raston CL, Atwood JL, Nichols PJ, Sudria IBN (1996) Chem Commun: 2615 b) Haino T, Yanase M, Fukazawa Y (1997) Angew Chem 109:288; Angew Chem Int Ed Engl 36:259

83. Yang Y, Arias F, Echegoyen L, Chibante LPF, Flanagan S, Robertson A, Wilson LJ (1995) J Am Chem Soc 117:7801

84. Matsuzawa N, Dixon DA, Fukunaga T (1992) J Phys Chem 96:7594

85. Austin SJ, Fowler PW, Sandall JPB, Birkett PR, Avent AG, Darwish AD, Kroto HW, Taylor R, Walton DRM (1995) J Chem Soc Perkin Trans 2:1027

86. Darwish AD, Avent AG, Taylor R, Walton DRM (1996) J Chem Soc Perkin Trans 2:2079

87. Haddon RC (1993) Science 261:1545

88. Haddon RC, Scuseria GE, Smalley RE (1997) Chem Phys Lett 272:38

89. Hawkins JM, Meyer A, Solow MA (1993) J Am Chem Soc 115:7499
90. Karfunkel HR, Hirsch A (1992) Angew Chem 104:1529; Angew Chem Int Ed Engl 31:1468
91. Henderson CC, Rohlfing CM, Cahill PA (1993) Chem Phys Lett 213:383
92. Henderson CC, Rohlfing CM, Gillen KT, Cahill PA (1994) Science 264:397
93. Balch AL, Catalano VJ, Costa DA, Fawcett WR, Frederco M, Ginwalla AS, Lee JW, Olmstead MM, Noll BC, Winkler K (1997) J Phys Chem Solids 58:1633
94. Henderson CC, Rohlfing CM, Assink RA, Cahill PA (1994) Angew Chem 106:803; Angew Chem Int Ed Engl 33:786
95. Hirsch A, Lamparth I, Grösser T, Karfunkel HR (1994) J Am Chem Soc 116:9385
96. Djojo F, Herzog A, Lamparth I, Hampel F, Hirsch A (1996) Chem Eur J 2:1537
97. Grösser T, Prato M, Lucchini V, Hirsch A, Wudl F (1995) Angew Chem 107:1462; Angew Chem Int Ed Engl 34:1343
98. Baker J, Fowler PW, Lazzeretti P, Malagoli M, Zanasi R (1991) Chem Phys Lett 184:182
99. Rathna A, Chandrasekhar J (1995) Fullerene Sci Technol 3:681
100. Taylor R (1993) J Chem Soc Perkin Trans 2:813
101. Note on the use of the expression "addend": As bridging (divalent) addends are among the most common in fullerene chemistry, the simple expression "addend" is often used in the sense of "bridging addend" and thus implies bonding to two carbon atoms on the fullerene surface. This situation can be compared to the addition of two monovalent addends like H or Cl, and, in terms of adduct symmetry, a C_{2v}-symmetrical (C_s-symmetrical) bridging addend corresponds to two identical (different) monovalent addends. It is important to be aware of this use of the language for avoiding confusion in discussing fullerene adduct isomers. Furthermore, talking about the symmetry of addends in this review, we mean the symmetry of the bound addend moiety which may differ from that of the reagent molecule prior to addition
102. Smith AB, Strongin RM, Brard L, Furst GT, Atkins JH, Romanow WJ, Saunders M, Jiménez-Vázquez HA, Owens KG, Goldschmidt RJ (1996) J Org Chem 61:1904
103. Bezmelnitsin VN, Eletskii AV, Schepetov NG, Avent AG, Taylor R (1997) J Chem Soc Perkin Trans 2:683
104. Balch AL, Catalano VJ, Lee JW, Olmstead MM, Parkin SR (1991) J Am Chem Soc 113:8953
105. Iyoda M, Ogawa Y, Matsuyama H (1995) Fullerene Sci Technol 3:1
106. Avent AG, Darwish AD, Heimbach DK, Kroto HW, Meidine MF, Parsons JP, Remars C, Roers R, Ohashi O, Taylor R, Walton DRM (1994) J Chem Soc Perkin Trans 2:15
107. a) Meidine MF, Avent AG, Darwish AD, Kroto HW, Ohashi O, Taylor R, Walton DRM (1994) J Chem Soc Perkin Trans 2:1189 b) Becker H, Javahery G, Petrie S, Bohme DK (1994) J Phys Chem 98:5591 c) Pang LSK, Wilson MA (1993) J Phys Chem 97:6761 d) Nie B, Rotello VM (1996) J Org Chem 61:1870
108. Wang Y, Schuster DI, Wilson SR, Welch CJ (1996) J Org Chem 61:5198
109. Gareis T, Köthe O, Beer E, Daub J (1996) In: Kadish KM, Ruoff RS (eds) Fullerenes – recent advances in the physics and chemistry of fullerenes and related materials, vol 3. The Electrochemical Society Inc, Pennington, NJ, p 1244
110. a) Sliwa W (1995) Fullerene Sci Technol 3:243 b) Eguchi S, Ohno M, Kojima S, Koide N, Yashiro A, Shirakawa Y, Ishida H (1996) Fullerene Sci Technol 4:303 c) Sliwa W (1997) Fullerene Sci Technol 5:1133
111. Suzuki T, Li QC, Khemani KC, Wudl F (1992) J Am Chem Soc 114:7301
112. Diederich F, Isaacs L, Philp D (1994) Chem Soc Rev 23:243
113. Bellavia-Lund C, Wudl F (1997) J Am Chem Soc 119:943
114. a) Hummelen JC, Prato M, Wudl F (1995) J Am Chem Soc 117:7003 b) Hummelen JC, Knight B, Pavlovich J, González R, Wudl F (1995) Science 269:1554
115. Schick G, Kampe K-D, Hirsch A (1995) J Chem Soc Chem Commun:2023
116. Lamparth I, Nuber B, Schick G, Skiebe A, Grösser T, Hirsch A (1995) Angew Chem 107:2473; Angew Chem Int Ed Engl 34:2257
117. Nuber B, Hirsch A (1996) Chem Commun:1421

118. Meier MS, Poplawska M, Compton AL, Shaw JP, Selegue JP, Guarr TF (1994) J Am Chem Soc 116:7044
119. Irngartinger H, Kohler C-M, Baum G, Fenske D (1996) Liebigs Ann: 1609
120. a) Yamago S, Nakamura E (1996) Chem Lett: 395 b) Irie K, Nakamura Y, Ohigashi H, Tokuyama H, Yamago S, Nakamura E (1996) Biosci Biotechnol Biochem 60:1359
121. Meier MS, Wang G-W, Haddon RC, Pratt Brock C, Lloyd MA, Selegue JP (1998) J Am Chem Soc 120:2337
122. Seiler P, Herrmann A, Diederich F (1995) Helv Chim Acta 78:344
123. Wilson SR, Lu Q (1995) J Org Chem 60:6496
124. Birkett PR, Avent AG, Darwish AD, Kroto HW, Taylor R, Walton DRM (1995) J Chem Soc Chem Commun: 683
125. Darwish AD, Abdul-Sada AK, Langley GJ, Kroto HW, Taylor R, Walton DRM (1994) J Chem Soc Chem Commun: 2133
126. a) Bürgi HB, Venugopalan P, Schwarzenbach D, Diederich F, Thilgen C (1993) Helv Chim Acta 76:2155 b) Roth G, Adelmann P (1992) J Phys I Fr 2:1541 c) Nikolaev AV, Dennis TJS, Prassides K, Soper AK (1994) Chem Phys Lett 223:143
127. Akasaka T, Mitsuhida E, Ando W, Kobayashi K, Nagase S (1994) J Am Chem Soc 116:2627
128. Boudon C, Gisselbrecht J-P, Gross M, Isaacs L, Anderson HL, Faust R, Diederich F (1995) Helv Chim Acta 78:1334
129. Cardullo F, Seiler P, Isaacs L, Nierengarten JF, Haldimann RF, Diederich F, Mordasini-Denti T, Thiel W, Boudon C, Gisselbrecht JP, Gross M (1997) Helv Chim Acta 80:343
130. Bingel C (1993) Chem Ber 126:1957
131. Bingel C, Schiffer H (1995) Liebigs Ann: 1551
132. Isaacs L, Diederich F (1993) Helv Chim Acta 76:2454
133. Wilson SR, Wu Y, Kaprinidis NA, Schuster DI, Welch CJ (1993) J Org Chem 58:6548
134. Prato M, Bianco A, Maggini M, Scorrano G, Toniolo C, Wudl F (1993) J Org Chem 58:5578
135. Toniolo C, Bianco A, Maggini M, Scorrano G, Prato M, Marastoni M, Tomatis R, Spisani S, Palú G, Blair ED (1994) J Med Chem 37:4558
136. Maggini M, Scorrano G, Bianco A, Toniolo C, Prato M (1995) Tetrahedron Lett 36:2845
137. Bianco A, Maggini M, Scorrano G, Toniolo C, Marconi G, Villani C, Prato M (1996) J Am Chem Soc 118:4072
138. Shen CK-F, Chien K-M, Juo C-G, Her G-R, Luh T-Y (1996) J Org Chem 61:9242
139. Novello F, Prato M, Da Ros T, De Amici M, Bianco A, Toniolo C, Maggini M (1996) Chem Commun: 903
140. Zhang X, Foote CS (1995) J Am Chem Soc 117:4271
141. Anderson JL, An Y-Z, Rubin Y, Foote CS (1994) J Am Chem Soc 116:9763
142. a) Balch AL, Lee JW, Olmstead MM (1992) Angew Chem 104:1400; Angew Chem Int Ed Engl 31:1356 b) Balch AL, Hao L, Olmstead MM (1996) Angew Chem 108:211; Angew Chem Int Ed Engl 35:188
143. Hsu H-F, Wilson SR, Shapley JR (1997) Chem Commun: 1125
144. Balch AL, Costa DA, Olmstead MM (1996) Chem Commun: 2449
145. Hirsch A, Lamparth I, Karfunkel HR (1994) Angew Chem 106:453; Angew Chem Int Ed Engl 33:437
146. Avent AG, Birkett PR, Darwish AD, Kroto HW, Taylor R, Walton DRM (1996) Tetrahedron 52:5235
147. Birkett PR, Avent AG, Darwish AD, Kroto HW, Taylor R, Walton DRM (1996) Chem Commun: 1231
148. Avent AG, Birkett PR, Crane JD, Darwish AD, Langley GJ, Kroto HW, Taylor R, Walton DRM (1994) J Chem Soc Chem Commun: 1463
149. Avent AG, Birkett PR, Darwish AD, Kroto HW, Taylor R, Walton DRM (1997) Fullerene Sci Technol 5:643
150. Heymann D, Chibante LPF (1993) Recl Trav Chim Pays-Bas 112:639
151. Darwish AD, Kroto HW, Taylor R, Walton DRM (1996) J Chem Soc Perkin Trans 2:1415
152. a) Boltalina OV, Sidorov LN, Bagryantsev VF, Seredenko VA, Zapol'skii AS, Street JM, Taylor R (1996) J Chem Soc Perkin Trans 2:2275 b) Boltalina OV, Sidorov LN, Sukhanova

EV, Sorokin ID (1994) Chem Phys Lett 230:567 c) Boltalina OV, Ponomarev DB, Borschevskii AY, Sorokin ID, Sidorov LN (1996) In: Kadish KM, Ruoff RS (eds) Fullerenes – recent advances in the physics and chemistry of fullerenes and related materials, vol 3. The Electrochemical Society Inc, Pennington, NJ, p 108 d) Borschevskii AY, Boltalina OV, Sidorov LN, Markov VY, Ioffe IN (1996) In: Kadish KM, Ruoff RS (eds) Fullerenes – recent advances in the physics and chemistry of fullerenes and related materials, vol 3. The Electrochemical Society Inc, Pennington, NJ, p 509 e) Boltalina OV, Dashkova EV, Sidorov LN (1996) Chem Phys Lett 256:253

153. Birkett PR, Darwish AD, Kroto HW, Langley GJ, Taylor R, Walton DRM (1995) J Chem Soc Perkin Trans 2:511
154. a) Hao CY, Liu ZY, Guo XH, Liu ZQ, Xu WG, Sun YQ, Liu SY (1997) Rapid Commun Mass Spectrom 11:1677 b) Kato T, Akasaka T, Kobayashi K, Nagase S, Kikuchi K, Achiba Y, Suzuki T, Yamamoto K (1997) J Phys Chem Solids 58:1779 c) Nagase S, Kobayashi K, Akasaka T (1997) Theochem – J Mol Struct 398:221
155. Guldi DM, Hungerbühler H, Asmus K-D (1995) J Phys Chem 99:9380
156. A hint for a removal of a Bingel addend at high temperature is found in ref. 95
157. Keshavarz-K M, Knight B, Haddon RC, Wudl F (1996) Tetrahedron 52:5149
158. Boudon C, Gisselbrecht JP, Gross M, Herrmann A, Rüttimann M, Crassous J, Cardullo F, Echegoyen L, Diederich F (1998) J Am Chem Soc 120:7860
159. Stone AJ, Wales DJ (1986) Chem Phys Lett 128:501
160. Harada N, Nakanishi K (1983) Circular dichroism spectroscopy – exciton coupling method in organic stereochemistry. Oxford University Press, Oxford
161. Harada N, Koumura N, Feringa BL (1997) J Am Chem Soc 119:7256, and references cited therein

Fullerene Materials

Maurizio Prato

Dipartimento di Scienze Farmaceutiche, Piazzale Europa 1, I-34127 Trieste, Italy.
E-mail: prato@univ.trieste.it

The range of potential applications of fullerenes and fullerene derivatives in materials science is becoming broader in virtue of the increased number of derivatives that are continuously produced. New opportunities arise from the combination of the fullerene properties with those of other classes of materials, such as polymers, electro- or photoactive units, liquid crystals, etc. In this article we will review the most recent achievements in this field.

Keywords: Polymers, Nonlinear optics, Electrochemistry, Photophysics, Photoconductors, Dyads.

1
Introduction

When the fullerenes stopped being a mere scientific curiosity and became a totally new research field, a great deal of effort was devoted to materials science applications [1]. Several innovative uses were envisioned [2]. The fullerenes were thus considered the ultimate superconductors, ferromagnets, lubricants, photoconductors, catalysts, etc. The high potential for the practical utilization of fullerenes has been mainly connected to the spectacular properties of the carbon cages. [60]Fullerene or C_{60}, the most common and abundant representative of the fullerene family, becomes a high T_c superconductor when treated with alkali metals [3–6]. In agreement with the postulated existence of three low-lying degenerate LUMO orbitals, cyclic voltammetric studies have demonstrated that in solution C_{60} takes reversibly up to six electrons [7, 8]. Due to a broad absorption range in the UV and the visible region, C_{60} provides a ready access to its excited states, singlets and triplets, which exhibit very rich chemistry and

Topics in Current Chemistry, Vol. 199
© Springer Verlag Berlin Heidelberg 1999

physics [9–15]. A charge transfer complex between C_{60} and the strong donor TDAE (tetrakis-diethylaminoethylene) gave rise to the highest T_c organic ferromagnet. Interesting nonlinear optical properties were found for C_{60} and C_{70} [16, 17].

Transferring the properties of the fullerene to bulk materials which exhibit the original features of pristine C_{60} has for years been a common task in the field. In general, however, C_{60}, the most common member and therefore the fullerene of choice for large scale applications, suffers several drawbacks. C_{60}, in fact, is insoluble or only sparingly soluble in most solvents, and it aggregates very easily, becoming even less soluble [18]. This problem could be partly surmounted with the help of the functionalization chemistry of the fullerenes [19–28].

In this article we discuss the most promising applications of the fullerenes in materials science, with particular emphasis to C_{60} derivatives. It should be pointed out that the derivatives may offer a high number of opportunities, as compared to pristine C_{60}. In most instances, a handle is provided for improving the properties or to broaden the spectrum of applications. A recent review describes materials with pristine C_{60} [29].

2
Fullerene-Based Materials

A schematic summary of the more promising applications of the fullerenes developed in this article includes:

1. fullerene-polymer combinations;
2. thin fullerene films;
3. electro-optical devices;
4. miscellaneous uses.

Some aspects of these application fields have been reviewed [28–31].

2.1
Polymers

The synthesis of fullerene-containing polymers is noteworthy for several reasons. On the one hand, once C_{60} is attached to a polymer, most of the fullerene properties are transferred to the polymer. Thus, for instance, electroactive and photoactive polymers or polymers with nonlinear optical properties can be prepared. On the other hand, hardly processible fullerenes embedded in highly soluble polymers may become more easily amenable to further treatments. The resulting materials might eventually be used for surface coating, photoconducting devices, or to create new molecular networks.

Generally speaking, the structural combination of fullerenes and polymers can lead to a few categories, depending on the relative spatial arrangement (Scheme 1) [25, 28–30, 32].

Type 1: fullerenes in the backbone of the polymer (pearl necklace or in-chain type);

Type 1: "main chain fullerenes"

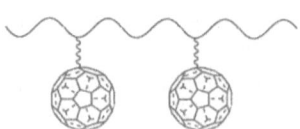

Type 2: "pendant fullerenes"

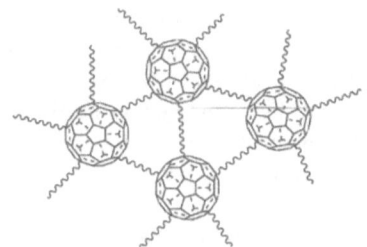

Type 3: starburst fullerenes

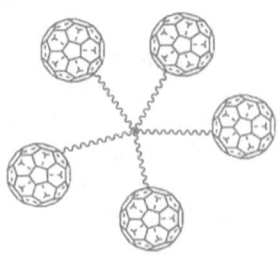

Type 4: fullerene end-capped polymers

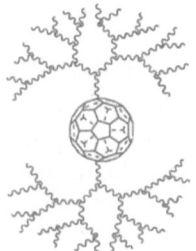

Type 5: dendrimer

Scheme 1

Type 2: polymers bearing pendant fullerenes (pendant charm bracelet or on-chain type);
Type 3: random polymers irradiating from a fullerene spheroid (starburst type or cross-link type);
Type 4: polymers terminated by a fullerene unit (end-chain type);
Type 5: dendrimers.

The chemical grafting of polymers to fullerenes can be achieved in several ways. If the polymerization process is carried out in the presence of fullerenes, the final polymer may contain fullerene units [33–42]. Typically this will happen when the polymerization conditions produce nucleophilic or free radical species, since the fullerene double bonds are highly responsive to this ambient. This methodology is useful as it represents a direct way for accessing C_{60} polymers. However, often, but not always (see below), multiple additions to the fullerene double bonds occur. This results in lack of strict control of the fullerene structure. Since the electrochemical and photophysical properties of the fullerenes depend heavily on the number of additions [43–45], a careful control of the addition to fullerenes is required. This control can be achieved by using a fullerene-containing monomer and polymerization conditions that are inert to the fullerene reactivity [46, 47]. Since a high number of fullerene units in a polymer causes low solubility, resulting in low polymerization degree, co-polymers are usually prepared [46, 47]. A disadvantage of this approach is that the preparation of the monomer can be painstaking. Other ways to fullerene-polymer combinations include reaction of C_{60} with polymers that contain reactive groups [48–56]. Synthesis of dendrimers [57–61] as well as starburst polymers [62, 63] has also been achieved (Scheme 2). In particular, the C_{60} core has been used as a building block for dendrimer synthesis. By successful regioselective additions, a T_h-symmetrical C_{60} core with an octahedral pattern was obtained, producing a dendrimer with a core branching multiplicity of 12 [59].

All-carbon fullerene polymers of type 1 (Scheme 1) can be obtained by irradiation of oxygen-free films or solutions of C_{60} [64–67] as well as by heating AC_{60} crystals (A = K, Rb, Cs) [6]. A quasi-linear structure, derived from [2+2] cycloadditions of C_{60} double bonds leading to four-membered rings, has been proposed for these polymers.

Living anionic polymerizations seem to be more easily controllable than radical polymerizations.

In a very elegant approach, Okamura et al. attached C_{60} to a range of styrene polymers with different molecular weights (from 1000 to 10,000) [32]. The attachment was provided by 1,4-radical addition to C_{60}, producing narrow-dispersity polystyryl adducts (Scheme 3). The electrochemical and ground state absorption properties of the polymers were identical to those of a low-mass, fully characterized model (Scheme 3). This indicates that the reported procedure is useful for the preparation of high-mass polymers, with retained fullerene properties.

Ford and collaborators reported a comparison between polymethyl methacrylate (PMMA)/C_{60} and polystyrene (PS)/C_{60} systems, generated by radical-chain polymerization [68]. Whereas high molecular PS contains up to one hundred fullerene units, high molecular PMMA contains only one C_{60} unit/polymer.

Addition of "living" anionic polymers to C_{60} may offer unique opportunities to control the number of chains grafted on the fullerene, the molecular mass, and the polymolecularity. Ederlé and Mathis showed that, when conducted under high purity conditions, the additon of "living" anionic polymers can be controlled and used to prepare adducts with well defined structures [69]. In toluene, star-shaped polymers with up to six branches can be produced using reactive carbanions like styryl or isoprenyl from low to high mass. The number of additions to C_{60} (three to six) could be controlled by stoichiometry. In THF two electrons are transferred from the carbanion to C_{60}, followed by addition. In another contribution, the same group studied the influence of C_{60} anions to initiate living polymerization of PS or PMMA. $C_{60}^{3-}(K^+)_3$ did not initiate polymerization of PS or PMMA [70]. C_{60}^{6-} did not initiate polymerization of PS but it did initiate polymerization of PMMA. However, no C_{60} units resulted attached to the polymer, which was taken as proof of an electron-transfer mechanism.

Scheme 2

"Living" star molecules with a C_{60} core and a well-defined number of arms bearing negative charges were prepared from addition of PS-Li to C_{60} in toluene. The living star molecules initiate anionic polymerization of monomeric styrene. Six carbanions are needed to initiate the anionic polymerization of styrene. Five carbanions are necessary for the polymerization of MMA. Heterostars containing six PS branches and two PMMA branches were also obtained.

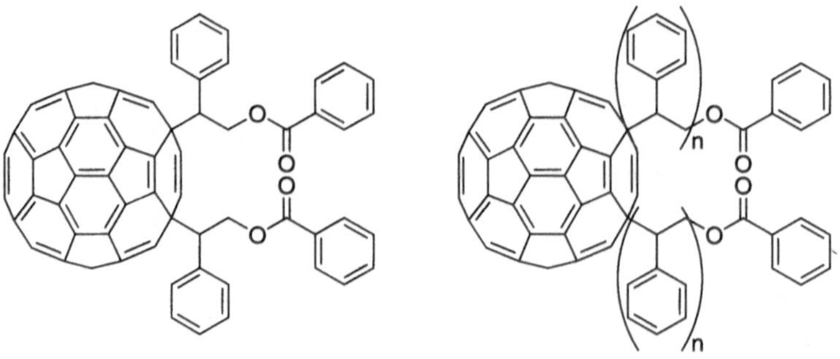

Scheme 3

The miscibility of different C_{60}-containing polystyrenes with other polymers was studied in attempts to transfer the fullerene properties to the resulting polymer blends [71]. It was found that poly(2,6-dimethyl-1,4-phenylene oxide), PPO, is miscible with all the PS/C_{60} samples investigated, whereas in the case of poly (vinyl methyl ether), PVME, PS/C_{60} samples were miscible only for lower contents of C_{60} [71].

Rotello and co-workers took advantage of the reversibility of the addition of cyclopentadiene to C_{60} for the temporary attachment of the fullerene to a modified Merrifield resin [72]. Addition of the cyclopentadiene-modified resin to C_{60} was achieved at room temperature, whereas the fullerene was released at 180 °C upon addition of maleic anhydride as a cyclopentadiene trap. Rotello and Nie proposed the methodology for a non-chromatographic purification of the fullerenes [73]. The same authors reported the preparation of thermoreversible polymer [74]. A bisanthracene derivative was added to C_{60} producing a polymeric substance with an estimated molecular range of 3000–25,000 or higher (Scheme 4). The addition is reversed at 75 °C, providing a thermoreversible material which, according to the authors, might be used for the creation of recyclable and thermally processible polymers.

Electrochemical polymerization of C_{60} derivatives is possible. Starting from a dialkynylated methanofullerene, Diederich et al. observed formation of

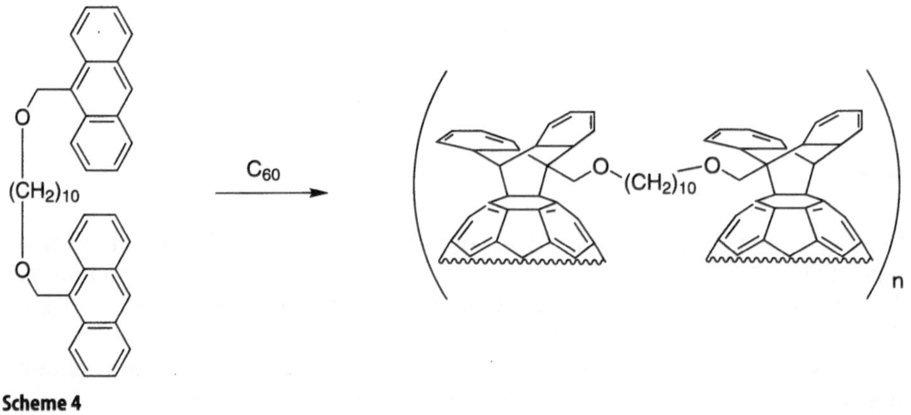

Scheme 4

an electrically conducting film on the surface of the platinum cathode [75]. A redox-active fullerene polymer with interesting mechanical and electrical properties was also obtained by electrochemically polymerizing the fullerene oxide $C_{60}O$ [76, 77]. A monomer unit having a cyclopentadithiophene moiety attached to C_{60} was electrochemically polymerized, leading to a conjugated polymer that contains C_{60} covalently attached [78]. Though some solubility problems arose, leading to a low polymerization degree, the authors reported that both components, namely the conjugated polymer and C_{60}, retain their original electrochemical properties, and that some new properties may be expected from their interactions.

2.2
Thin Films

Fullerene-based thin films are of high interest because of the possibility of transferring the interesting fullerene properties to bulk materials by surface coating. In this respect, self-assembled monolayers (SAM) [29, 79, 80] and Langmuir films are being increasingly used due to the interesting organized structures that can be obtained.

Self-assembled monolayers of C_{60} derivatives have been successfully obtained using thiol-functionalized fullerenes [29, 79–82]. In a recent report, Echegoyen and collaborators used a 1,10-phenanthroline-C_{60} (phen-C_{60}) adduct to form a self-assembled monolayer on Au(111) (Scheme 5) [83]. However, whereas phen-C_{60} alone did not form stable monolayers, very well ordered stacks were obtained using 1,10-phenanthroline as a template, within which the fullerene derivative intercalates very efficiently. The geometry of the packing seems to be controlled by the dimensions of the two partners, as the fullerene diameter is about 10 Å and the phen-phen distance is approximately 1/3 (3.3 Å).

Echegoyen et al. used molecular recognition to induce the formation of molecular monolayers of an 18-crown-6 functionalized fullerene [84]. A gold surface was modified using a thiol-terminated ammonium salt. When the modified gold layer was immersed into a CH_2Cl_2 solution of the fullerene, surface coverage was obtained which corresponds to a compact monolayer of C_{60}, as found by OSWV

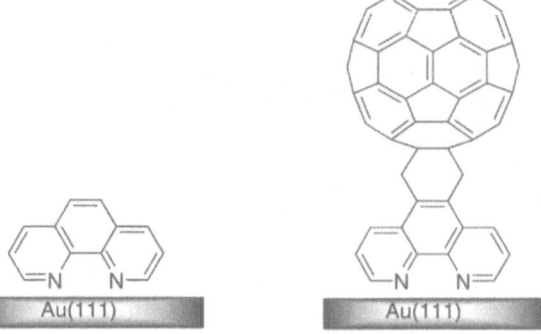

Scheme 5

measurements. The attachment of the fullerene to the ammonium salt, and thus to the gold surface, was demonstrated to be reversible in a CH_2Cl_2 solution.

A major problem encountered during the preparation of Langmuir films of fullerenes is related to the high hydrophobicity of the carbon cage campounds and to aggregation problems related to the strong intermolecular forces. Efforts have been made aimed at the preparation of fullerene derivatives which present a hydrophilic end [60, 85–98]. In these cases, monomolecular layers with areas per molecule of approximately 100 Å have often been obtained. Langmuir-Blodgett transfers to solid substrates, however, proved very difficult, and only a few successful cases have been reported.

The two fullerene-dendrimer conjugates (Scheme 6) constitute one of the rare examples of fullerene derivatives able to give stable monomolecular Langmuir films, which also do not show aggregation at the air-water interface. The monolayers were transferred onto quartz slides with relatively good transfer ratios for the up cycles [60].

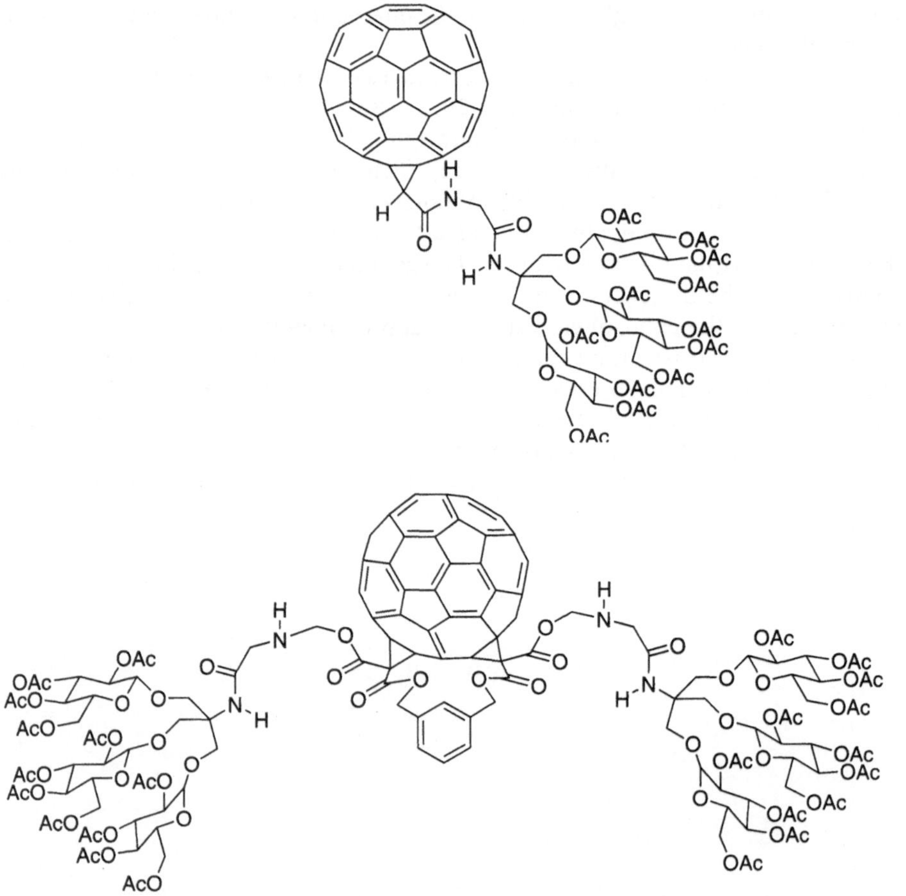

Scheme 6

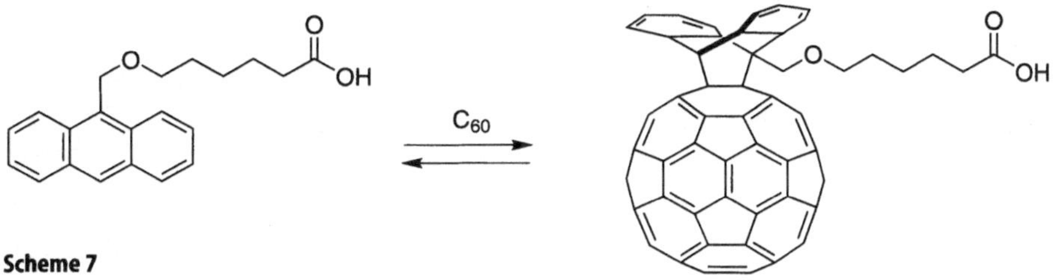

Scheme 7

In a clever attempt to form Langmuir monolayers of pure C_{60}, Metzger and collaborators prepared an amphiphilic derivative by a thermally reversible Diels-Alder reaction (Scheme 7) [99]. The temporary introduction of the carboxylic acid functionality guarantees an efficient anchoring of the fullerene spheroid to the water layer. The monolayers could be successfully transferred onto solid substrates to form LB monolayers and multilayers. Upon heating, the Diels-Alder adduct decomposes, giving back C_{60}. Thus, the final film consists of mixed layers of pure C_{60} and C_{60}/adduct.

Thin fullerene films for laser protection can be obtained by incorporation or covalent attachment of fullerenes to transparent solid matrices. The optical limiting (OL) properties of C_{60}, originally detected in toluene solutions [16], can be transferred to solid substrates without significant activity loss [100]. The solid substrates so far employed are organic polymeric [34, 100–102] and organic-inorganic hybrids [103–108]. These latter materials offer the advantage of better resistance against high power laser pulses, with very high damage thresholds, which makes them ideal for OL purposes. The best way to pursue the preparation of homogeneous thin glassy films of optical quality, with tunable amounts of fullerene for applications in the OL field, would then be to attach covalently the monofunctionalized fullerene to the silicon matrix [109, 110].

2.3
Electro-Optical Devices

Due to its rich electronic and electrochemical behavior, and to its versatile chemical reactivity, fullerene C_{60} has been considered as the ideal partner in photo-induced processes [28, 111, 112]. C_{60}, in fact, is a good electron-acceptor, and has a low reorganization energy [113]. For these reasons, an increasingly high number of donors have been covalently linked to C_{60}, for potential use as novel electronic materials and for applications in artificial photosynthesis. Many classes of donor units have been attached to C_{60}, including aromatics [12, 13, 114–118], porphyrins [11, 119–125] and phthalocyanins [126, 127], rotaxanes [128, 129], tetrathiafulvalenes [130–133], carotenes [125], Ru-bipy- [134, 135] and Ru-terpy- [135, 136] complexes, as well as ferrocene [130, 137].

A few triads containing C_{60} have been synthesized (Scheme 8) and studied in terms of electron and energy transfer [138–140]. The results have confirmed the ability of C_{60} to slow down back electron transfer in the charge separated states.

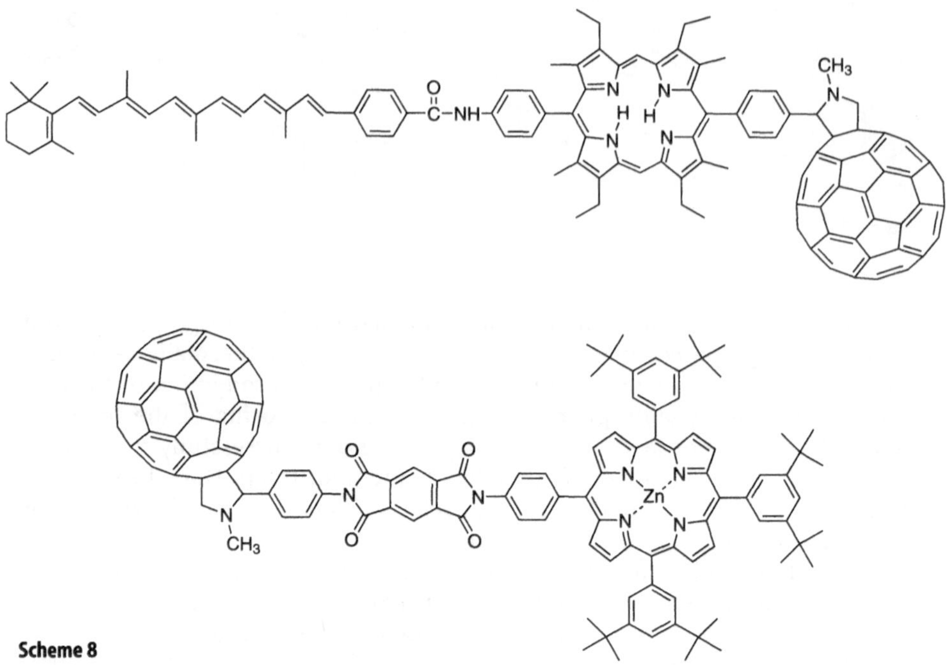

Scheme 8

The heterogeneous mixing of fullerenes and fullerene derivatives with π-conjugated polymer generates excellent materials for photovoltaic devices [141 – 143]. Upon irradiation of the fullerene/polymer blends, charge-transfer from the polymer to C_{60} occurs, resulting in efficient photoconductivities. Better behavior of the fullerene derivatives as compared to pristine C_{60} has been observed, and attributed to the better miscibility of the functionalized species [144 – 147].

Photo-induced changes in the complex index of refraction in conjugated polymer/fullerene blends have been reported [148].

2.4
Miscellaneous Uses

A thermotropic liquid crystal containing two cholesterol units attached to a C_{60} has been synthesized and its mesomorphic behavior was investigated [149]. A high melting point was obtained, and attributed to the fullerene moiety. A very interesting liquid crystalline material, containing a ferrocene unit attached to C_{60} has been prepared [150]. This compound is particularly appealing for its potential use in photo-induced processes. Other mesogenic groups were covalently linked to C_{60} [151].

Silicon alkoxide functionalized fullerenes can be profitably used in the preparation of HPLC stationary phases [152]. The silicon alkoxide group, in fact, guarantees the chemical grafting to silica (Scheme 9). Standard chromatographic tests with simple aromatic compounds revealed that the new phase works with high efficiency both in organic and water-rich media. Specific inter-

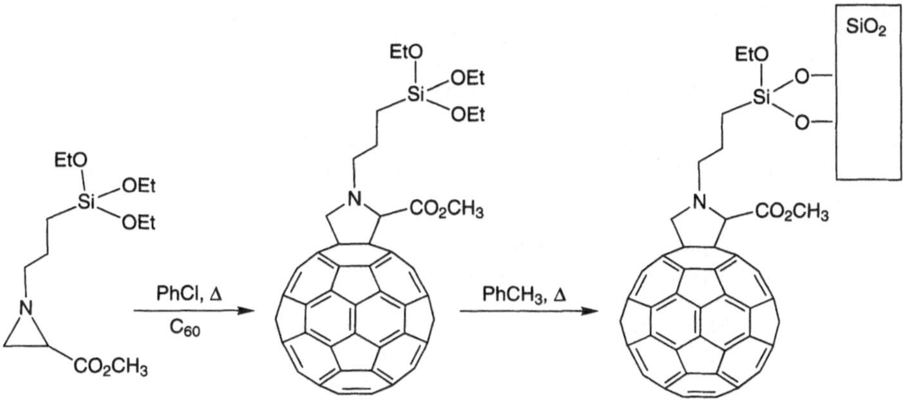

Scheme 9

actions towards complex solutes capable of establishing multipoint contacts with the spheroidal shape of the fullerene are expected. Accordingly, exceptionally high size-selectivities were obtained for cyclic oligomeric compounds like calixarenes and cyclodextrins in organic and water-rich media, respectively. A number of helical-shaped peptides, containing hydrophobic cavities complementary in size to C_{60}, bind selectively to the grafted fullerene.

3
References

1. Baum RM (1993) Chem Eng News 71:8
2. Bachmann PK, Messier R (1990) Chem Eng News 68:24
3. Hebard AF, Rosseinski MJ, Haddon RC, Murphy DW, Glarum SH, Palstra TTM, Ramirez AP, Kortan AR (1991) Nature 350:600
4. Holczer K, Klein O, Huang S-M, Kaner RB, Fu K-J, Whetten RL, Diederich F (1991) Science 252:1154
5. Haddon RC (1992) Acc Chem Res 25:127
6. Rosseinsky MJ (1995) J Mater Chem 5:1497
7. Xie Q, Pérez-Cordero E, Echegoyen L (1992) J Am Chem Soc 114:3978
8. Ohsawa Y, Saji T (1992) J Chem Soc, Chem Commun 781
9. Foote CS (1994) Top Curr Chem 169:347
10. Anderson JL, An Y-Z, Rubin Y, Foote CS (1994) J Am Chem Soc 116:9763
11. Liddell PA, Sumida JP, Macpherson AN, Noss L, Seely GR, Clark KN, Moore AL, Moore TA, Gust D (1994) Photochem Photobiol 60:537
12. Williams RM, Zwier JM, Verhoeven JW (1995) J Am Chem Soc 117:4093
13. Nakamura Y, Minowa T, Tobita S, Shizuka H, Nishimura J (1995) J Chem Soc, Perkin Trans 2:2351
14. Guldi DM, Hungerbühler H, Asmus K-D (1995) J Phys Chem 99:9380
15. Bensasson RV, Bienvenue E, Janot J-M, Leach S, Seta P, Schuster DI, Wilson SR, Zhao H (1995) Chem Phys Lett 245:566
16. Tutt LW, Kost A (1992) Nature 356:225
17. Kajzar F, Taliani C, Danieli R, Rossini S, Zamboni R (1994) Chem Phys Lett 217:418
18. Ruoff RS, Tse DS, Malhotra R, Lorents DC (1993) J Phys Chem 97:3379
19. Wudl F (1992) Acc Chem Res 25:157

20. Taylor R, Walton DRM (1993) Nature 363:685
21. Hirsch A (1993) Angew Chem, Int Ed Engl 32:1138
22. Hirsch A (1994) The chemistry of the fullerenes. Thieme, Stuttgart
23. Diederich F, Isaacs L, Philp D (1994) Chem Soc Rev 23:243
24. Hirsch A (1995) Synthesis 895
25. Taylor R (1995) The chemistry of fullerenes .World Scientific, Singapore
26. Smith AB (ed) (1996) Fullerene chemistry. Tetrahedron Symposia-in-Print Number 60, p 4925
27. Diederich F, Thilgen C (1996) Science 271:317
28. Prato M (1997) J Mater Chem 7:1097
29. Mirkin CA, Caldwell WB (1996) Tetrahedron 52:5113
30. Hirsch A (1993) Adv Mater 5:859
31. Sun Y-P (1997) In: Ramamurthy V, Schanze KS (eds) Molecular and supramolecular photochemistry, vol 1. Marcel Dekker, New York, p 325
32. Okamura H, Terauchi T, Minoda M, Fukuda T, Komatsu K (1997) Macromolecules 30:5279
33. Loy DA, Assink RA (1992) J Am Chem Soc 114:3977
34. Kojima Y, Matsuoka T, Takajashi H, Kurauchi T (1995) Macromolecules 28:8868
35. Nigam A, Shecharam T, Bharadwaj T, Giovanola J, Narang S, Malhotra R (1995) J Chem Soc, Chem Commun 1547
36. Samulski ET, DeSimone JM, Hunt MO, Menceloglu YZ, Jarnagin RC, York GA, Labat KB, Wang H (1992) Chem Mater 4:1153
37. Cao T, Webber SE (1995) Macromolecules 28:3741
38. Camp AG, Lary A, Ford WT (1995) Macromolecules 28:7959
39. Ma B, Lawson GE, Bunker CE, Kitayforodskiy A, Sun Y-P (1995) Chem Phys Lett 247:51
40. Olah GA, Bucsi I, Lambert C, Aniszfeld R, Trivedi NJ, Sensharma DK, Prakash GKS (1991) J Am Chem Soc 113:9387
41. Sun Y-P, Liu B, Lawson G (1997) Photochem Photobiol 66:301
42. Sun Y-P, Lawson G, Bunker C, Johnson R, Ma B, Farmer C, Riggs J, Kitaygorodskiy A (1996) Macromolecules 29:8441
43. Suzuki T, Li Q, Khemani KC, Wudl F, Almarsson Ö (1991) Science 254:1186
44. Boudon C, Gisselbrecht J-P, Gross M, Isaacs L, Anderson HL, Faust R, Diederich F (1995) Helv Chim Acta 78:1334
45. Hirsch A, Lamparth I, Grösser T, Karfunkel HR (1994) J Am Chem Soc 116:9385
46. Shi S, Li Q, Khemani KC, Wudl F (1992) J Am Chem Soc 114:10,656
47. Zhang N, Schricker SR, Wudl F, Prato M, Maggini M, Scorrano G (1995) Chem Mater 7:441
48. Geckeler KE, Hirsch A (1993) J Am Chem Soc 115:3850
49. Bergbreiter DE, Gray HN (1993) J Chem Soc, Chem Commun 645
50. Manolova N, Rashkov I, Beguin F, van Damme H (1993) J Chem Soc, Chem Commun 1725
51. Liu B, Bunker CE, Sun Y-P (1996) Chem Commun 1241
52. Hawker CJ (1994) Macromolecules 27:4836
53. Bunker CE, Lawson GE, Sun Y-P (1995) Macromolecules 28:3744
54. Weis C, Friedrich C, Mülhaupt R, Frey H (1995) Macromolecules 28:403
55. Dai L, Mau AWH, Griesser HJ, Spurling TH, White JW (1995) J Phys Chem 99:17,302
56. Wignall GD, Affholter KA, Bunick GJ, Hunt MO, Menceloglu YZ, DeSimone JM, Samulski ET (1995) Macromolecules 28:6000
57. Wooley KL, Hawker CJ, Fréchet JMJ, Wudl F, Srdanov G, Shi S, Li C, Kao M (1993) J Am Chem Soc 115:9836
58. Hawker CG, Wooley KL, Fréchet JM (1994) J Chem Soc, Chem Commun 925
59. Camps X, Schönberger H, Hirsch A (1997) Chem Eur J 3:561
60. Cardullo F, Diederich F, Echegoyen L, Habicher T, Jayaraman N, Leblanc R, Stoddart J, Wang S (1998) Langmuir 14:1955
61. Brettreich M, Hirsch A (1998) Tetrahedron Lett 39:2731
62. Chiang LY, Wang LY, Tseng S-M, Wu J-S, Hsieh K-H (1994) J Chem Soc, Chem Commun 2675

63. Chiang LY, Wang LY, Kuo C-S (1995) Macromolecules 28:7574
64. Rao AM, Zhou P, Wang K-A, Hager GT, Holden JM, Wang Y, Lee W-T, Bi X-X, Ecklund PC, Cornett DS, Duncan MA, Amster IJ (1993) Science 259:955
65. Stephens PW, Bortel G, Faigel G, Tegze M, Jánossy A, Pekker S, Oszlanyi G, Forró L (1994) Nature 370:636
66. Fischer JE (1994) Science 264:1548
67. Sun Y-P, Ma B, Bunker CE, Liu B (1995) J Am Chem Soc 117:12,705
68. Ford WT, Graham TD, Mourey TH (1997) Macromolecules 30:6422
69. Ederlé Y, Mathis C (1997) Macromolecules 30:2546
70. Ederlé Y, Mathis C (1997) Macromolecules 30:4262
71. Zheng J, Go SH, Lee SY (1997) Macromolecules 30:8069
72. Guhr KI, Greaves MD, Rotello VM (1994) J Am Chem Soc 116:5997
73. Nie B, Rotello VM (1996) J Org Chem 61:1870
74. Nie B, Rotello VM (1997) Macromolecules 39:3949
75. Anderson HL, Boudon C, Diederich F, Gisselbrecht J-P, Gross M, Seiler P (1994) Angew Chem, Int Ed Engl 33:1628
76. Fedurco M, Costa DA, Balch AL, Fawcett WR (1995) Angew Chem, Int Ed Engl 34:194
77. Winkler K, Costa DA, Balch AL, Fawcett WR (1995) J Phys Chem 99:17,431
78. Benincori T, Brenna E, Sannicolò F, Trimarco L, Zotti G, Sozzani P (1996) Angew Chem, Int Ed Engl 35:648
79. Chupa JA, Xu S, Fischetti RF, Strongin RM, McCauley JP, Smith AB, Blasie JK, Peticolas LJ, Bean JC (1993) J Am Chem Soc 115:4383
80. Shi X, Caldwell WB, Chen K, Mirkin CA (1994) J Am Chem Soc 116:11,598
81. Caldwell WB, Chen K, Mirkin CA, Babinec SJ (1993) Langmuir 9:1945
82. Akiyama T, Imahori H, Ajawakom A, Sakata Y (1996) Chem Lett 907
83. Domínguez O, Echegoyen L, Cunha F, Tao N (1998) Langmuir 14:821
84. Arias F, Godínez LA, Wilson SR, Kaifer AE, Echegoyen L (1996) J Am Chem Soc 118:6086
85. Maliszewskyi NC, Heiney PA, Jones DH, Strongin RM, Cichy MA, Smith AB III (1993) Langmuir 9:1439
86. Diederich F, Jonas U, Gramlich V, Herrmann A, Ringsdorf H, Thilgen C (1993) Helv Chim Acta 76:2445
87. Goldenberg LM, Williams G, Bryce MR, Monkman AP, Petty MC, Hirsch A, Soi A (1993) J Chem Soc, Chem Commun 1310
88. Williams G, Soi A, Hirsch A, Bryce MR, Petty MC (1993) Thin Solid Films 230:71
89. Hawker CJ, Saville PM, White JW (1994) J Org Chem 59:3503
90. Li Y, Xu Y, Mo Y, Bai F, Li Y, Wu Z, Han H, Zhu D (1994) Solid State Commun 92:185
91. Maggini M, Karlsson A, Pasimeni L, Scorrano G, Prato M, Valli L (1994) Tetrahedron Lett 35:2985
92. Maggini M, Pasimeni L, Prato M, Scorrano G, Valli L (1994) Langmuir 10:4164
93. Wang JY, Vaknin D, Uphaus RA, Kiaer K, Lösche M (1994) Thin Solid Films 242:40
94. Jonas U, Cardullo F, Belik P, Diederich F, Gügel A, Harth E, Herrmann A, Isaacs L, Müllen K, Ringsdorf H, Thilgen C, Uhlmann P, Vasella A, Waldraff CAA, Walter M (1995) Chem Eur J 1:243
95. Guldi DM, Tian Y, Fendler JH, Hungerbülher H, Asmus K-D (1995) J Phys Chem 99:17,673
96. Matsumoto M, Tachibana H, Azumi R, Tanaka M, Nakamura T, Yunome G, Abe M, Yamago S, Nakamura E (1995) Langmuir 11:660
97. Ravaine S, Le Peq F, Mingotaud C, Delhaes P, Hummelen JC, Wudl F, Patterson LK (1995) J Phys Chem 99:9551
98. Patel HM, Didymus JM, Wang KKW, Hirsch A, Skiebe A, Lamparth I, Mann S (1996) Chem Comm 611
99. Kawai T, Scheib S, Cava MP, Metzger RM (1997) Langmuir 13:5627
100. Kost A, Tutt L, Klein MB, Dougherty TK, Elias WE (1993) Opt Lett 18:334
101. Sun Y-P, Riggs J (1997) J Chem Soc, Faraday Trans 66:301

102. Tang B, Peng H, Leung S, Au C, Poon W, Chen H, Wu X, Fok M, Yu N, Hiraoka H, Song C, Fu J, Ge W, Wong G, Monde T, Nemoto F, Su K (1998) Macromolecules 31:103
103. Bentivegna F, Canva M, Georges P, Brun A, Chaput F, Malier L, Boilot J-P (1993) Appl Phys Lett 62:1721
104. Maggini M, Scorrano G, Prato M, Brusatin G, Innocenzi P, Guglielmi M, Renier A, Signorini R, Meneghetti M, Bozio R (1995) Adv Materials 7:404
105. Brunel M, Canva M, Brun A, Chaput F, Malier L, Boilot J-P (1995) In: Crane R, Lewis K, Stryland EV, Khoshnevisan M (eds) Materials for optical limiting materials. Research Society, Pittsburgh, p 281
106. Bozio R, Meneghetti M, Signorini R, Maggini M, Scorrano G, Prato M, Brusatin G, Guglielmi M (1996) In: Kajzar F, Agranovich VM, Lee CY-C (eds) Photoactive organic materials. Kluwer, Dordrecht, p 159
107. Signorini R, Zerbetto M, Meneghetti M, Bozio R, Maggini M, Scorrano G, Prato M, Brusatin G, Menegazzo E, Guglielmi M (1996) In: Kafafi ZH (ed) Fullerenes and photonics III. SPIE, Bellingham, p 130
108. McBranch DW, Klimov V, Smilowitz LB, Grigorova M, Robinson JM, Koskelo A, Mattes BR, Wang H, Wudl F (1996) In: Kafafi ZH (ed) Fullerenes and photonics III. SPIE, Bellingham, p 140
109. Signorini R, Zerbetto M, Meneghetti M, Bozio R, Maggini M, De Faveri C, Prato M, Scorrano G (1996) Chem Commun 1891
110. Kraus A, Schneider M, Gügel A, Müllen K (1997) J Mater Chem 7:763
111. Imahori H, Sakata Y (1997) Adv Mater 9:537
112. Martin N, Segura J, Seoane C (1997) J Mater Chem 7:1661
113. Imahori H, Hagiwara K, Aoki M, Akiyama T, Taniguchi S, Okada T, Shirakawa M, Sakata Y (1996) Chem Phys Lett 263:545
114. Khan SI, Oliver AM, Paddon-Row MN, Rubin Y (1993) J Am Chem Soc 115:4919
115. Lawson JM, Oliver AM, Rothenfluh DF, An Y-Z, Ellis GA, Ranasinghe MG, Khan MG, Franz AG, Ganapathi PS, Shephard MJ, Paddon-Row MN, Rubin Y (1996) J Org Chem 61:5032
116. Belik P, Gügel A, Kraus A, Walter M, Müllen K (1995) J Org Chem 60:3307
117. Matsubara Y, Tada H, Nagase S, Yoshida Z (1995) J Org Chem 60:5372
118. Williams RM, Koeberg M, Lawson JM, An Y-Z, Rubin Y, Paddon-Row MN, Verhoeven JW (1996) J Org Chem 61:5055
119. Drovetskaya T, Reed CA, Boyd P (1995) Tetrahedron Lett 36:7971
120. Imahori H, Hagiwara T, Akiyama T, Taniguchi S, Okada T, Sakata Y (1995) Chem Lett 265
121. Imahori H, Sakata Y (1996) Chem Lett 199
122. Ranasinghe MG, Oliver AM, Rothenfluh DF, Salek A, Paddon-Row MN (1996) Tetrahedron Lett 37:4797
123. Bell T, Smith T, Ghiggino K, Ranasinghe M, Shephard M, Paddon-Row M (1997) Chem Phys Lett 268:223
124. Kuciauskas D, Lin S, Seely GR, Moore AL, Moore TA, Gust D, Drovetskaya T, Reed CA, Boyd PDW (1996) J Phys Chem 100:15,926
125. Imahori H, Cardoso S, Tatman D, Lin S, Noss L, Seely G, Sereno L, Chessa de Silber J, Moore TA, Moore AL, Gust D (1995) Photochem Photobiol 62:1009
126. Linssen TG, Dürr K, Hirsch A, Hanack M (1995) J Chem Soc, Chem Commun 103
127. Durr K, Fiedler S, Linssen T, Hirsch A, Hanack M (1997) Chem Ber 130:1375
128. Diederich F, Dietrich-Buchecker C, Nierengarten J-F, Sauvage J-P (1995) J Chem Soc, Chem Commun 781
129. Armaroli N, Diederich F, Dietrich-Buchecker C, Flamigni L, Marconi G, Nierengarten J-F, Sauvage J-P (1998) Chem Eur J 4:406
130. Prato M, Maggini M, Giacometti C, Scorrano G, Sandonà G, Farnia G (1996) Tetrahedron 52:5221
131. Martin N, Sánchez L, Seoane C, Andreu R, Garín J, Orduna J (1996) Tetrahedron Lett 37:5979

132. Martin N, Pérez I, Sánchez L, Seoane C (1997) J Org Chem 62:5690
133. Llacay J, Mas M, Molins E, Veciana J, Powell D, Rovira C (1997) Chem Commun 659
134. Maggini M, Donò A, Scorrano G, Prato M (1995) J Chem Soc, Chem Commun 845
135. Armspach D, Constable EC, Diederich F, Housecroft CE, Nierengarten J-F (1998) Chem Eur J 4:723
136. Armspach D, Constable EC, Diederich F, Housecroft CE, Nierengarten J-F (1996) Chem Commun 2009
137. Maggini M, Karlsson A, Scorrano G, Sandonà G, Farnia G, Prato M (1994) J Chem Soc, Chem Commun 589
138. Liddell PA, Kuciauskas D, Sumida JP, Nash B, Nguyen D, Moore AL, Moore TA, Gust D (1997) J Am Chem Soc 119:1400
139. Imahori H, Yamada K, Hasegawa M, Taniguchi S, Okada T, Sakata Y (1997) Angew Chem, Int Ed Engl 36:2626
140. Carbonera D, Di Valentin M, Corvaja C, Agostini G, Giacometti G, Liddell PA, Kuciauskas D, Moore AL, Moore TA, Gust D (1998) J Am Chem Soc 120:4398
141. Wang Y (1992) Nature 356:585
142. Sariciftci NS, Smilowitz L, Heeger AJ, Wudl F (1992) Science 258:1474
143. Morita S, Zakhidov AA, Yoshino K (1992) Solid State Commun 82:249
144. Janssen RAJ, Christiaans MPT, Pakbaz K, Moses D, Hummelen JC, Sariciftci NS (1995) J Chem Phys 102:2628
145. Janssen RAJ, Hummelen JC, Lee K, Pakbaz K, Sariciftci NS, Heeger AJ, Wudl F (1995) J Chem Phys 103:788
146. Yu G, Gao J, Hummelen JC, Wudl F, Heeger AJ (1995) Science 270:1789
147. Kraabel B, Hummelen JC, Vacar D, Moses D, Sariciftci NS, Heeger AJ, Wudl F (1996) J Chem Phys 104:4267
148. Miller E, Lee K, Hasharoni K, Hummelen J, Wudl F, Heeger A (1998) J Chem Phys 108:1390
149. Chuard T, Deschenaux R (1996) Helv Chim Acta 79:736
150. Deschenaux R, Even M, Guillon D (1998) Chem Commun 537
151. Ravaine S, Vicentini F, Mauzac M, Delhaes P (1995) New J Chem 19:1
152. Bianco A, Gasparrini F, Maggini M, Misiti D, Polese A, Prato M, Scorrano G, Toniolo C, Villani C (1997) J Am Chem Soc 119:7550

Nanotubes: A Revolution in Materials Science and Electronics

Mauricio Terrones[1] · Wen Kuang Hsu[2] · Harold W. Kroto[3] · David R. M. Walton[4]

School of Chemistry Physics and Environmental Science, University of Sussex, Brighton BN1 9QJ, Great Britain. [1]E-mail: kapa4@sussex.ac.uk · [2]E-mail: kapa5@sussex.ac.uk · [3]E-mail: h.w.kroto@sussex.ac.uk · [4]E-mail: d.walton@sussex.ac.uk

Nanotube theoretical and experimental research has developed very rapidly over the last seven years, following the bulk production of C_{60} and structural identification of carbon nanotubes in soot deposits formed during plasma arc experiments. This review summarises achievements in nanotube technology, in particular various routes to carbon nanotubes and their remarkable mechanical and conducting properties. The creation of novel nanotubules, nanowires and nanorods containing other elements such as B, N, Si, O, Mo, S and W is also reviewed. These advances are paving the way to nanoscale technology and promise to provide a wide spectrum of applications.

Keywords: Nanotubes, Fullerenes, Nanofibres, BN, BC_2N, Nanowires, Encapsulates.

Topics in Current Chemistry, Vol. 199
© Springer Verlag Berlin Heidelberg 1999

1
Introduction

Fullerenes consist of closed (sp^2 hybridized) carbon networks, organised on the basis of 12 pentagons and any number of hexagons except one. This definition encompasses elongated fullerenes, known as nanotubes [1] (Fig. 1). In 1991, Iijima (NEC) reported the existence of such structures, consisting of concentric graphite tubes, produced in the Krätschmer/Huffman fullerene reactor [2]. A few months after Iijima's publication, Ebbesen and Ajayan described the bulk synthesis of nanotubes [3] formed as an inner core cathode deposit generated by arcing graphite electrodes in an inert atmosphere; a similar procedure to that used for fullerenes.

Nowadays, carbon nanotubes can be produced by diverse techniques such as arc discharge [4–8], pyrolysis of hydrocarbons over catalysts [9–11], laser vaporisation of graphite [12, 13], and by electrolysis of metal salts using graphite electrodes [14, 15]. The products exhibit various morphologies (e.g. straight, curled, hemitoroidal, branched, spiral, helix-shaped, etc.).

It was predicted from the outset that carbon nanotubes would behave as metallic, semiconducting or insulating nanowires depending upon their diameter and the hexagonal pattern along the direction of the tube axis (helicity) [16, 17]. Recent transport measurements on bulk nanotubes [18, 19], individual multi-layered tubes [20, 21] and single-walled tubules [22, 23] have revealed that their conducting properties depend markedly on the degree of graphitisation, helicity and diameter. Young's moduli measurements also show that multi-layered nanotubes are mechanically much stronger than conventional carbon fibres [24, 25] and are extraordinarily flexible when subjected to large strain [26].

Soon after carbon nanotubes had been prepared by the arc discharge techniques [3], the suggestion was made that it might be possible to introduce metals into the inner core of the nanotubes. Such a proposal followed logically from the

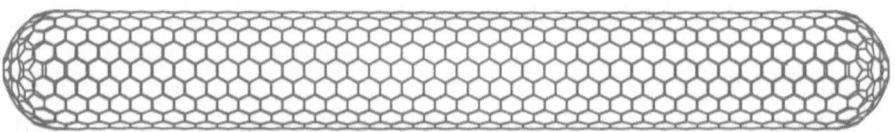

Fig. 1. Representation of a nanotube, which can be regarded as an elongated fullerene

successful preparation of fullerenes containing endohedral metals [27, 28]. Elements such as Pb, Bi, Cs, S and Se were first introduced by heating with nanotubes (opened by selective oxidation of the end caps) [29–31].

In addition to graphite, other layered materials such as BN [32], BC_2N [33] and BC_3 [34], were predicted to form fullerene-like structures. The synthesis of these materials soon followed and, recently, new families of cages (e.g. BN [35–37], BC_2N [38–42], BC_3 [40–42], MoS_2 [43], WS_2 [44] and VS_2 [45]) were made. It is noteworthy that related B/N/C structures were described in 1986 by Bartlett et al. [46, 47] as the outcome of thermolysis of various precursors (e.g. $C_2H_2 + BCl_3$). The structure of these materials, as well as possible growth mechanisms associated with their production, is discussed in the following sections.

At present, theoretical and experimental research continues to accelerate, and it is likely that the proportion of nanotube publications devoted to applications (broadly termed nano-engineering) will increase. Recent examples include: (i) gas storage of Ar [48] and H_2 [49]; (ii) STM probes [50] and field emission sources [51–54]; (iii) high power electrochemical capacitors [55–56]; and (iv) the production of nanorods (e.g. TiC, NbC, Fe_3C, GaN, SiC and BC_x) using carbon nanotubes as reacting templates [57, 58]. However, in order to employ such nanotubes or nanorods on a commercial base, it is necessary to control their growth, length, diameter and crystallinity.

2
Carbon Nanotubes

2.1
Structure of Carbon Nanotubes

In 1991, Iijima [2] found, by electron diffraction, that the interlayer spacing within multi-walled nanotubes ($d_{(002)} = 3.4$ Å) was slightly greater than that of graphite ($d_{(002)} = 3.35$ Å). This fact was ascribed to a combination of tubule curvature and van der Waals force interactions between successive graphene layers. The X-ray diffraction pattern is consistent with this spacing and provides information about crystallinity and structural dimensions in the bulk phase.

2.1.1
Straight Nanotubes

Carbon nanotubes consist of concentric hexagon-rich cylinders, made up of sp^2 hybridised carbon, as in graphite, and terminated by end-caps arising from the presence of 12 pentagons (six per end). It is possible to construct a cylinder by rolling up a hexagonal graphene sheet in different ways. Two of these are "non-helical" in the sense that the graphite lattices at the top and the bottom of the tube are parallel. These arrangements are termed "armchair" and "zig-zag". In the armchair structure, two C-C bonds on opposite sides of each hexagon are perpendicular to the tube axis, whereas in the zig-zag arrangement, these bonds are parallel to the tube axis (Fig. 2a, b). In all other conformations, the C-C bonds lie at an angle to the tube axis and a helical structure results (Fig. 2c, d).

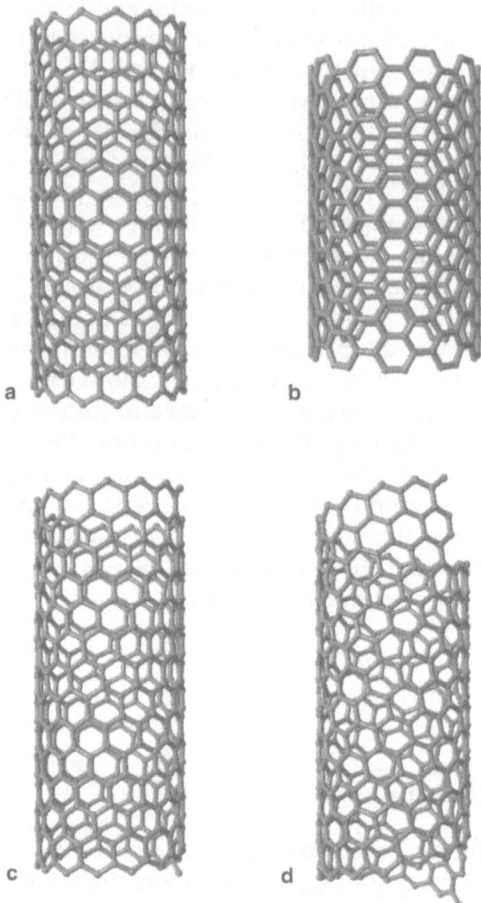

Fig. 2a–d. Molecular models of single-walled nanotubes with different helicities: **a** zig-zag arrangement; **b** armchair configuration; **c, d** two different helicities

Theoretical calculations indicate that the electronic properties for a single-walled nanotube will vary as a function of its diameter and helicity [59–61], and that the tube may behave as a semiconductor or metallic conductor, giving rise to the possibility of novel applications in solid state chemistry and to the development of nano-scale electronic devices.

2.1.2
Tubule Helicities and their Implications

In order to create a tubule, one first selects the origin *(0,0)* and then a lattice point in the graphene sheet given by *(m, n)* (see Fig. 3). By rolling up the sheet so that the chosen lattice point is superposed in the origin, the desired tubule is generated (Fig. 3).

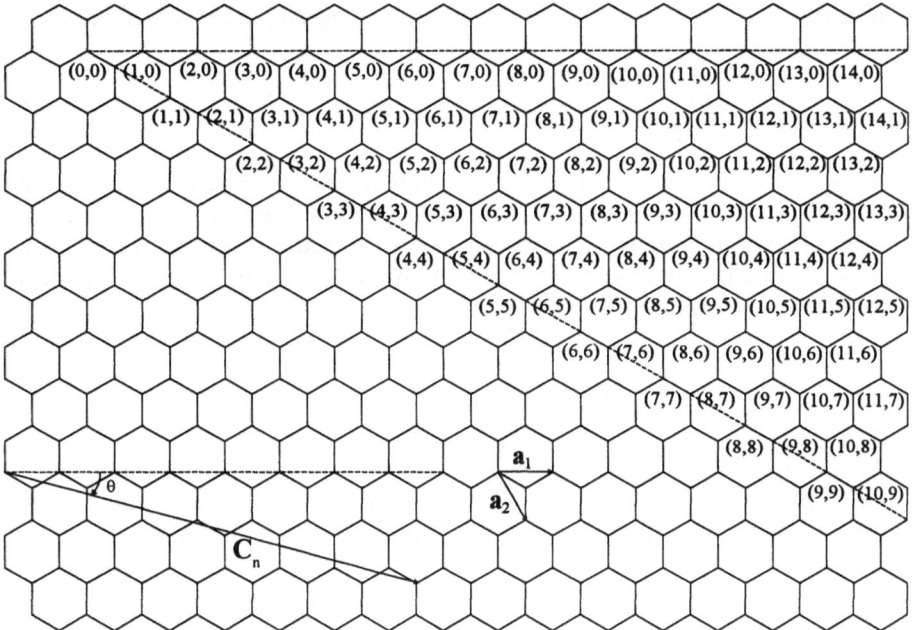

Fig. 3. Graphene sheet. There are many ways to roll it over and therefore, different types of tubules. This vector convention is used to define each point on the lattice. Unitary vectors a_1 and a_2 are necessary to determine the rolling direction expressed by vector C_n

The index (m,n) represents the lattice point which can be used to determine tube diameter and helicity. There are two ways of rolling the graphite sheet (from back to front or vice versa) resulting in chiral tubules.

The vector $C_n = m a_1 + n a_2$ determines the direction of rolling and therefore, the diameter can be expressed as

$$d = \frac{a\sqrt{m^2 + mn + n^2}}{\pi} \tag{1}$$

where $a = 1.42 \times \sqrt{3}$ Å corresponds to the lattice constant in the graphite sheet. Note that the C-C distance is 1.42 Å.

The generalised description of chiral tubules includes the range of orientations for C_n. To determine the armchair and zigzag structure in terms of (m,n) and the inclination angle θ, it is necessary to have the following conditions:

$$\theta = 0, (m, n) = (p, 0), \text{ where } p \text{ is integer (zigzag)} \tag{2}$$

$$\theta = \pm 30°, (m, n) = (2p, -p) \text{ or } (p, p) \text{ (armchair)} \tag{3}$$

The chiral angle θ (angle between C_n and the zigzag direction) in defined as

$$\theta = \arctan\left[-\frac{\sqrt{3}n}{2m+n}\right] \qquad (4)$$

Theoretical studies suggest that all armchair fibres are metallic, as are zigzag tubules when m is a multiple of three [60]. However, one may obtain a larger fraction of metallic tubes if the seed caps are centred about a pentagon, which yields an armchair arrangement. The metallic conductivity condition for these structures can be written

$$\frac{(2m+n)}{3} = \text{integer} \qquad (5)$$

If Eq. (5) is not satisfied, it is hypothesised that the tube may exhibit semiconducting properties [59–61]. Therefore, one third of these tubules may be one-dimensional metals and the other two thirds one-dimensional semiconductors.

2.1.3
Electron Diffraction of Nanotubes

Electron diffraction (ED) is a powerful tool for determining the fine structure of nanotubes (e.g. helicity, interlayer spacing, morphology, etc.). However, when dealing with multi-walled nanotubes, several helicities and complex configurations may arise due to various stacking and orientations within the graphene shells.

Liu and Cowley studied electron diffraction patterns generated by multi-walled nanotubes, and concluded that up to nine helix angles may appear, depending upon the number of walls [62]. They also noticed that the helix angle changes after three to five graphitic sheets, and that for most of the tubes these angles increase successively by regular increments. Furthermore, some of the tubular structures possess a polygonal (instead of a perfectly circular) cross section, reflected in the ED patterns.

2.1.4
Nanotube Caps

Nanotube end-caps provide a rationale for the growth and helicity of a tube. As noted already, six pentagons (regularly or randomly distributed) are necessary and sufficient to close one end of a nanotube. However, five pentagons are capable of closing a nanotube in a cone-like shape [63]. Figures 4 and 5 show different types of nanotube caps and their associated morphologies, depending upon the pentagonal arrangement within the cap. The strain introduced by the pentagons renders the caps chemically reactive. Confirmation of this is provided by oxidation, which takes place in air at ca. 750 °C [64] (see also Sect. 2.2.2).

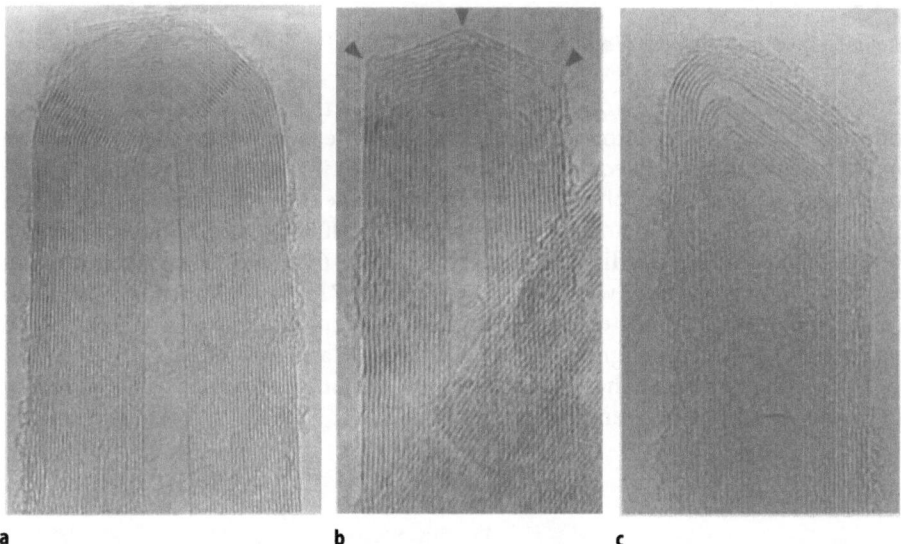

a					b					c

Fig. 4a–c. Various caps for carbon nanotubes: **a, b** hemispherical – *headed arrows* denote likely pentagon sites; **c** triangular cap (interlayer spacing ca. 0.34 nm)

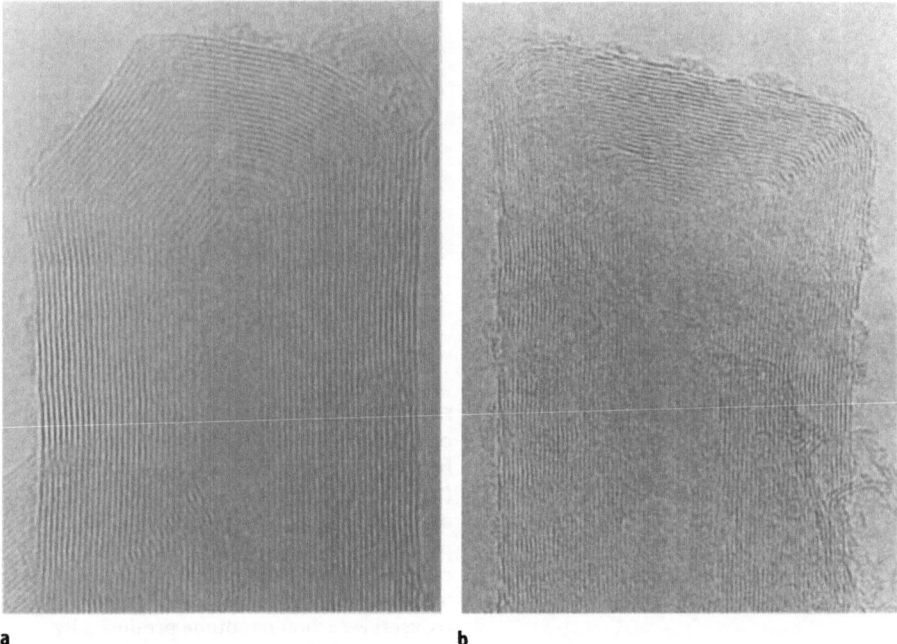

a					b

Fig. 5a,b. Unsymmetrical nanotube caps: **a** irregular cap; **b** square-like cap (interlayer spacing ca. 0.34 nm)

2.1.5
Helix-Shaped and Hemitoroidal Nanotubes

In 1991 Mackay and Terrones [65] proposed, for the first time, that the introduction of heptagons into the hexagonal graphite network would give rise to negative curvature. Subsequently, Iijima et al. [66] introduced this concept with respect to carbon nanotubes on the basis of one extra pentagon per heptagon (see Fig. 6a). Helicoidal and toroidal graphite with 5-, 6- and 7-membered carbon rings was also predicted theoretically [67–69] and, later, experimental evidence for its existence was obtained [4, 10, 70–72]. As a result, it is possible to "curl" and "twist" graphite by regular introduction of pentagon and heptagon pairs into the hexagonal graphite network (Figs. 6 and 7).

Helicoidal nanotubes have been observed among the products obtained by pyrolysis of acetylene over a cobalt catalyst [10, 11, 70, 73, 74]. During pyrolysis

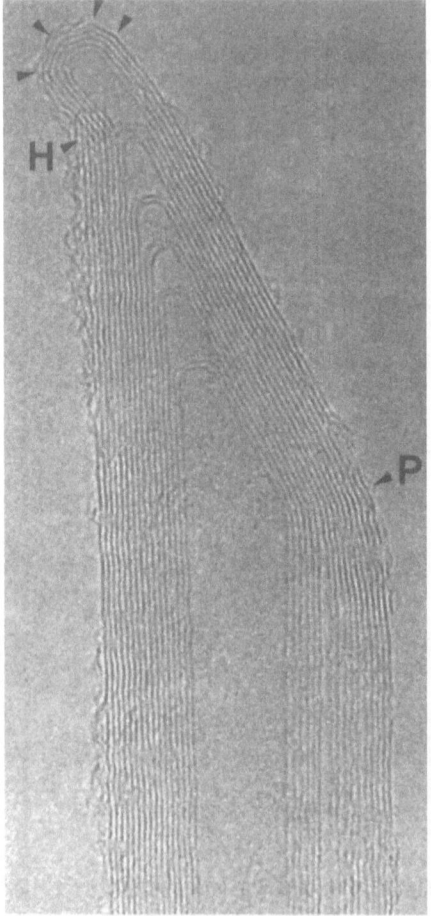

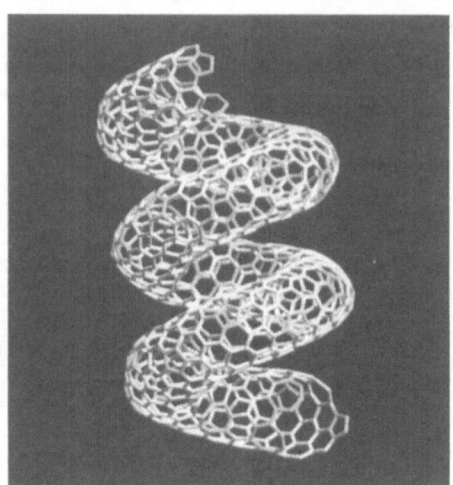

Fig. 6. a Nanotube cap exhibiting negative curvature due to the presence of a pentagon-heptagon pair; the 30°-pentagon declination (*right hand bend*) cancels the 30°-heptagon inclination (*left top bend*), ending with six pentagons at the hemispherical cap (interlayer spacing, ca. 0.34 nm). **b** Helcoidal graphite, simulated single-wall corkscrew carbon nanotube produced by interspersing five and seven-membered rings judiciously within the mainly hexagonal network

(see Sect. 2.3), the catalytic particle evidently plays an important role in nanotube growth, since the accretion of carbon atoms appears to occur on the metal surface. As a result, different rates of carbon agglomeration may arise within the nanotube walls and are responsible for the bent structures (e.g. spirals and helices; see Sect. 2.3). These helices are multi-walled and complex however, and the simple archetypal single wall helix (Fig. 6b) [69, 75] has yet to be reported.

Electron diffraction and HRTEM studies on these coiled nanotubes reveal the presence of successive offset 30° bends at regular intervals, which cause the tube to coil systematically [76]. Recent calculations on these archetypal helices suggest that they may possess metallic properties, determined by the distributions of five- and seven-membered rings [77, 78]. The synthesis of such "perfect" helicoidal graphite (Fig. 6b) and its use in nanoscale engineering and in electronics represents an exciting challenge for the future.

Hemitoroidal nanotube tips (axially-elongated concentric doughnuts; Fig. 7) have been produced by passage of an arc between graphite rods [4, 79] (Fig. 7b, c). In addition, the formation of hemi-toroidal-linked structures between adjacent concentric walls in pyrolytically produced carbon nanotubes has been observed and discussed in detail [63, 71]. These structures appear to form as a result of optimal graphitisation of two or more adjacent concentric graphene tubules.

The growth mechanism for these hemi-toroidal structures remains unclear. It is noteworthy that the arrangement for these "sock-like" (inwardly-folded; Fig. 7a) structures can be envisaged as cauterised rim-seals, which link the inner or the

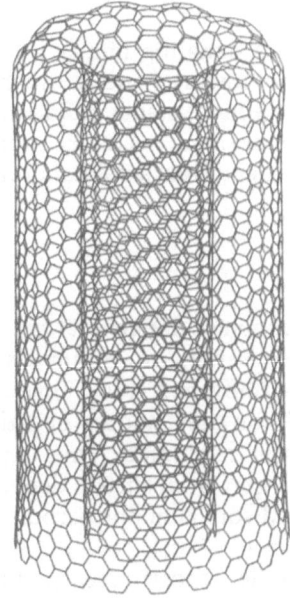

a

Fig. 7. a Molecular model of a hemi-toroidal nanotube cap, consisting of two concentric nanotubes joined together at their top rims

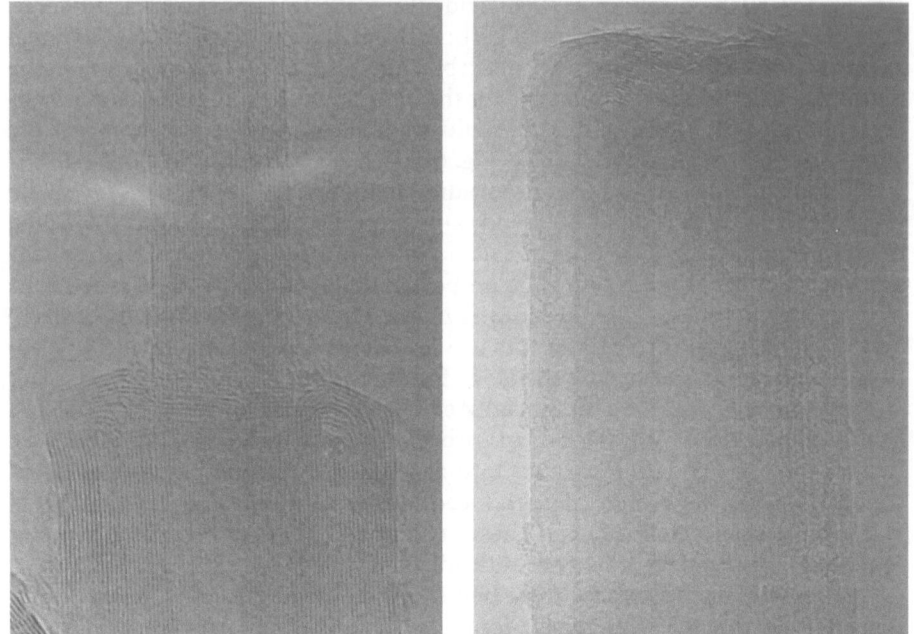

b   c

Fig 7. b, c Hemi-toroidal nanotube caps found in the cathode deposit after arcing hexagonal BN/graphite mixtures (interlayer spacing ca. 0.34 nm)

outer tubes, thereby avoiding dangling bonds [71]. These structures can be thought of as concentrically growing nanotubes which have been forced into U-turns in order to generate a more stable graphene configuration. Because of the insertion of non-hexagonal rings, electronic properties for these structures may well diverge from those of conventional nanotubes.

2.2
Arc Discharge Techniques

The optimum conditions for nanotube generation involve passage of a direct current (100 A) through two graphite (6–8 mm diameter) electrodes (separation ca. 1 mm) in an He atmosphere (500 Torr). During arcing, a deposit forms at a rate of 1 mm/min on the cathode (negative electrode), whereas the anode (positive electrode) is consumed. This deposit exhibits a cigar-like structure (Fig. 8), in which a grey hard shell is deposited on the periphery. The inner core, which is dark and soft, contains carbon nanotubes and polyhedral particles.

SEM studies on the cathode deposit reveal multiple features. Curved and compact graphite layers are deposited on the outer shell. Inner core deposits contain bundle-like structures, which contain randomly distributed nanotubes and polyhedral particles (Fig. 9). It is possible that, as a result of carbon vapour

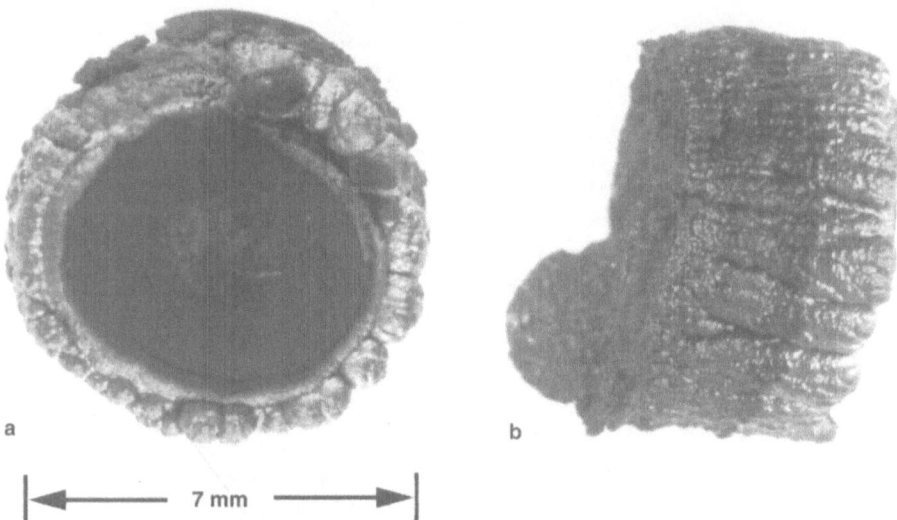

a

b

|←————— 7 mm —————→|

Fig. 8 a, b. Two views of a typical cathode deposit after arcing two graphite electrodes in an He atmosphere: **a** cross-section of the deposit exhibiting a cigar-like structure with a black inner core (containing nanotubes and nanoparticles) and a hard outer grey shell (graphite layered material); **b** side view of the same deposit only showing the grey outer shell

condensation generated by the arc and by temperature gradients (in the neighbourhood of the reaction zone) caused by the He gas, these two different micromorphologies (outer grey shell and inner core bundles) form within the deposit. On the nanometer scale, carbon ion aggregation in the high temperature zone may result in the formation of anisotropic structures (tubules) on the cathode. It is believed that an open-end mechanism (see below) may play a role in the generation of elongated structures. Polyhedral particles may arise from frustration associated with axial open-growth, possibly caused by arc instabilities such as current and localised abrupt pressure changes. When a thin film of Ni metal is deposited on the surface of the graphite anode, a more stable arc is produced and, consequently, the number of polyhedral particles is reduced considerably (by 50%); in addition, longer nanotubes are formed. This result is consistent with the mechanism outlined above.

2.2.1
Deposit Formation and Growth Mechanisms

Two main "atomistic" growth mechanisms have been proposed to account for the formation of individual carbon nanotubes. The first [4, 5] states that carbon atoms (e.g. C, C_2, C_3 units) attach to the edges of a growing carbon cylinder, which closes when conditions are not suitable for growth. An alternative scheme [1] proposes that nanotubes are essentially elongated giant fullerenes which grow by direct insertion of carbon species, accreted from the vapour phase, into the closed network.

Fig. 9 a–c. SEM image sequence of inner core deposit bundles: **a** low resolution of a bundle consisting of smaller aligned bundles; **b** higher magnification of the aligned bundles; **c** single bundle exhibiting randomly distributed nanotubes

It has recently been found that if hydrogen is used as a buffer gas in arc discharge experiments, the formation of C-H bonds at the point of tube growth inhibits nanotube closure [80]. Additionally, the observation that small diameter carbon tubules are contained within larger diameter tubules [81] provides evidence that the inner tube is denied access to a carbon source during growth. These two results support the open-ended mechanism for nanotube formation proposed by Iijima [5].

Bulk nanotube growth mechanisms have also been proposed [82–85], one of the latest being due to Gamaly and Ebbesen [84]. Their model deals with the carbon velocity distribution in the reaction zone near the cathode surface and contains two groups of carbon species, designated fast and slow. The fast carbons are thought to be anisotropic (one symmetry axis) and are held to be responsible for the nanotube length. The second (slow) group exhibits an isotropic carbon velocity distribution (Maxwellian), which is possibly responsible for nanotube width and, hence, multilayered nanostructures. This second group may be responsible for the formation of polyhedral nanoparticles, and may favour nanotube capping after arcing is discontinued.

Ajayan et al. analysed cathode deposits by SEM [85], and obtained results similar to those described in Sect. 2.2 above. The authors state that nucleation and growth of nanostructures can occur only in a high density carbon vapour, and that porous-like microstructures form as part of the inner core cathode deposit [85].

Ebbesen et al. found that aligned nanotubes could be grown within the inner core bundles as in a fractal configuration [86]. In most cases such nanotube alignment has never been observed [19, 85]. It is possible that, when cutting and/or manipulating bundles, unintentional alignment arises [18, 85]. It is important to bear in mind that the porous microstructure, which constitutes the cathode core, is a sign of random deposition [85]. To sum up, the growth of nanotubules in well-organised patterns during arc discharge seems unlikely.

2.2.2
Nanotube Purification

Carbon nanotubes produced by arc discharge techniques can be freed from polyhedral particles by differential oxidation [64; see Fig. 10]. This method suffers from the disadvantage that more than 95% of the original material is destroyed, and the remaining nanotubes lose their end caps [64]. As a result, further annealing at high temperatures (3000 K) is necessary in order to eliminate dangling bonds at the edges of the open tubules.

In the presence of air, the onset of nanotube weight loss occurs at ca. 700 °C (Fig. 11). The mass then decreases rapidly and, at ca. 860 °C, the nanotubes are completely oxidised to CO and CO_2. By comparison, C_{60} is less resistant to oxidation. Figure 11 compares the weight loss vs temperature for inner core deposits (containing nanotubes and polyhedral particles) and C_{60}. Thus nanotubes constitute one of the forms of carbon most resistant towards oxidation.

Fortunately, non-destructive methods for separating carbon nanotubes from polyhedral particles have been reported [87, 88]. These involve the use of well-

Fig. 10 a – c. TEM images: **a** bulk material containing nanotubes and polyhedral particles, 1:2 ratio; **b** oxidised sample (10.5 % of the original mass remaining), exhibiting a decrease in the number of polyhedral particles; **c** oxidised sample (1 % of the mass remaining), showing essentially pure nanotubes

Thermogravimetric Analysis

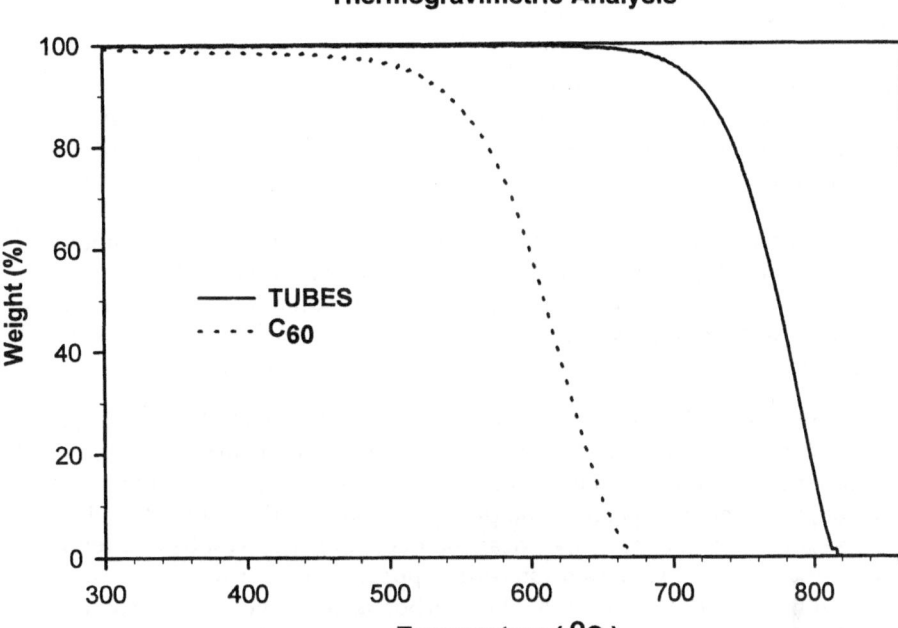

Fig. 11. Thermogravimetric (TGA) analysis of carbon nanotubes and C_{60}. The nanotubes are clearly more resistant than pure C_{60} towards oxidation

dispersed colloidal suspensions of tubes/particles, coupled with surface-active agents (surfactants, polymers or other colloidal particles), which prevent aggregation. Subsequent filtration of the suspensions, using porous filters [87] and size exclusion chromatography [88] yield size-graded almost pure nanotubes.

2.3
Hydrocarbon Pyrolysis

Pyrolysis of hydrocarbons (e.g. benzene, acetylene, naphthalene, ethylene, etc.) in the presence of catalysts (e.g. Co, Ni and Fe deposited on substrates such as silicon, graphite or silica) provides an additional route to fullerenes and carbon nanotubes. Prior to the discovery of fullerenes in 1985, pyrolytically grown nanofibres/nanotubes had actually been observed and structurally identified by several authors [74, 89–91]. Even at that early stage, a growth mechanism was postulated involving metal (catalyst) particles, which were held to be responsible for the agglomeration of carbon and subsequent axial growth of the fibre.

2.3.1
Filament Formation

The formation of carbon filaments was first described over a century ago by Schützenberger and Schützenberger [89], who produced microfilaments by passing cyanogen over porcelain at "cherry red" heat. In those days it was impossible to observe nanometer-scale structures, although it is highly likely that they may have been formed. It is now well recognised that pyrolysis of hydrocarbons and of carbon monoxide over catalysts (e. g. Co, Ni, Fe, Pt and Cu) at high temperatures ($> 600\,°C$) yields graphite filaments and nanotubes [92].

Several authors have proposed mechanisms to account for the formation of carbon filaments and nanotubes by pyrolysing hydrocarbons over catalysts. At present, the following mechanisms have been widely recognised.

1. Carbon diffusion "through" catalytic particles: Baker et al. [92–96] postulated that decomposition of acetylene on the exposed surfaces of the metal catalyst resulted in gaseous hydrogen and carbon. The carbon then diffuses through the catalytic particle and precipitates at the other end of the filament. The exothermic nature of hydrocarbon decomposition results in a temperature gradient across the catalyst, which appears to be a crucial factor associated with the model in question. Thus, carbon is precipitated at the colder zone of the particle, allowing the filament to grow (Fig. 12 c). This process continues until the catalytic activity of the leading particle is "neutralised" (Fig. 12 d).

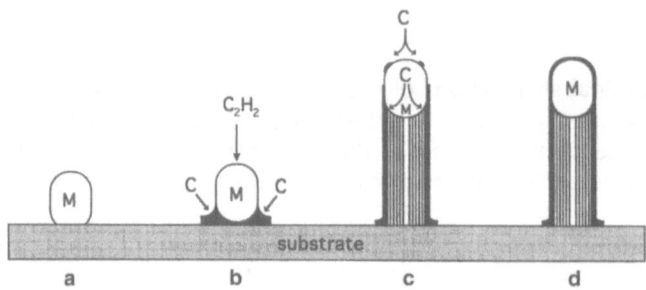

Fig. 12. Growth mechanism postulated by Baker et al. for the formation of carbon filaments by pyrolysis of acetylene (C_2H_2) on a metal particle (M); (C) denotes carbon. (From Baker et al.)

2. Carbon diffusion "on" catalytic particles: Baird and Fryer in 1974 [97] and Oberlin et al. in 1976 [98] postulated that carbon filaments are formed by surface diffusion on the particle (Fig. 13). This process should be easier if the catalyst particles are liquid [74].

Carbon nanotubes can be regarded as hollow carbon filaments, which may account for preferential carbon growing on selected faces of the catalytic particle [99] [i.e. the Fe particle is oriented with its (100) axis along the filament axis, whereas Fe/Co or Fe/Ni alloys prefer the (110) orientation].

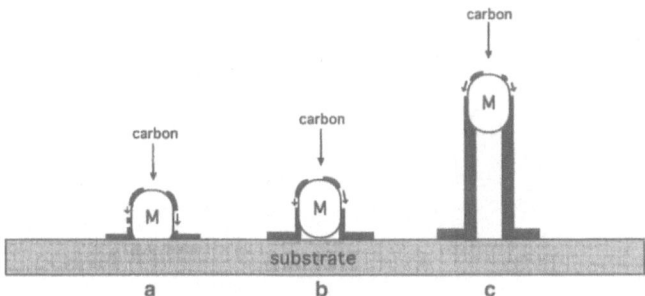

Fig. 13. Schematic illustration of the growth mechanism proposed for carbon fibres and filaments formed by pyrolysing benzene over catalytic particles. (From Oberlin et al.)

It is important to note that diffusion parameters can also depend critically on the dimensions of small particles. A recent model accounts for the formation of hollow filaments, based upon the precipitation of carbon on the catalytic particle. As a result, graphite planes grow parallel to the exterior fibre planes and a hollow core is formed [100]. However, other studies by Boellard et al. [101] on the filament tips where the metal particle is located suggested that the basal planes were positioned at an angle to the filament axis. In this case, the Ni particles were aligned with respect to its [112] axis, parallel to the filament axis (cone-like structures).

From the above argument, it is clear that there may not be a unique growth mechanism for the production of carbon filaments and nanotubes. Changing parameters, such as type of hydrocarbon, catalysts, particle size and temperature, leads to different filaments possessing various morphologies and degrees of graphitisation. However, any pyrolysis experiment in the presence of catalysts results in the formation of elongated structures. Therefore, deposition of "uniform catalyst particles" must play a vital role in preferential nanotube generation, thus controlling diameter and length.

2.3.2
Aligned Nanotubes

In 1994, Ajayan and co-workers generated aligned nanotube arrays by cutting thin slices (50–200 nm thick) of a nanotube-composite polymer [102]. The tubes appeared to be well separated as a result of mechanical deformation suffered by their being embedded in the polymer. However, the alignment strongly depended upon the thickness of the composite slice and the process becomes impractical for larger areas [102] (Fig. 14). Therefore, in order to generate large areas of aligned nanotubes, alternative bulk growth methods should be considered.

In this context, laser etching of cobalt thin films provides a novel route to catalysts which, in conjunction with the pyrolysis of organic precursors (e.g. 2-amino-4,6-dichloro-*S*-triazine, etc.) yields aligned nanotubes [103]. These tubes are of uniform length (≤ 100 μm) and diameter (30–50 nm o.d.) and grow perpendic-

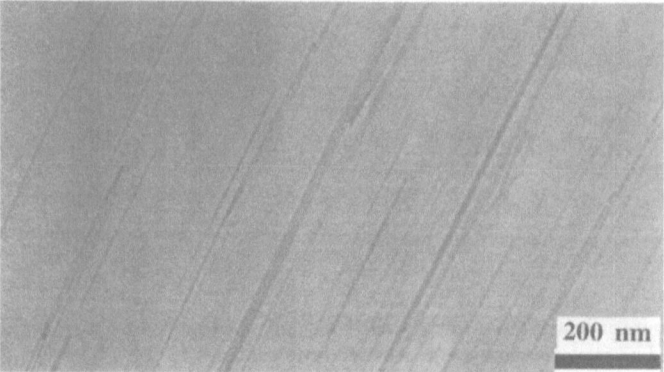

Fig. 14. TEM image of a nanotube-polymer slice (80 nm thick) exhibiting alignment of perfectly parallel multi-walled carbon nanotubes (Courtesy of P.M. Ajayan)

ularly to the catalytic substrate "only" in the etched regions. The orientation of the catalyst particles, ablated during etching, appears to be crucial, nanotubes with preferred helicity being favoured (e.g. armchair 25–30%). It is possible that matrices consisting of aligned nanotube bundles may be useful as novel mechanically strong composite materials and as ultra-fine field emission sources [18, 104–106] (Fig. 15).

Another recent report describes the large scale synthesis of aligned carbon nanotubes, of uniform length and diameter, by passage of acetylene over iron nanoparticles embedded in mesoporous silica [107]. The latter two methods, based on the pyrolysis of organic precursors over templated/catalysts supports, are by far superior by comparison with plasma arcs, since other graphitic structures such as polyhedral particles, encapsulated particles and amorphous carbon are notably absent (Fig. 16).

2.4
Laser Vaporisation

Laser vaporisation of graphite at high temperature (1200°C) in an inert atmosphere (Ar) generates multi-walled carbon nanotubes [12, 108]. The success of this technique was explained in terms of a gas-phase scheme, in which carbon atoms attach to the adjacent edges of the growing multi-walled graphite sheets, prolonging the life-time of the open structure, finally resulting in a multi-layered nanotube [12]. This mechanism states that a graphitic flake, with at least one pentagon, is first created from the carbon vapour formed by graphite vaporisation. Under specified high temperature conditions this flake may anneal, thus creating fullerenes, which may nucleate and form anions. If the carbon density is high enough, successive concentric graphite layers form around the template and multi-walled nanotubes are produced.

It is also possible that the tubes were formed by self-assembly caused by the carbon vapour interacting with the surface of the silica tube [108]. Additionally, condensation originating from temperature gradients around the hot area,

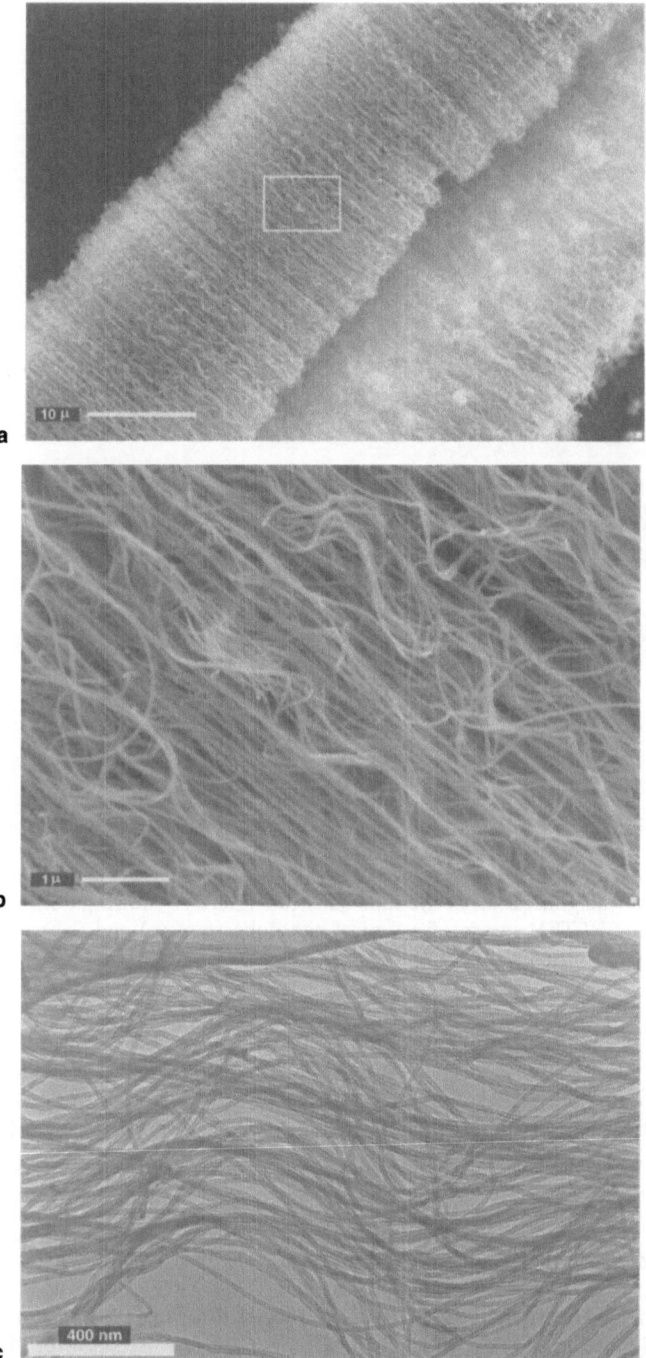

Fig. 15 a–c. SEM images of aligned nanotube bundles: **a** low magnification of adjacent bundles in which nanotubes appear fairly aligned; **b** higher magnification of one bundle showing aligned nanotubes [uniform length (20 μm) and diameters (30–50 nm)]; **c** TEM image of a typical region full of pure nanotubes

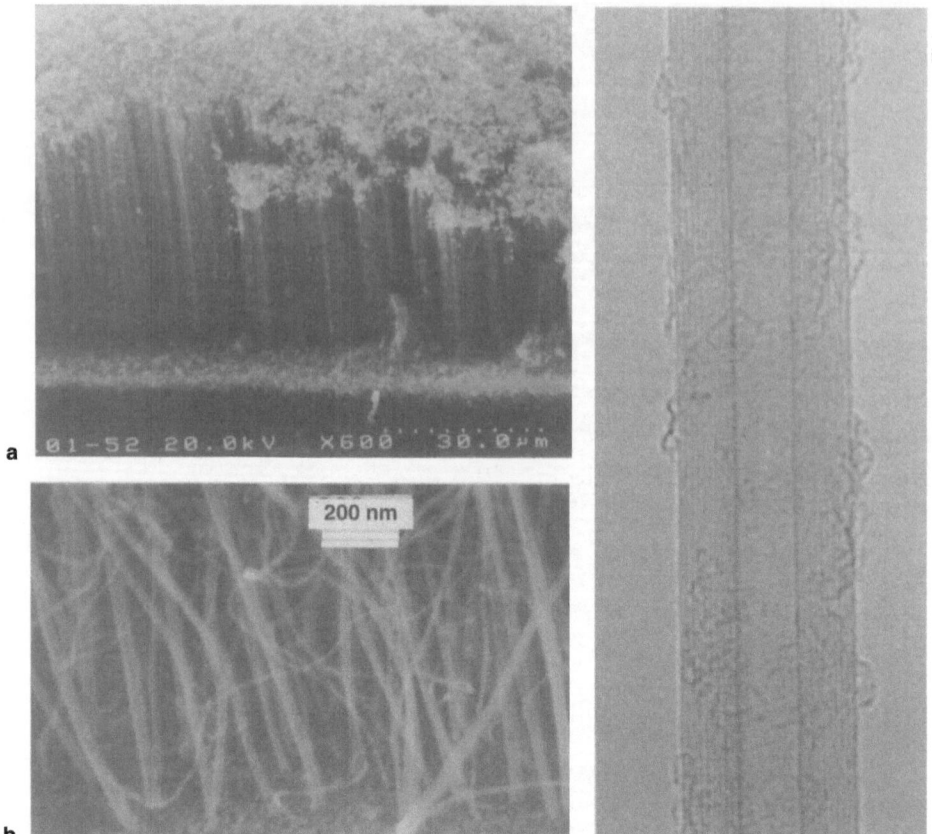

Fig. 16. a SEM image (thin film) exhibiting aligned nanotubes grown perpendicular to the substrate. **b** Higher magnification of the tube-substrate interface showing uniform diameter within the tubes. **c** HRTEM image of a typical nanotube from (a) and (b) with interlayer spacings of ca. 0.34 nm (images courtesy of S. S. Xie and W. Z. Li)

where a laser beam vaporises the graphite powder, may be crucial for nanotube growth and subsequent closure. In other words, carbon may start to agglomerate on the silica surface. Further aggregation (due to the high vapour pressure) may then result in the formation of elongated structures [108]. Round graphitic morphologies were also observed [108].

2.4.1
Crystalline Bundles of Single-Walled Nanotubes

If graphite/Co-Ni mixtures are used as a target during laser vaporisation at 1200 °C, bundles or ropes of single-walled nanotubes are produced [13]. These usually appear to contain mainly (10,10) nanotubes (armchair tubes 13.8 Å diameter; see section 2.1.2) packed in a crystalline form. X-ray powder diffraction reveals that these ropes or bundles consist of 100 – 500 single-walled tubes packed in a two dimensional triangular constant lattice of 17 Å [13]. A "scooter" growth mechanism was proposed to account for the generation of these ropes. Initially, Ni or Co chemisorbs to the open end of a graphene sheet, binding to its edge and catalysing rearrangements of carbon rings (generated in the gas phase), thus favouring the axial growth of the tubes [13]. It is not obvious why single atoms would scoot around the edge of these tubules. Therefore, it is more likely that Co and/or Ni clusters are responsible for the growth of such tubular structures. In this context, a novel growth mechanism, based on the absorption of Ni/Co atoms on the surface of C_{60} [109], was proposed. This model suggests that a metal-coated fullerene acts as a growing template and, once growth has been initiated, nanotube propagation may occur without the particle been involved. If the resulting nanotube is conical, rather than cylindrical, growth will cease rapidly, especially if the propagating walls converge.

Bundles of single-walled nanotubes can also be produced using arc discharge techniques in presence of Co or Ni [110–113] (Fig. 17). These nanotubes are to be found in the soot collected from the arc discharge generator, and it is believed that metal clusters are responsible for their production. Very recently, bundles of single-walled nanotubes were obtained in high yield by plasma arcs, using

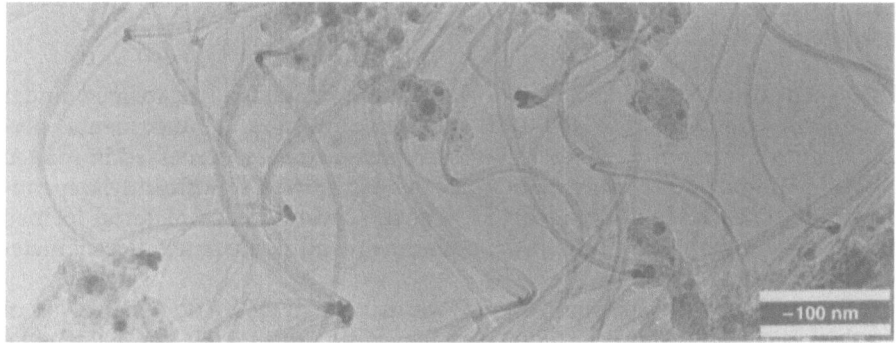

a

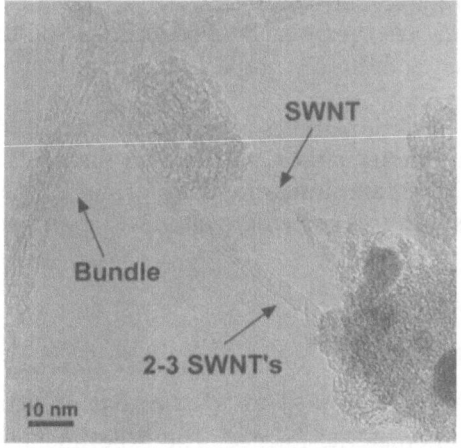

Fig. 17. **a** TEM image showing single-walled nanotube bundles, with encapsulated metal particles, produced by arc discharge techniques using Ni-Y/graphite mixtures; *scale bar* 100 nm (courtesy of P.M. Ajayan). **b** HRTEM image showing single walled nanotubes (SWNT's) and bundles. (Courtesy of J.P. Zhang)

b

graphite anode mixtures containing yttrium and nickel [114]. These bundles are deposited on a "collaret", formed around the cathode, generated after arc vaporisation. They constitute the trigonal lattice reported by Thess et al. [13]. Thus, arc discharge routes are now able to generate bundles of single-walled nanotubes in gram quantities per day.

A disadvantage, associated with the methods described above, arises from the presence of catalytic particles. However, recent reports have shown that separation of single-walled nanotube bundles from the catalytic particles can be achieved by chemical methods. For example, Tohji et al. [115] described a technique involving hydrothermal treatment of the material, followed by fullerene extraction, oxidation and metal particle dissolution. This process led to a 95% pure sample of separated single-walled nanotubes, accompanied by polyhedral particles and small concentrations of encapsulated metal particles. Almost simultaneously, Bandow et al. [116] reported a technique for separating single-walled nanotubes using cationic surfactant suspensions. The tubes were then trapped on a membrane filter without resorting to the use of acid, heat or oxidative treatment. Similar approaches, based upon size exclusion chromatography of single-walled nanotubes, have been performed successfully [117].

2.5
Electrolysis

The recent discovery that fullerene-related nanomaterials are formed under electrolytic conditions [14, 15] adds further technical and fundamental perspectives to the nanotube field. Using graphite electrodes immersed in molten LiCl under argon, nanotube production can be improved significantly. Depending upon the conditions, some 20–30% of the carbonaceous material formed may consist of graphitic nanotubes. Under other conditions quite different material is produced.

The condensed-phase apparatus [15] was used with LiCl. After electrolysis of graphite electrodes immersed in the molten ionic salt (e.g. LiCl at 600°C), the equipment was allowed to cool and carbonaceous materials were separated by dissolving the ionic salt in distilled water. After removal by filtration, the solids revealed the presence of carbon nanotubes and other graphite-like nanostructures [15].

TEM and HRTEM images of the electrolytic samples reveal the presence of nanotubes, encapsulated particles, amorphous carbon and carbon filaments. Nanotubes produced in this way (see Fig. 18) are similar to those formed in the carbon arc processes (i.e. multi-walled structures possessing interlayer spacings of ca. 0.34 nm; Fig. 18b).

2.5.1
Growth Mechanism for Condensed-Phase Processes

It is clear that the cathode immersion depth, as well as the current, play important roles in controlling the type of nanomaterial formed. The experimental

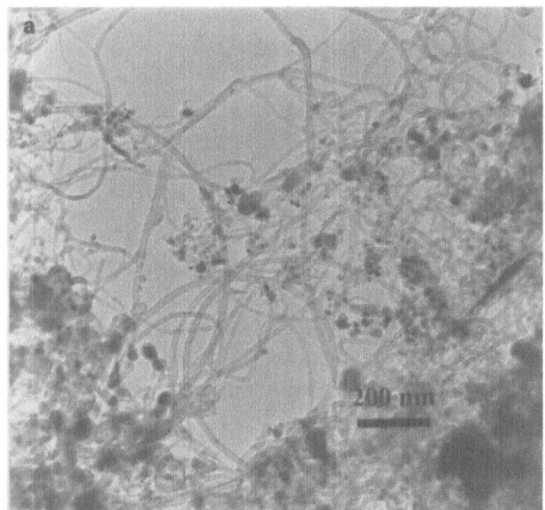

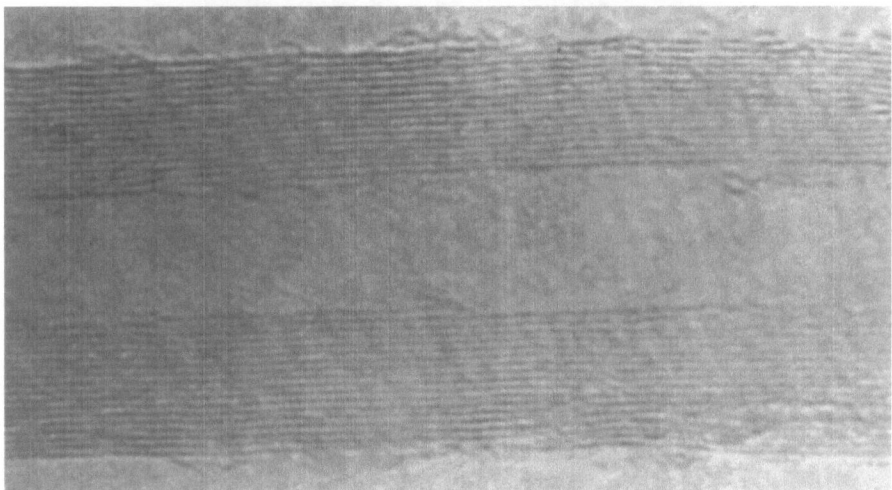

Fig. 18 a, b. TEM images: **a** nanotube sample generated by electrolysis; **b** a multi-wall nanotube showing D-spacing, ca. 0.34 nm

temperatures are much too low to disrupt completely the strong carbon bonds which make up the graphite sheets. Nevertheless, it is possible that some combination of the electrical potential and metal atom/cluster, or radical reactions (involving lithium metal atoms or ions) with the carbon electrode at the surface of the cathode may be involved in network fabrication, leading to nanomaterials. In other words, molten droplets of Li may form by localised electrolysis at prominent points on the carbon cathode surface. These droplets may dissolve carbon, which is subsequently transformed into nanotubes. As a consequence, these nanotubes, may concentrate the electric field of the cathode, serving as an enhanced site for electrolysis of molten LiCl.

3
Mechanical and Conducting Properties

Direct HRTEM observations reveal that nanotubes are remarkably flexible (see Fig. 19) [102, 118–120]. For instance, they can be bent mechanically by sonication, grinding [108, 119] or by being embedded in a polymer resin [102]. Theoretical calculations predicted these properties [121–124], noting that the tubes would soften with decreasing radius. It was also predicted that the tube stiffness would be dependent upon chirality (e.g. zig-zag tubes would be less stiff than the armchair tubes) [122].

The first attempt to determine Young's modulus for individual multi-walled carbon nanotubes was made by Treacy et al. [24] who measured, in the transmission electron microscope, the amplitudes of thermal vibrations in single

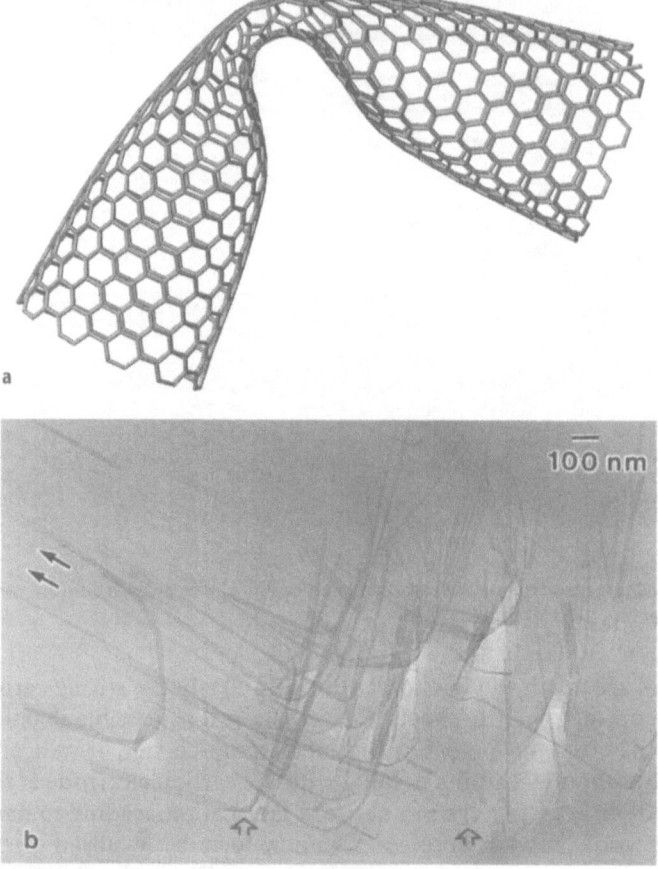

Fig. 19. a Molecular model of a collapsed nanotube showing bending within the hexagonal network. **b** Aligned and deformed tubes embedded in a polymer resin [102]; *arrows* indicate the nature of bending within the tubes; the larger tube has been twisted but the smaller one shows no such defect. (Courtesy of P. M. Ajayan)

nanotubes. The results showed that the nanotubes possess an average Young's modulus of $Y = 1.8$ TPa, which is much higher than that of commercially available carbon fibres ($Y \sim 800$ GPa).

Recently, direct measurements of the bending force as a function of displacement, inside an Atomic Force Microscope (AFM), on individual carbon nanotubes revealed an average Young's modulus of ca. 1.28 TPa, independent of tube diameter [25]. In this context, Falvo et al. [26] reported that carbon nanotubes can be repeatedly bent through large angles using the tip of an AFM, without undergoing catastrophic failure [26]. The latter measurements confirm that carbon nanotubes are extraordinarily flexible under large strain, which makes them good candidates for super-strong carbon-fibre-reinforced materials.

The first transport measurements on single bundles of carbon nanotubes, using two gold contacts attached by lithographic techniques, revealed a temperature-dependent resistance, which fitted well (2–300 K) a simple two-band semi-metal model [125]. Subsequent measurements on carbon nanotubes lying between coplanar electrodes showed the presence of thermal activated electrical behaviour with band gaps ranging from 6 to 260 meV [126]. More recent conductivity measurements on "aligned" bundles of carbon nanotubes [18, 19, 127, 128] showed that the material behaved as conducting rods and exhibited anisotropy in respect of transport properties for different alignment configurations [19, 126]. For instance, measurements on a single bundle-like particle taken from the inner core cathode deposits ($0.5 \times 0.6 \times 1.1$ mm^3) and consisting of smaller bundles composed of randomly distributed nanotubes (Fig. 9) revealed a nearly temperature-independent conductivity along the bundle axis. However, further microwave contactless conductivity measurements showed that inner core cathode deposits (nanotubes/nanoparticles generated by arc discharge methods) exhibited thermal activated behaviour with activation energies of ca. 70 meV [19] (somewhat smaller than those reported for individual nanotubes) [20]. Due to the variation in transport response among different bulk samples, it was concluded that the best way to determine the conductivity mechanism of carbon nanotubes would be direct four-probe measurements on well graphitised single tubes, thus avoiding contact problems and inhomogeneities within the nanostructures contained in the measured samples.

Single multi-layered nanotube conductivity measurements (Fig. 20) revealed that each nanotube exhibits unique conductivity properties [20, 21], leading to both metallic and non-metallic behaviour (resistivities at 300 K of ca. 1.2×10^{-4} to 5.1×10^{-6} ohm cm; activation energies < 300 meV for semiconducting tubes) [20]. This suggests that the geometric differences (e.g. bends, helicity, diameter, etc.) and graphitisation of the structures play important roles in the electronic behaviour. In addition, bundles of single-walled nanotubes have been shown to behave as metals with resistivities in the 0.34×10^{-4} to 1.0×10^{-4} ohm cm range [13].

The most recent achievement in this respect, involves transport measurements on a single-walled nanotube (1 nm in diameter; Fig. 21), showing that it behaves as a quantum wire, in which electrical conduction seems to occur via well-separated, discrete electron states that are quantum-mechanically coherent over long distances [22].

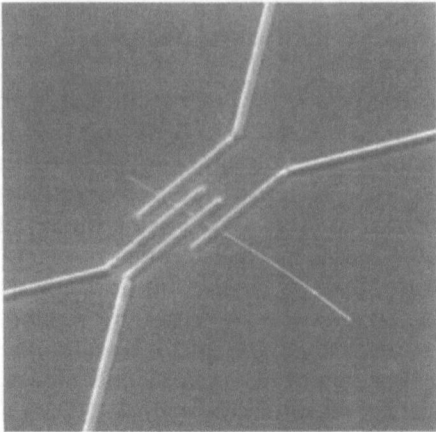

Fig. 20. Multi-walled nanotube connected to four tungsten wires (diam. 80 nm). (Courtesy of T. W. Ebbesen)

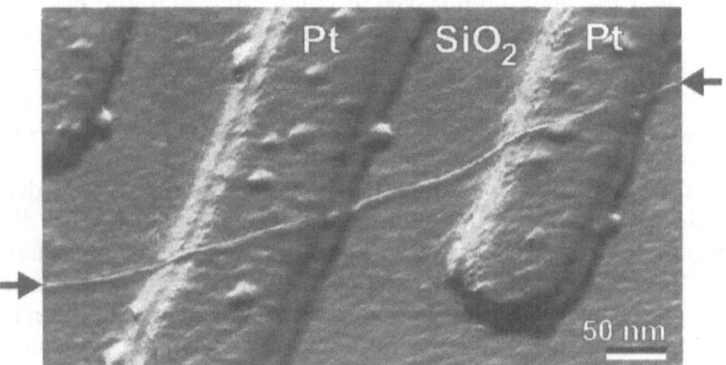

Fig. 21. Single-walled nanotube on nanofabricated electrodes. (Courtesy of C. Dekker)

Using a scanning tunnelling microscope (STM), Wildöer et al. [22] and Odom et al. [23] imaged single-walled carbon nanotubes generated by laser vaporisation techniques [13] with atomic resolution, and observed a chiral winding pattern of hexagons along the tube (Fig. 22). Further local density of states measurements on different tubes revealed for the first time clear experimental verification of the theoretical band structure predictions [22, 25] (see Fig. 22). Basically, these results show that nanotubes can be either metals or semiconductors depending on small variations in the chiral angle or diameter.

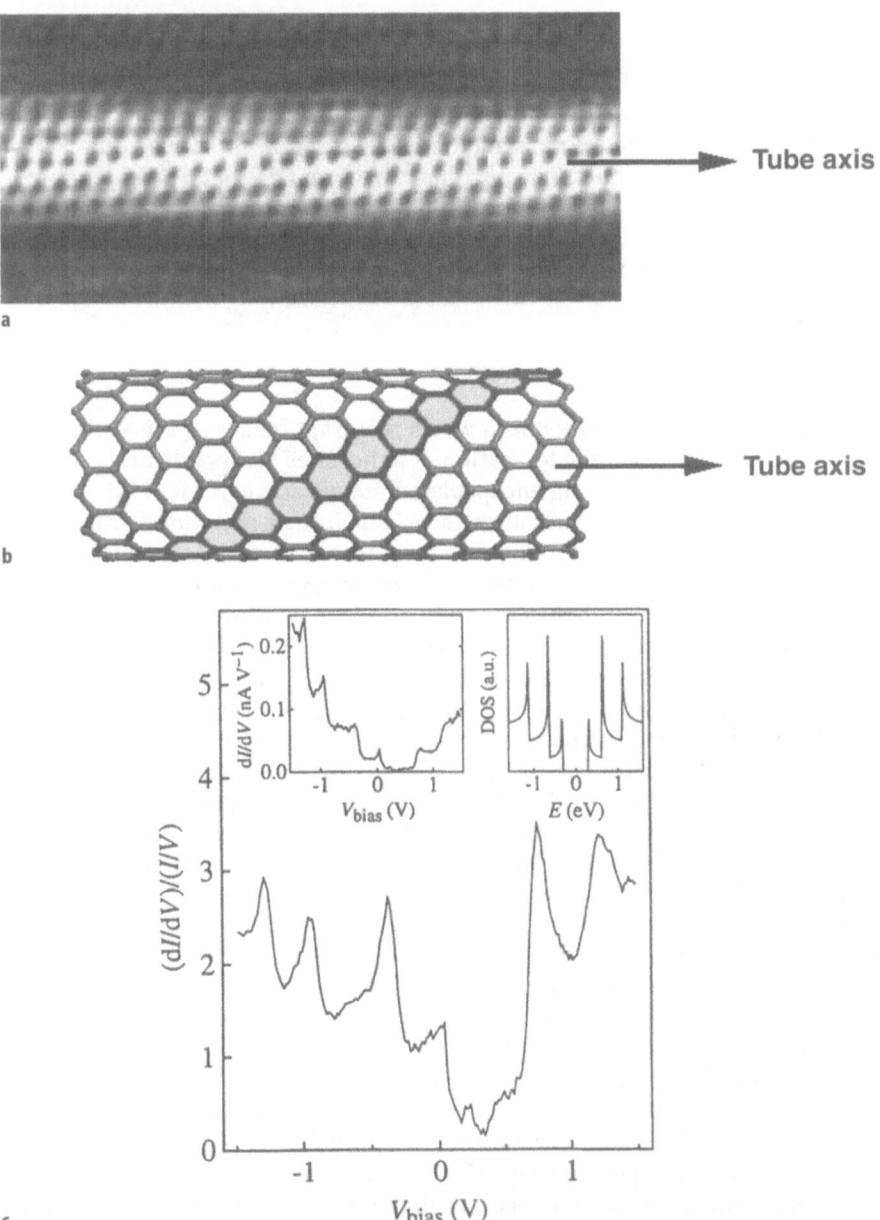

Fig. 22. a Atomically resolved STM image of an individual single-walled carbon nanotube. The lattice on the cylinder surface allows clear identification of tube chirality by measuring the angle between the tube axis and the hexagon rows (*black dots*) (Courtesy of C. Dekker). **b** Molecular model of a chiral single-walled nanotube highlighting a hexagonal row (represented by the *black dots* in a). **c** (dI/dV)/(I/V) the measure of the density of states (DOS) vs. V for a semiconducting nanotube. The *left inset* displays the raw dI/dV data, while the *right inset* displays the calculated DOS for a (16,0) semiconducting tube. The overall shape of the experimental peaks resembles that predicted by theory. (Courtesy of C. Dekker)

4
Encapsulates

Encapsulation by carbon nanotubes and the fabrication of nanocomposite materials using nanotubes as reacting templates can now be carried out using a range of disparate techniques. The introduction of metals, metal carbides or oxides into nanotubes may alter significantly their conducting and electronic properties caused by the internal framework within the structures. The following section discusses methods for generating filled nanotubes.

4.1
Metal Nanowires and Encapsulated Particles

Typical arc discharge experiments, in which the graphite anode was drilled and packed with a metal, led to the formation of nanotubes and polyhedral particles at the cathode, both containing encapsulated metals and carbides [129–134]. A problem arises from the arc discharge technique in that nanotubes are thereby only partly filled. This fact may be disadvantageous if nanoscale applications (e.g. conduction) are to be considered. Recently, Loiseau and Pascard found optimal conditions for more-or-less complete encapsulation of Se, S, Sb and Ge using arc discharge techniques [135]. It was concluded from allied experiments [136] that elements with incomplete electronic shells stood the best chance of being encapsulated. However, the growth mechanism associated with these filled nanotubes (termed nanowires) is not well understood. At present it has proved possible to encapsulate by plasma arcs LaC_2 [137], YC_2 [130, 138], LaC_2 [139], CeC_2 [134, 139], Gd_2C_3 [136, 140], TiC [136, 141], V_4C_3 [142], ZrC [142], TaC [19, 134, 137], MoC [19, 143], NbC [19], HfC [144], Fe_3C [136], Ni_3C [136] and other carbides of Cr [136], Dy [136] and Yb [136]. Only a few metals such as Mn, Co and Cu, and semiconducting elements like Se, Ge and Sb have been inserted into nanotubes without carbide formation taking place.

Chemical techniques have also been used to prepare nanotube-encapsulated materials. In this context, carbon nanotubes were oxidatively opened by boiling aqueous nitric acid, then filled with oxides of Ni [145], Co [145], Fe [145], U [145], Mo [146], Sn [147, 148], Nd [149], Sm [149], Eu [149], La [149], Ce [149], Pr [149], Y [149], Zr [149], Cd [149], pure metals such as Pd [150], Ag [151], Au [151], proteins or enzymes (e.g. Zn2Cd5-metallothionein, cytochrome c(3), β-lactamase I, and DNA complexes) [152, 153], and Au_2S_3 [149], AuCl [151], CdS [149] by wetting techniques (see Fig. 23; also [148, 149]). In addition, other capillarity wetting methods, such as heating the elements with open-ended nanotubes, have led to the introduction of the elements Pb, Bi, Cs, S and Se into nanotubes [29–31]. From these studies, it was concluded that only low surface tension substances can be introduced into nanotubes.

Carbon nanotubes also react with metal oxides above 1200 °C to yield carbide nanorods [57, 58] (i.e. filaments consisting of pure metal carbides which are not encapsulated, i.e. TiC, NbC, Fe_3C, GaN, SiC and BC_x). Ajayan and co-workers reported that it is also possible to coat carbon nanotubes with a single layer of metal oxide (e.g. V_2O_5) in addition to inserting the oxides into the core [154].

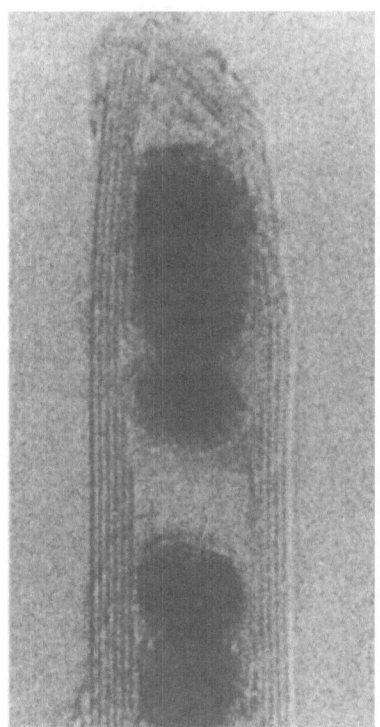

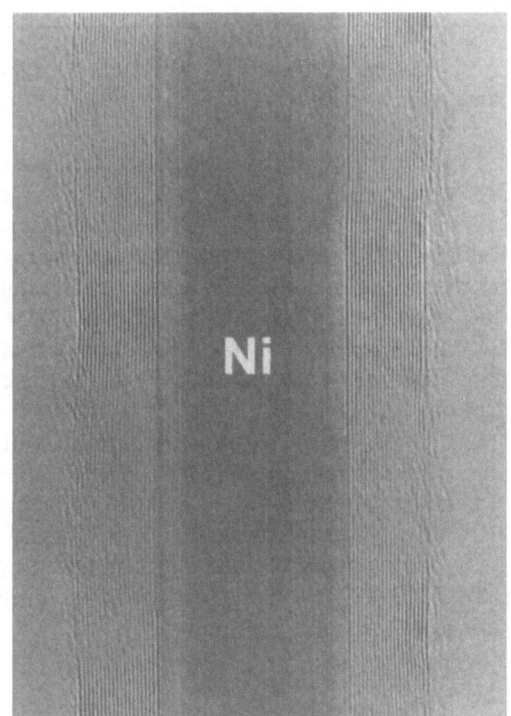

Ni

Fig. 23. a UO$_{2-x}$ filled nanotube generated by oxidative opening of carbon nanotubes in boiling nitric acid, then filled with metal oxide, interlayer spacing, ca. 0.34 nm (Courtesy of S. C. Tsang and P. J. F. Harris [145]). **b** Ni-filled nanotube grown during pyrolysis of solid organic precursors (interlayer spacing, ca. 0.34 nm; courtesy of N. Grobert and H. Terrones). **c** PbO filled nanotube generated by capillarity techniques (interlayer spacing, ca. 0.34 nm; courtesy of P. M. Ajayan)

These authors produced so-called nanocomposites by annealing partly oxidised nanotubes with oxide mixtures in air above the melting point of the corresponding oxide [154].

In the case of encapsulated ferromagnetic nanoparticles (e.g. Fe, Ni and Co), it may prove possible to develop magnetic data storage devices, toners and inks for xerography [155]. Pyrolysis of solid organic precursors [156, 157] in the presence of Co, Ni and Fe leads to highly graphitic nanotubes containing these metals (Fig. 23). Refractory metal carbides such as TaC, NbC and MoC exhibit superconducting transitions at low temperatures (10–14 K) [19, 134, 143]. Therefore, carbide nanorods or nanotubes containing encapsulated metal carbides may be useful in constructing superconducting nanoscale circuits.

Condensed-phase processes offer alternative routes to nanotubes containing encapsulated metals [14, 15, 158, 159]. For example, the passage of an electric current through graphite electrodes immersed in molten mixtures of LiCl and $SnCl_2$ under argon has been shown to yield carbon-coated Sn nanowires [19, 158], which behave as metals [19]. Interestingly, low melting point metals such as Sn, Bi and Pb generate nanowires by this method when using metal+LiCl mixtures as electrolytes [159] (Fig. 24).

4.2
Non-Carbon Nanorods

The synthesis of crystalline Si and Ge semiconducting nanowires [160, 161] may lead to improved electrical and optical devices, although this goal is proving

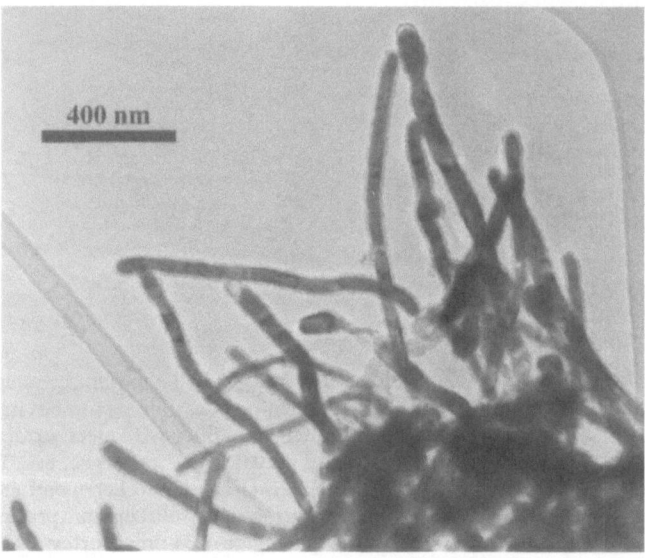

Fig. 24. TEM image of fully-filled tin nanotubes (nanowires) produced by electrolysis of graphite electrodes in $SnCl_2$ and LiCl mixtures (0.5 : 99.5 by weight respectively)

difficult to achieve. Alternative methods such as template-mediated processes using carbon nanotubes or zeolites [162, 163] can result in a measure of controlled growth, but in general polycrystalline nanostructures result. Novel approaches using laser vaporisation techniques involving vapour-liquid-solid growth have been developed [164] and a recently gas-solid phase reaction involving CO, catalysed by Co particles, led to the formation of novel flower-like nanostructures consisting of silicon oxide nanofibres, radially attached to a single catalytic particle [165] (Fig. 25). These structures may prove to be useful in nanoscale devices, optoelectronic sensors, 3-dimensional composite materials, novel catalytic supports and biological microfilters and, in particular, in situations where network growth is important.

5
$B_xC_yN_z$ Nanomaterials

This section describes the production and characterisation of BN and $B_xC_yN_z$ nanostructures generated by arc discharge techniques and by pyrolysis of $CH_3CN \cdot BCl_3$ over cobalt at 1000 °C. The resulting graphite-like nanostructures show remarkable conducting and mechanical properties.

Fig. 25. SEM image of flower-like SiO_x nanofibers generated by gas-solid reactions

5.1
BN Nanotubes

h-BN can be considered to be a non-carbon analogue of graphite, due to similarities in crystal structure (h-BN: a = 2.50 Å, c = 6.66 Å; graphite: a = 2.46 Å, c = 6.71 Å). A remarkable characteristic of this binary system, compared to graphite, is the different layer-by-layer stacking (aA, aA, ...), in which the B_3N_3 hexagons eclipse and alternate with N_3B_3 hexagons (see Fig. 26). Furthermore, h-BN is an insulator (band gap = 5.8 eV) [166] whereas graphite is a semi-metal (band overlap = 0.04 eV) [167].

Theoretical studies suggest [32] that BN nanotubes should be semiconductors with a band gap of 5.5 eV lower than in bulk BN. This property is independent of tube diameter, chirality and the number of walls. However, recent ab intio calculations show that under polygonisation, BN nanotubes reduce their band gap to ca. 1.5–2 eV [168]. They also appear to be resistant to oxidation and may therefore find use in materials science. In fact, experimental determination of the Young's modulus in BN tubes (Y = 1.22 ± 0.24 TPa) [169] confirms various theoretical elastic calculations on nanotubes [170–172] and reveals that BN tubes are highly crystalline and may be the strongest insulating nanofibres [169, 170].

With BN it is not possible to generate defects containing odd membered rings (e.g. pentagons or heptagons). This is because B-B and N-N bonds within rings

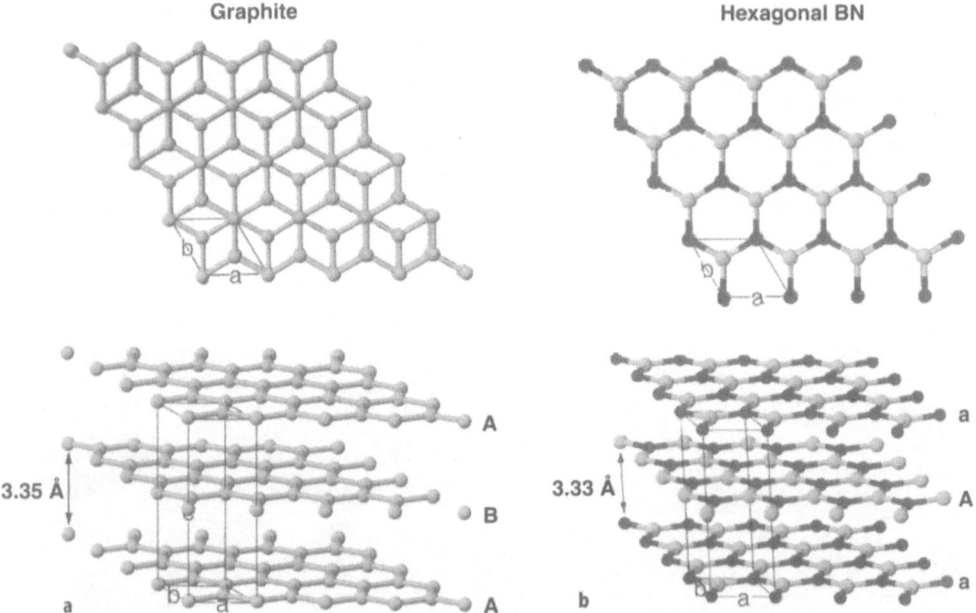

Fig. 26 a, b. Crystal structure similarities between h-BN and graphite: **a** the aAaA... stacking sequence in h-BN, showing h-BN sheets eclipsing each other (i.e. B eclipses N and vice versa), interlayer spacing 3.33 Å; **b** perfect graphite crystal, showing an ABAB... stacking, interlayer distances 3.35 Å

are less stable (e.g. the presence of N-N bonds leads to the formation of stable N_2 molecules at high temperatures). Therefore, it is necessary to introduce even-membered BN rings, such as squares or octagons, in order to achieve closure within the BN hexagonal sheet. Euler's Law for even-membered rings can be written as $12 = 2N_4 - 0N_6 - 2N_8$, where N_4 and N_8 equal the number of squares and octagons respectively. It is thus clear that the introduction of six squares will close a BN structure if the number of octagons is zero.

In particular, BN nanotubes and encapsulated polyhedral nanoparticles have been produced by arc discharge techniques employing hexagonal boron nitride and transition metals (i.e. W [35], Ta [36] and Hf [37]). Some of the nanotubes show metal/nitride/boride particles at their tips depending on the metal involved in the arc process. In other cases, nanotubes with flat caps (no metal particles) are found (Fig. 27). A general result from these experiments is that the number of layers (ca. 2–7) in these tubules tends to be lower than in carbon nanotubes. In particular, HfB_2 and Ta, used in conjunction with the arc apparatus, generate square-capped nanotubes, possibly due to the production of square B_2N_2 rings in the predominantly hexagonal network (Fig. 27). Electron Energy Loss Spectroscopy (EELS) measurements also showed that the nanotubes exhibit ca. 1:1 B:N ratios [35–37].

Laser vaporisation of h-BN under nitrogen at high pressure also leads to BN nanotubes, which apparently grow from disordered BN material [173]. In this case the tubes are poorly crystalline, possibly due to dislocations and defects within the hexagonal framework. However, further high electron irradiation is capable of annealing such a defect [173].

5.2
BC₂N Nanofibres and Tubes

$B_xC_yN_z$ materials constitute another possible layer-by-layer system, composed of h-BN and graphite hybrids. They may be semiconductors, and be used as photoluminescent materials (light sources), transistors working at high temperatures, lightweight electrical conductors or high temperature lubricants [174]. These applications depend not only on the composition but also on the arrangement of the B, C and N atoms.

BC₂N nanotubes, in particular, have been proposed theoretically [33, 34] and their successful synthesis described elsewhere [38–42]. These tubular structures are predicted to be semiconducting [33].

Pyrolytic methods can be used to generate BC_xN nanotubes. These techniques, when compared with arc discharge methods, produce a rather uniform range of BC_xN nanostructures, preserving stoichiometry [41]. In particular, pyrolysis of $CH_3CN \cdot BCl_3$ over cobalt powder at 950–1000 °C leads to ca. $[BC_2N_z]_n$ (z = 0.3–0.6) nanofibres and nanotubes possessing different morphologies as compared with nanotubes produced in the plasma arc.

The $[BC_2N_z]_n$ nanofibres/nanotubes, produced by pyrolysis, generate *non-cylindrical* layer-by-layer tubes, most of which (> 90%) appeared to consist of stacked cones in which the general cone-cone surface separation is close to that of graphite. The cone axes appear to be roughly aligned with the overall nano-

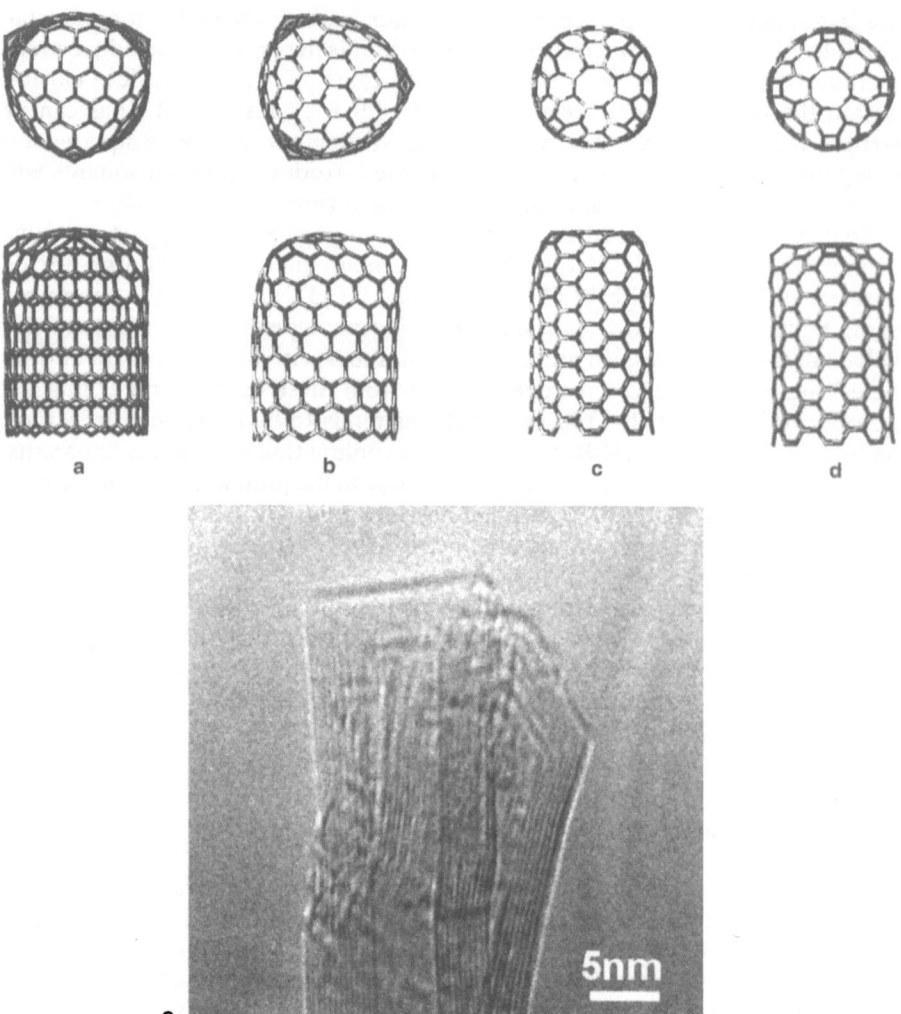

Fig. 27 a – e. Molecular model of a BN tubule cap: **a, b** three squares are equilaterally distributed over a zigzag type hexagonal tube; rotations along the tubule axis show that the square shape varies; **c, d** another BN armchair nanotube containing four squares and 1 octagon contributing to a perfect flat cap; **e** HRTEM image of BN nanotube produced by arcing h-BN/Ta mixtures. The flat cap may arise from possible introduction of four-membered rings into the hexagonal network

tube/nanofibre axis. Individual cones tend to adopt the shape and size of one end of the encapsulated catalytic nanoparticle (Fig. 28), indicating that the shapes of these particles are intimately responsible for structural features of the resulting tubules and fibres [41]. At the outset, carbon starts to agglomerate around the metal particles, thus creating a carbon (graphitic) coating with a metal/carbide interface, which allows carbon species to travel as in solid solution, forming graphitic layers and adopting the shape of the metal/carbide particle. As single graphitic cones form, strains are expected which can be relaxed by sliding the cones slightly apart, thus generating the BC_xN nanofibres.

Recently, a laser vaporisation method has provided a synthesis of heterogeneous B-C-N nanotubes by ablating C/BN/metal pellets under a nitrogen flow [175]. However, the B-C-N tubular structures possess large outer diameters, are of non-uniform morphology, and vary in chemical composition [175]. Other arc discharge routes involving HfB_2 anodes in conjunction with graphite cathodes and a nitrogen atmosphere yield nanotubes consisting of coaxial (sandwich-like) BN and C domains (e.g. C-BN-C nanotubes, in which carbon shells make up the inner core and the outer region of the tubular structure, BN shells appearing in between) [176].

5.3
B/N Doped Carbon Nanotubes

Submission of h-BN/graphite mixtures to a plasma arc generates nanotubes of varying B:C stoichiometries, long carbon tubules containing boron at their tips [38–40, 42, 177], and enhanced yields of higher fullerenes (e.g. C_{70}, C_{76} and C_{78}) compared to C_{60} [39, 42].

Some of the graphitic products exhibit anomalous interlayer spacing (ca. 3.45–3.55 Å) [42, 108]. In some cases, these spacing irregularities alternate periodically. These irregularities can be attributed to defects caused by the generation of non-cylindrical tubules (e.g. polygonal cross section [62]). Microwave

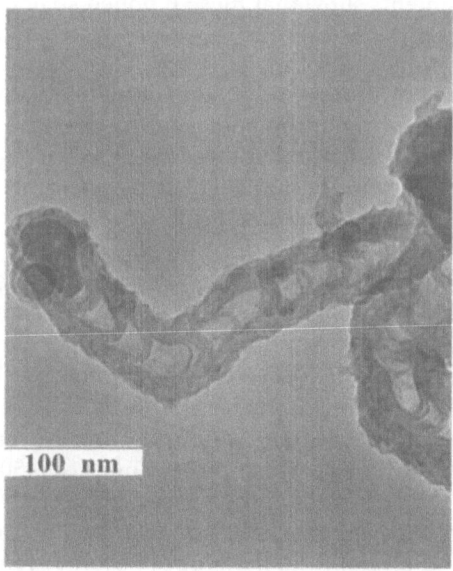

100 nm

Fig. 28. a $BC_2N_{0.6}$ nanofibre (exhibiting a stacked cone graphitic-like arrangement) produced by pyrolysis of $CH_3CN·BCl_3$ at 950 °C over Co

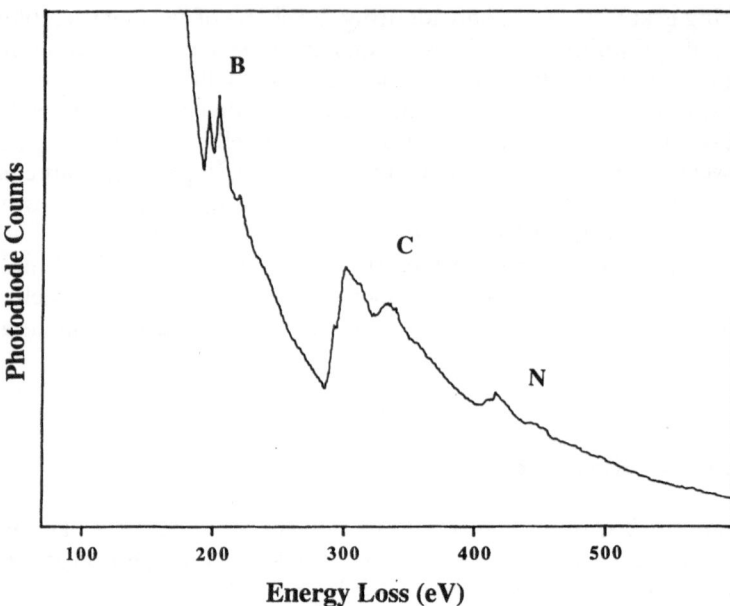

Fig. 28. b EELS spectrum near a nanofibre tip showing the ionisation edges at ca. 188 eV, 284 eV and 399 eV, corresponding to the K-shell edges of B, C and N respectively. The B and C regions shows sharply defined π^* and σ^* pre-ionisation-edge fine structural features, which are well-known indicators of sp^2 hybridisation (seen in graphitic structures)

conductivity measurements show that these B doped structures are intrinsically metallic (Fig. 29) [19], thus differing from standard pure carbon nanotubes which show thermally-activated transport (see Sect. 3). EELS analysis reveals that a small percentage of tubes exhibit stoichiometries close to BC_3 [42, 108]. Similar arc experiments using boron and graphite generate BC_3 islands in carbon nanotubes [177]. It is believed that these islands alter significantly the local density of states (LDOS) from a semimetal to an intrinsic metal (Fig. 29), and are consistent with theoretical calculations [177].

Nanotubes can also be doped by introducing nitrogen into the graphitic network. For example, pyrolysis of triazines over Co catalysts [103, 156] generates carbon nanotubes containing nitrogen (< 5%). Pyrolysis experiments by Sen et al. using pyridine and methylpyrimidine also led to nanotubes containing nitrogen [178]. Recent accounts by Sen et al. [179] and by Terrones et al. [180] described nanotubes with stoichiometries close to $C_{12}N$, and confirmed that nitrogen is bonded in an sp^2 fashion within the carbon framework. The latter tubes are easily oxidised (e.g. combustion sets in at ca. 450 °C in air, whereas pure carbon tubes do not ignite in air below ca. 700°C) [179]. Graphitisation is very dependent upon nitrogen concentration (i.e. the less the nitrogen content the more graphitic and straight the nanotubes become) [156, 180]. Tunnelling conductance of these N-doped tubes is generally higher than that of multi-walled all-carbon nanotubes [179].

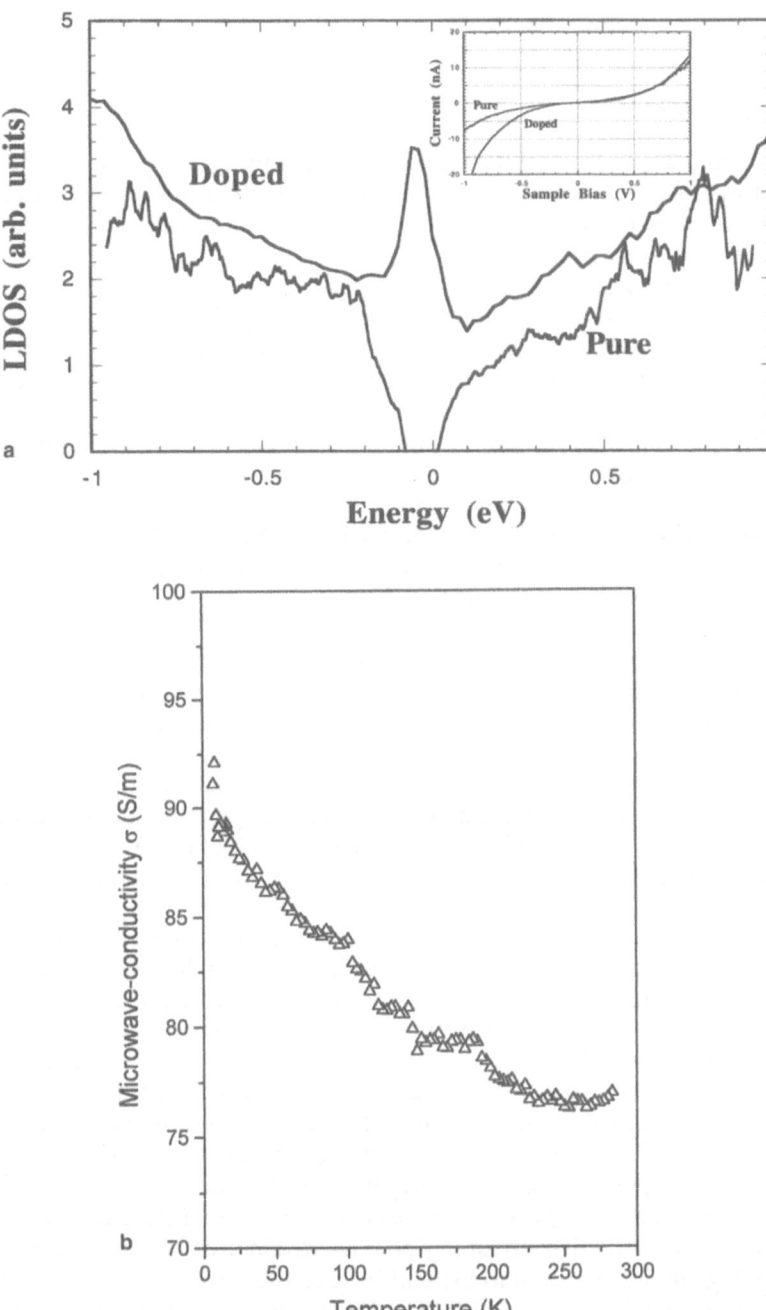

Fig. 29. a LDOS of carbon nanotubes and boron-doped carbon nanotubes [177]. It is interesting to note the intrinsic metallic behaviour of the doped tubes. A possible explanation is based upon the existence of BC_3 islands within the carbon hexagonal network (courtesy of P.M. Ajayan). **b** Microwave conductivity measurements of B-doped nanotubes obtained by arcing a mixture of h-BN and graphite in an He atmosphere exhibiting the unusual metallic behaviour [19]

6
Future Applications

Fullerene-related materials science has evolved rapidly over the past decade, and a fascinating field of nanotechnology has now opened up. Many applications of carbon nanotubes are envisaged, among which nano-scale engineering and electronics are prominent.

For example, very recently Dekker et al. fabricated a three-terminal switchable device based upon a single nanotube molecule [181]. This transistor, which operates at room temperature, consists of a semiconducting single-walled nanotube connected to metal nanoelectrodes. The performance, in terms of switching speed due to low capacitance, is excellent [181].

An inherent problem associated with nanotubes lies in the difficulty in manipulating them. In this context, a method devised to control the length and, consequently, the electronic properties of individual nanotubes by STM nanostructuring has been reported [182]. This technique allows one to cut nanotubes into shorter sections, which may be useful for future construction of molecular machinery and/or nano-scale circuits.

Integration of multiple devices into circuits may be feasible in future if molecular self assembly methods can be controlled, thus producing nanotubes with desired dimensions and properties. In this context, TEM images – of inner core cathode deposits produced in the arc experiment – show the existence of 30°-bent multi-walled nanotubes [79, 108, 183] (Fig. 30). These bends arise as a result of pentagon insertion (outer rim with positive curvature) matched by the inclusion of a heptagon (inner rim with negative curvature) on opposite sides of the tubule (Fig. 30). Further electron diffraction studies [79, 108] of these bent structures have shown that the hexagon chirality changes by 30° consequent upon the 30° angle bend.

Helicity changes by 30° may cause significant variations in the electronic properties (e.g. from metallic conductor to semiconductor because a bend can transform a zig-zag configuration into an armchair arrangement). The properties of these bent structures have been widely investigated by Lambin et al. [184–186] and others [187]. The results show that metal-metal and semiconductor-metal nano-scale junctions arise from the presence of a single pentagon-heptagon pair within the tubule. Therefore, these structures may behave as nano-switches. Collins et al. observed abrupt transport changes (from graphite-like response to highly non-linear) along single walled nanotubes [188]. These changes may be due to the presence of defects such as pentagon-heptagon pairs [189–191]. However, further investigation is needed in order to confirm this and other topological defect calculations within graphite networks (e. g. pentagon-heptagon, pentagon-octagon, etc.) [189–193]. Additionally, nanowires or metal encapsulated nanotubes and nanorods may also have a significant role to play in nanoscale electronic circuit construction.

Novel structures based upon sp^2 carbon, such as concentric giant fullerenes [194], graphite cones (see Fig. 31) [195, 196], graphitic cubes [197] and discs [195], bundles of tori (see Fig. 32) [72], helices [10–11] and metal/graphite needles [157] have been successfully generated, and await transport and me-

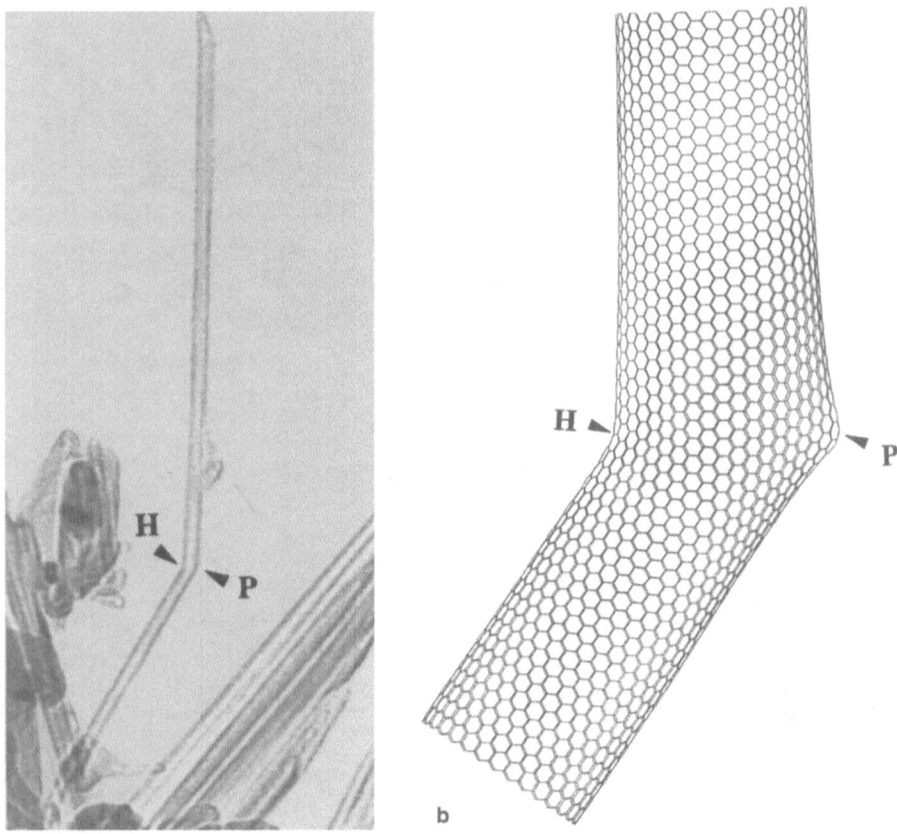

Fig. 30. a TEM image of a 30°-bent nanotube. **b** Molecular simulation model showing the change in helicity by 30° (from zigzag to armchair) – two different electronic properties (e. g. semiconductor and metallic conductor) are possible

chanical studies. Predicted structures remained to be constructed [65, 75, 198, 199].

Encapsulation of metals within nanotubes may be of interest, especially if the metal is ferromagnetic (e. g. data storage devices and stiffer STM tips). Investigation on nanotubes and related fullerene-like structures for hydrogen storage is a necessary precursor to the development of fuel-cell electric vehicles [49].

It may prove possible to develop matrices of well-aligned and highly graphitic nanotubes, in addition to carbon-tubes/Fe/S alloys which will be stronger than steel. The bulk generation and separation of carbon onions (giant concentric fullerenes) may be of interest as solid-state lubricants, as are the existing fullerene-like cages WS_2 and MoS_2 [200].

A wide variety of routes to nanotubes, including pyrolytic, laser beam, plasma arc and electrolytic techniques, have been reviewed. However, superior nanotube yield and structural quality is needed if electronic and optical applications

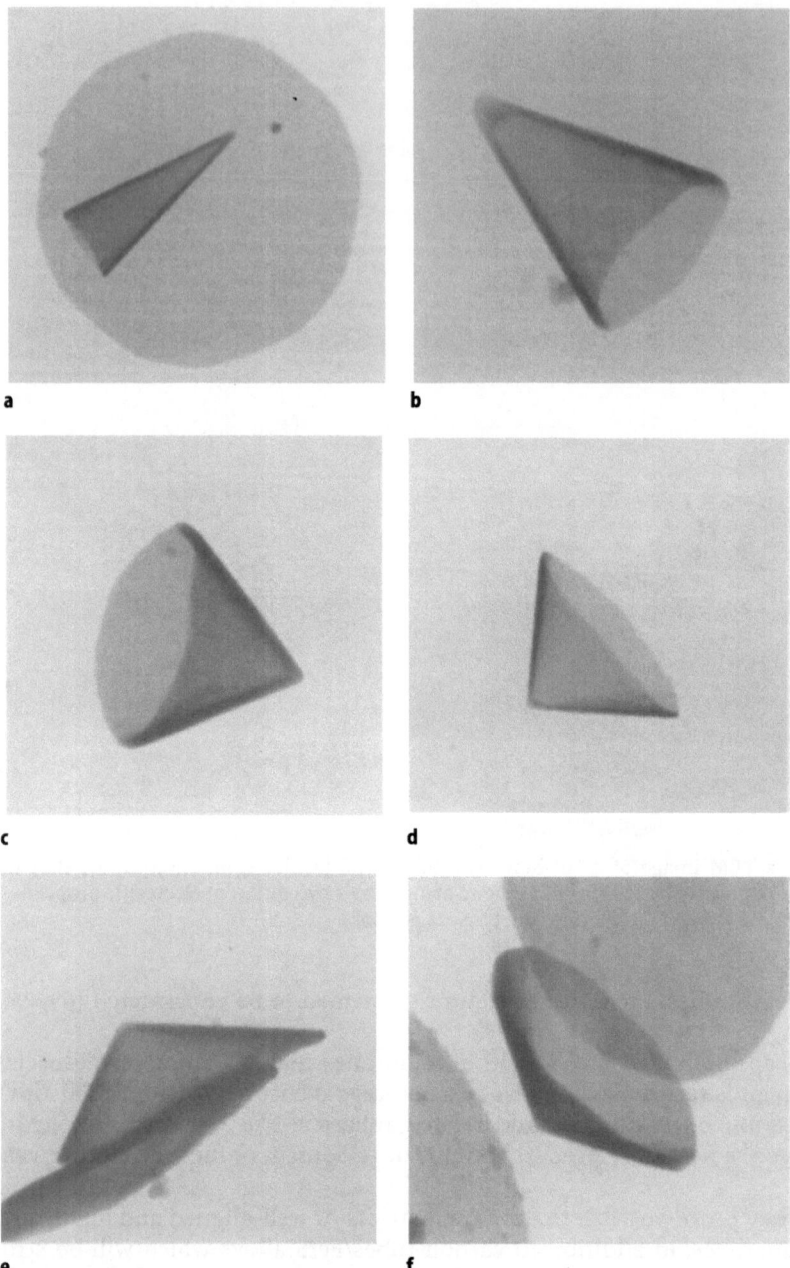

Fig. 31. Graphitic cones, possessing different angles, obtained by pyrolytic techniques. (Courtesy of T.W. Ebbesen)

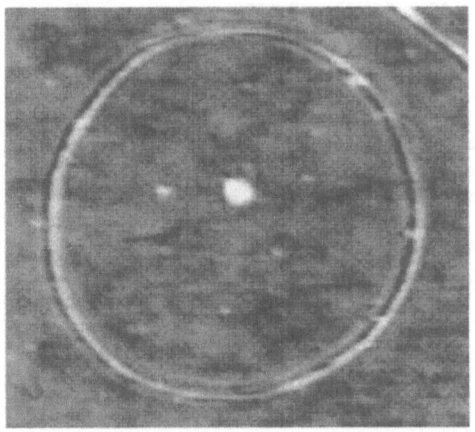
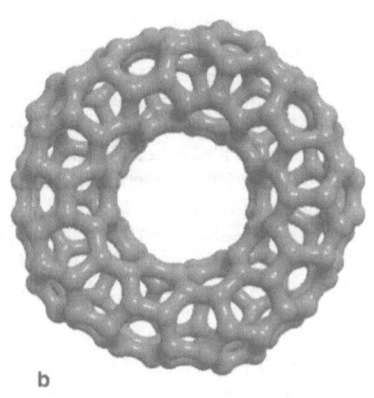

a

b

Fig. 32. **a** Scanning force image of a bundle of tori (bundle of toroidal single walled nanotubes; radius ca. 300 nm) generated by laser vaporisation of graphite [72]. The structure exhibits an apparent height of ca. 1.5 nm and width of 4–8 nm (Courtesy of C. Dekker). **b** Model of a toroidal 240-atom graphite structure (Courtesy of H. Terrones)

are to be developed. It is likely that, in the near future, novel nano-scale materials including nanofibres, nanowires and nanorods may also prove to be useful in optoelectronic sensors, three-dimensional composite materials, novel catalytic supports and biological microfilters.

7
References

1. Endo M, Kroto HW (1992) J Phys Chem 96:6491
2. Iijima S (1991) Nature 56:354
3. Ebbesen TW, Ajayan PM (1992) Nature 358:220
4. Iijima S, Ajayan PM, Ichihashi T (1992) Phys Rev Lett 69:3100
5. Iijima S (1993) Mat Sci Eng B 19:172
6. Ajayan PM, Iijima S, Ichihashi T (1993) Phys Rev B 47:6859
7. Zhang XF, Zhang XB, Van Tendeloo G, Amelinckx S, Op de Beck M, Van Landuyt J (1993) J Cryst Growth 130:368
8. Ebbesen TW (1994) Annu Rev Mater Sci 24:235
9. Endo M, Takeuchi K, Igarashi S, Kobori K, Shiraishi M, Kroto H W (1993) J Phys Chem Solids 54:1841
10. Amelinckx S, Zhang XB, Bernaerts D, Zhang XF, Ivanov V, Nagy JB (1994) Science 265:635
11. Bernaerts D, Zhang XB, Zhang XF, Amelinckx S, Van Tendeloo G, Van Landuyt J, Ivanov V, Nagy JB (1995) Phil Mag A 71:605
12. Guo T, Nikoleav P, Rinzler AG, Tománek D, Colbert DT, Smalley RE (1995) J Phys Chem 99:10,694
13. Thess A, Lee R, Nikolaev P, Dai H, Petit P, Robert J, Xu C, Lee YH, Kim SG, Rinzler AG, Colbert DT, Scuseria GE, Tománek D, Fischer JE, Smalley RE (1996) Science 273:483
14. Hsu WK, Hare JP, Terrones M, Harris PFJ, Kroto HW, Walton DRM (1995) Nature 377:687
15. Hsu WK, Terrones M, Hare JP, Terrones H, Kroto HW, Walton DRM (1996) Chem Phys Lett 261:161

16. Hamada N, Sawada S, Oshiyama A (1992) Phys Rev Lett 68:1579
17. Saito R, Fujita M, Dresselhaus G, Dresselhaus MS (1992) Phys Rev B 46:1804
18. De Heer WA, Chatelain A, Ugarte D (1995) Science 270:1179
19. Terrones M, Hsu WK, Schilder A, Terrones H, Grobert N, Hare JP, Zhu YQ, Schwoerer M, Prassides K, Kroto HW, Walton DRM (1998) Appl Phys A 66:307
20. Dai H, Wong EW, Lieber CM (1996) Science 272:52
21. Ebbesen TW, Lezec HJ, Hiura H, Bennett JW, Ghaemi HF, Thio T (1996) Nature 382:54
22. Wildöer JWG, Venema LC, Rinzler AG, Smalley RE, Dekker C (1998) Nature 391:59
23. Odom TW, Huang JL, Kim P, Lieber CM (1998) Nature 391:62
24. Treacy M, Ebbesen TW, Gibson JM (1996) Nature 381:678
25. Wong EW, Sheehan PE, Lieber CM (1997) Science 277:1971
26. Falvo M R, Clary GJ, Taylor RM II, Chi V, Brooks FP Jr, Washburn S, Superfine R (1997) Nature 389:582
27. Heath JR, O'Brien SC, Zhang Q, Liu Y, Curl RF, Kroto HW, Tittel FK, Smalley RE (1985) J Am Chem Soc 107:7779
28. O'Brien SC, Heath JR, Curl RF, Smalley RE (1988) J Chem Phys 88:220
29. Ajayan PM, Iijima S (1993) Nature 361:333
30. Ajayan PM, Ebbesen TW, Ichihashi T, Iijima S, Tanigaki K (1993) Nature 362:522
31. Dujardin E, Ebbesen TW, Hiura H, Tanigaki K (1994) Science 265:1850
32. Blase X, Rubio A, Louie SG, Cohen ML (1994) Europhys Lett 28:335
33. MiyamotoY, Rubio A, Cohen ML, Louie SG (1994) Phys Rev B 50:4976
34. MiyamotoY, Rubio A, Cohen ML, Louie SG (1994) Phys Rev B 50:18,360
35. Chopra NG, Luyken RJ, Cherrey K, Crespi VH, Cohen ML, Louie SG, Zettl A (1995) Science 269:966
36. Terrones M, Hsu WK, Terrones H, Zhang JP, Ramos S, Hare JP, Castillo R, Prassides K, Cheetham AK, Kroto HW, Walton DRM (1996) Chem Phys Lett 259:568
37. Loiseau A, Willaime F, Demoncy N, Hug G, Pascard H (1996) Phys Rev Lett 76:4737
38. Stephan O, Ajayan PM, Colliex C, Redlich Ph, Lambert JM, Bernier PM, Lefin P (1994) Science 266:1683
39. Redlich Ph, Loeffler J, Ajayan PM, Bill J, Aldinger F, Rühle M (1996) Chem Phys Lett 260:465
40. Wengsieh, Z, Cherrey K, Chopra NG, Blase X, Miyamoto Y, Rubio A, Cohen ML, Louie SG, Zettl A, Gronsky AR (1995) Phys Rev B 51:11,299
41. Terrones M, Benito AM, Manteca-Diego C, Hsu WK, Osman OI, Hare JP, Reid DG, Terrones H, Cheetham AK, Prassides K, Kroto HW, Walton DRM (1996) Chem Phys Lett 257:576
42. Terrones M, Hsu WK, Ramos S, Castillo R, Terrones H (1998) Full Sci Tech 6:787
43. Margulis L, Saltra G, Tenne R, Tallanker M (1993) Nature 365:113
44. Tenne R, Margulis L, Genut M, Hodes G (1992) Nature 360:444
45. Homyonfer M (1997) private communication
46. Kaner RB, Kouvetakis J, Warble CE, Sattler ML, Bartlett N (1986) Mat Res Bull 22:399
47. Kouvetakis J, Kaner RB, Sattler ML, Bartlett N (1986) Chem Soc Chem Commun 1758
48. Gadd GE, Blackford M, Moricca S, Webb N, Evans PJ, Smith AM, Jacobsen G, Leung S, Day A, Hua Q (1997) Science 277:933
49. Dillon AC, Jones KM, Bekkedahl TA, Kiang CH, Bethune DS, Heben MJ (1997) Nature 386:377
50. Dai HJ, Hafner JH, Rinzler AG, Colbert DT, Smalley RE (1996) Nature 384:147
51. de Heer WA, Chatelain A, Ugarte D (1995) Science 270:1179
52. Rinzler AG, Hafner JH, Nikolaev P, Lou L, Kim SG, Tomànek D, Nordlander P, Colbert DT, Smalley RE (1995) Science 269:1550
53. Collins PG, Zettl A (1996) Appl Phys Lett 69:1969
54. Saito Y, Hamaguchi K, Hata K, Uchida K, Tasaka Y, Ikazaki F, Yumura M, Kasuya A, Nishima Y (1997) Nature 389:555
55. Niu C, Sichel EK, Hoch R, Moy D, Tennent H (1997) Appl Phys Lett 70:1480

56. Britto PJ, Santhanam KSV, Ajayan PM (1996) Bioelectrochem Bioenergetics 41:121
57. Dai H, Wong EW, Lu YZ, Fan S, Lieber CM (1995) Nature 375:769
58. Han W, Fan S, Li Q, Hu Y (1997) Science 277:1287
59. Saito R, Fujita M, Dresselhaus G, Dresselhaus MS (1992) Phys Rev B 46:1804
60. Saito R, Fujita M, Dresselhaus G, Dresselhaus MS (1992) Appl Phys Lett 60:2204
61. Hamada N, Swada S, Oshiyama A (1992) Phys Rev Lett 68:1579
62. Liu M, Cowley JM (1994) Carbon 32:393
63. Endo M, Takeuchi K, Igarashi S, Kobori K, Shiraishi M, Kroto HW, Sarkar A (1995) Carbon 33:873
64. Ebbesen TW, Ajayan PM, Hiura H, Tanigaki K (1994) Nature 367:519
65. Mackay AL, Terrones H (1991) Nature 352:762
66. Iijima S, Toshinary I, Ando Y (1992) Nature 356:776
67. Chernozatonskii LA (1992) Phys Lett A 170:37
68. Itoh S, Ihara S, Kitakami J (1993) J Phys Rev B 47:1703
69. Ihara S, Itoh S, Kitakami J (1993) J Phys Rev B 47:12,908
70. Ivanov V, Nagy JB, Lambin Ph, Lucas A, Zhang XB, Zhang XF, Bernaerts D, Van Tendeloo G, Amelinckx S, Van Landuyt (1994) J Chem Phys Lett 223:329
71. Sarkar A, Kroto HW, Endo M (1995) Carbon 33:51
72. Liu J, Dai H, Hafner JH, Colbert DT, Smalley RE, Tans SJ, Dekker C (1997) Nature 385:780
73. Boehm HP (1973) Carbon 11:583
74. Dresselhaus MS, Dresselhaus G, Sugihara K, Spain IL, Goldberg HA (1988) Synthesis of graphite fibers and filaments. In: Graphite fibers and filaments, vol 3. Springer, Berlin Heidelberg New York, p 12
75. Kroto HW, Hare JP, Sarkar A, Hsu K, Terrones M, Abeysinghe JR (1994) MRS Bulletin 19:51
76. Bernaerts D, Zhang XB, Zhang XF, Amelinckx S, Van Tendeloo G, Van Landuyt J, Ivanov V, Nagy JB (1995) Phil Mag A 71:605
77. Akagi, K, Tamura R, Tsukada M, Itoh S, Ihara S (1995).Phys Rev Lett 74:2307
78. Ricardo-Chávez JL, Dorantes-Dávila J, Terrones H, Terrones M (1997) Phys Rev B 56:12,143
79. Terrones M, Hsu WK, Hare JP, Walton DRM, Kroto HW, Terrones H (1996) Phil Trans Roy Soc A 354:2055
80. Wang XK, Lin XW, Dravid VP, Ketterson JB, Chang RPH (1995) Appl Phys Lett 66:2430
81. Dravid VP, Lin X, Wang Y, Wang XK, Yee A, Ketterson JB, Chang RPH (1993) Science 259:1601
82. Saito Y, Yoshikawa M, Inagaki M, Tomita M, Hayashi T (1993) Chem Phys Lett 204:277
83. Colbert DT, Zhang J, McClure SM, Nikolaev P, Chen Z, Hafner JH, Owens DW, Kotula PG, Carter JH, Weaver AG, Rinzler AG, Smalley RE (1994) Science 266:1218
84. Gamaly EG, Ebbesen TW (1995) Phys Rev B 52:2083
85. Ajayan PM, Redlich Ph, Rühle M (1997) Mater Res 12:1
86. Ebbesen TW, Hiura H, Fujita J, Ochai Y, Matsui S, Tanigaki K (1993) Chem Phys Lett 209:83
87. Bonard JC, Stora T, Salvetat JP, Maier F, Stöckli T, Duschl C, Forró L, de Heer WA, Châtelain A (1997) Adv Mater 9:827
88. Duesberg GS, Burghard M, Muster J, Philipp G, Roth S (1998) J Chem Soc Chem Commun 435
89. Schützenberger P, Schützenberger LCR (1890) Acad Sci 111:774
90. Pebalon CHCR (1905) Acad Sci 137:706
91. Endo M, Katoh A, Sugiura T, Shiraishi M (1987) 18th beinnial conference on carbon, Worcester, MA
92. Baker RTK, Harris PS (1978) Chemistry and physics of carbon. Marcel Dekker, New York, p 83
93. Baker RTK, Braber MA, Harris PS, Feates FS, Waite RJ (1972) J Catal 26:51
94. Baker RTK, Chludzinski JJ Jr (1980) J Catal 64:464
95. Baker RTK, Harris PS, Thomas RB, Waite RJ (1973) J Catal 30:86
96. Baker RTK, Waite RJ (1975) J Catal 37:101
97. Baird T, Fryer JR (1974) Carbon 12:591

98. Oberlin A, Endo M, Koyama T (1976) J Cryst Growth 32:335
99. Audier M, Coulon M, Oberlin A (1980) Carbon 18:73
100. Tibbets G (1984) J Cryst Growth 66:632
101. Boellard E, de Bokx PK, Kock AJHM, Geus JW (1985) J Catal 96:481
102. Ajayan PM, Stephan O, Colliex C, Trauth D (1994) Science 265:1212
103. Terrones M, Grobert N, Olivares J, Zhang JP, Terrones H, Kordatos K, Hsu WK, Hare JP, Townsend PD, Prassides K, Cheetham AK, Kroto HW, Walton DRM (1997) Nature 388:52
104. Rinzler AG, Hafner JH, Nikolaev P, Lou L, Kim SG, Tomànek D, Nordlander P, Colbert DT, Smalley RE (1995) Science 269:1550
105. Collins PG, Zettl A (1996) Appl Phys Lett 69:1969
106. Saito Y, Hamaguchi K, Hata K, Uchida K, Tasaka Y, Ikazaki F, Yumura M, Kasuya A, Nishima Y (1997) Nature 389:555
107. Li WZ, Xie SS, Qian LX, Chang BH, Zou BS, Zhou WY, Zhao RA, Wang G (1996) Science 274:1701
108. Terrones M (1997) PhD Thesis, University of Sussex
109. Birkett PR, Cheetham AK, Eggen BR, Hare JP, Kroto HW, Walton DRM (1997) Chem Phys Lett 281:114
110. Iijima S, Ichihashi T (1993) Nature 363:603
111. Ajayan PM, Lambert JM, Bernier P, Barbedette L, Colliex C, Planeix JM (1993) Chem Phys Lett 215:509
112. Saito Y, Yoshikawa T, Okuda M, Ohkohchi M, Ando Y, Kasuya A, Nishina Y (1993) Chem Phys Lett 209:72
113. Zhou D, Seraphin S, Wang S (1994) Appl Phys Lett 65:1593
114. Journet C, Maser WK, Bernier P, Loisseau A, Lamy de la Chapelle M, Lefrant S, Deniard P, Lee P, Fischer JE (1997) Nature 388:756
115. Tohji K, Takahashi H, Shinoda Y, Shimizu N, Jeyadevan B, Matsuoka I, Saito Y, Kasuya A, Ito S, Nishina Y (1997) J Phys Chem B 101:1974
116. Bandow S, Rao AM, Williams KA, Smalley RE, Eklund PC (1997) J Phys Chem B 101:8839
117. Duesberg G (1998) personal communication
118. Iijima S, Brabec C, Maiti A, Bernholc J (1996) J Chem Phys 104:2089
119. Chopra NG, Benedict LX, Crespi VH, Cohen ML, Louie SG, Zettl A (1995) Nature 377:135
120. Ruoff RS, Lorents DC (1995) Carbon 33:925
121. Overney G, Zhong W, Tománek D (1993) Z Phys D 27:93
122. Robertson DH, Brener DW, Minmire JW (1992) Phys Rev B 4:12,592
123. Kelly BT (1981) Physics of Graphite Applied Science, London
124. Tersoff J (1992) Phys Rev B 46:15,546
125. Langer L, Stockman L, Heremans JP, Bayot V, Olk CH, Van Haesendonck C, Bruynseraede Y, Issi JP (1994) J Mater Res 9:927
126. Nakayama Y, Akita S, Shimada Y (1995) Jpn J Appl Phys 34:L10
127. De Heer WA, Bacsa WS, Châtelain A, Gerfin T, Humphrey-Baker R, Forró L, Ugarte D (1995) Science 268:845
128. Baumgartner G, Carrard M, Zuppiroli L, Bacsa W, de Heer WA, Forró L (1997) Phys Rev B 55:6704
129. Subramoney S, Kavelaar V, Ruoff RS, Lorents DC, Malhotra R, Kazmer AJ (1994) In: Kadish KM, Ruoff RS (eds) Chemistry and physics of fullerenes and related materials PV 94–24. The Electrochemistry Society, Pennington, NJ, p 1498
130. Ruoff RS, Lorents DC, Chan BC, Malhotra R, Subramoney S (1993) Science 259:346
131. Liu M, Cowley JM (1995) Carbon 33:225
132. Seraphin S (1994) In: Kadish KM, Ruoff RS (eds) Chemistry and physics of fullerenes and related materials PV 94–24. The Electrochemistry Society, Pennington, NJ, p 1433
133. Subramoney S, Ruoff RS, Lorents DC, Chan BC, Malhotra R, Dyer RMJ, Parvin K (1994) Carbon 32:507
134. Yosida Y (1993) Appl Phys Lett 62:3447
135. Loisseau A, Pascard H (1996) Chem Phys Lett 256:246

136. Guerret-Piécourt C, Le Bouar Y, Loisseau A, Pascard H (1994) Nature 372:761
137. Tomita M, Saito Y, Hayashi T (1993) Jpn J Appl Phys 32:L280
138. Saito Y, Yoshikawa M, Inagaki M, Tomita M, Hayashi T (1993) Chem Phys Lett 204: 277
139. Murakami Y, Shibata T, Okuyama K, Arai T, Suematsu H, Yoshida Y (1994) J Phys Chem Solids 54:1861
140. Majetich SA, Artman JO, McHenry ME, Nuhfer NT, Staley SW (1993) Phys Rev B 48:16,845
141. Seraphin S, Zhou D, Jiao J, Withers JC, Loufty R (1993) Appl Phys Lett 63:2073
142. Bandow S, Saito Y (1994) Jpn J Appl Phys 32:L1677
143. Hare JP, Hsu WK, Kroto HW, Lappas A, Prassides K, Terrones M, Walton DRM(1996) Chem Mater 8:6
144. Ata M, Yamaura K, Hudson AJ (1995) J Adv Mater 7:286
145. Tsang SC, Chen YK, Harris PJF, Green MLH (1994) Nature 372:159
146. Chen YK, Green MLH, Tsang SC (1996) J Chem Soc Chem Commun 2489
147. Sloan J, Cook J, Heesom JR, Green MLH, Hutchison JL (1997) J Cryst Growth 173:81
148. Sloan J, Cook J, Green MLH, Hutchison JL, Tenne R (1997) J Mater Chem 7:1089
149. Chen YK, Chu A, Cook J, Green MLH, Harris PJF, Heesom R, Humphries M, Sloan J, Tsang SC, Turner JFC (1997) J Mater Chem 7:545
150. Lago RM, Tsang SC, Lu KL, Chen YK, Green MLH (1995) J Chem Soc Chem Commun 1355
151. Chu A, Cook J, Heesom RJR, Hutchison JL, Green MLH, Sloan J (1996) J Chem Mater 8:2751
152. Tsang SC, Davis JJ, Green MLH, Allen H, Hill O, Leung YC, Sadler PJJ (1995) Chem Soc Chem Commun 1803
153. Tsang SC, Guo Z, Chen YK, Green MLH, Hill AO, Hambley TW, Sadler PJ (1997) Angew Chem Int Ed Engl 36:2198
154. Ajayan PM, Stephan O, Redlich Ph, Colliex C (1995) Nature 375:564
155. McHenry ME, Nakamura Y, Kirkpatrick S, Johnson F, Curtin S, DeGraef M, Uhfer NT, Majetich SA, Brunsman EM (1995) In: Kadish KM, Ruoff RS (eds) Chemistry and physics of fullerenes and related materials PV 95–26. The Electrochemistry Society, Pennington, NJ, p 1463
156. Terrones M, Grobert N, Zhang JP, Terrones H, Olivares J, Hsu WK, Hare JP, Cheetham AK, Kroto HW, Walton DRM (1998) Chem Phys Lett 285:299
157. Grobert N, Terrones M, Osborne OJ, Terrones H, Hsu WK, Trasobares S, Zhu YQ, Hare JP, Kroto HW, Walton DRM (1998) Appl Phys A (in press)
158. Hsu WK, Terrones M, Terrones H, Grobert N, Kirkland AI, Hare JP, Prassides K, Townsend PD, Kroto HW, Walton DRM (1998) Chem Phys Lett 284:177
159. Hsu WK, Terrones M, Terrones H, Grobert N, Zhu YQ, Hare JP, Kroto HW, Walton DRM (to be published)
160. Saunders GD, Chang YC (1992) Phys Rev B 45:9202
161. Brus LE (1994) J Phys Chem, 98:3575
162. Martin CR (1994) Science 266:1961
163. Mann S (1993) Nature 365:499
164. Morales AM, Lieber CM (1998) Science 279:208
165. Zhu YQ, Hsu WK, Terrones M, Grobert N, Terrones H, Hare JP, Kroto HW, Walton DRM (1998) J Mater Chem 8:1859
166. Zunger A, Katzir A, Halperin A (1976) Phys Rev B 13:5560
167. Dresselhaus MS, Dresselhaus G, Eklund PC (1996) Science of fullerenes and carbon nanotubes. Academic Press, New York
168. Charlier JC, Blase X (1998) personal communication
169. Chopra NG, Zettl A (1998) Solid State Commun 105:297
170. Hernández E, Goze C, Bernier P, Rubio A (1998) Phys Rev Lett 80:4502
171. Yakobson BI, Brabec CJ, Bernholc J (1996) J Phys Rev Lett 76:2411
172. Lu JP (1997) Phys Rev Lett 79:1297
173. Goldberg D, Bando Y, Eremets M, Takemura K, Kurashima K, Tamiya K, Yusa H (1997) Chem Phys Lett 279:191

174. Kawaguchi M, Bartlett N (1995) In: Nakajima T (ed) Fluorine-carbon and fluoride-carbon materials. Marcel Dekker Inc, New York, p 187
175. Zhang Y, Gu H, Suenaga K, Iijima S (1997) Chem Phys Lett 279:264
176. Suenaga K, Colliex C, Demoncy N, Loiseau A, Pascard H, Willaime F (1997) Science 278:653
177. Carroll DT, Curran S, Ajayan PM, Redlich Ph (1998) Phys Rev Lett 81:2332
178. Sen R, Satishkumar BC, Govindaraj S, Harikumar KR, Renganathan MK, Rao CNR (1997) J Mater Chem 7:2335
179. Sen R, Satishkumar BC, Govindaraj S, Harikumar KR, Raina G, Zhang JP, Chetham AK, Rao CNR (1998) Chem Phys Lett 287:671
180. Terrones M, Redlich Ph, Grobert N, Trasobares S, Terrones H, Hsu WK, Rühle M, Kroto HW, Walton DRM (to be published)
181. Sander SJ, Verschueren RM, Dekker C (1998) Nature 393:49
182. Venema L, Wildöer JW, Temminck Tuinstra HLJ, Dekker C, Rinzler AG, Smalley RE (1997) Appl Phys Lett (1997) 71:2629
183. Terrones H, Terrones M, Hsu WK (1995) Chem Soc Rev 24:341
184. Lambin Ph, Fonseca A, Vigneron JP, Nagy JB, Lucas AA (1995) Chem Phys Lett 245:85
185. Meunier V, Henrad L, Lambin Ph (1998) Phys Rev B 57:2586
186. Lambin Ph, Fonseca A, Nagy JB, Lucas AA (1996) Synth Met 77:249
187. Chico L, Crespi VH, Benedict LH, Louie SG, Cohen ML (1996) Phys Rev Lett 76:971
188. Collins PG, Zettl A, Bando H, Thess A, Smalley RE (1997) Science 278:100
189. Terrones M, Terrones H (1996) Full Sci Tech 4:517
190. Crespi VH, Cohen ML, Rubio A (1997) Phys Rev Lett 79:2093
191. Charlier JC, Ebbesen TW, Lambin Ph (1996) Phys Rev B 53:11,108
192. Charlier JC, Lambin Ph, Ebbesen TW (1996) Phys Rev B 54:R8377
193. Delaney P, Choi HJ, Ihm J, Louie SG, Cohen ML (1998) Nature 391:466
194. Ugarte D (1992) Nature 359:709
195. Krishnan A, Dujardin E, Treacy MMJ, Hugdahl J, Lynum S, Ebbesen TW (1997) Nature 388:451
196. Sattler K (1995) Carbon 33:920
197. Saito Y, Matsumoto T (1998) Nature 392:237
198. Lenoski T, Gonze X, Teter M, Elser V (1992) Nature 355:333
199. Terrones H, Terrones M (1997) Phys Rev B 55:9969
200. Rapoport L, Bilik Y, Feldman Y, Homyonfer M, Cohen SR, Tenne R (1997) Nature 387:791

Author Index Volumes 151–199

The volume numbers are printed in italics

Springer
and the
environment

At Springer we firmly believe that an international science publisher has a special obligation to the environment, and our corporate policies consistently reflect this conviction.
We also expect our business partners – paper mills, printers, packaging manufacturers, etc. – to commit themselves to using materials and production processes that do not harm the environment. The paper in this book is made from low- or no-chlorine pulp and is acid free, in conformance with international standards for paper permanency.